Metal Organic Frameworks

Metal Organic Frameworks

Synthesis and Application

Special Issue Editors

Victoria Samanidou
Eleni Deliyanni

MDPI • Basel • Beijing • Wuhan • Barcelona • Belgrade

Special Issue Editors
Victoria Samanidou Eleni Deliyanni
Aristotle University of Thessaloniki Aristotle University of Thessaloniki
Greece Greece

Editorial Office
MDPI
St. Alban-Anlage 66
4052 Basel, Switzerland

This is a reprint of articles from the Special Issue published online in the open access journal *Molecules* (ISSN 1420-3049) from 2018 to 2020 (available at: https://www.mdpi.com/journal/molecules/special_issues/MOFs).

For citation purposes, cite each article independently as indicated on the article page online and as indicated below:

LastName, A.A.; LastName, B.B.; LastName, C.C. Article Title. *Journal Name* **Year**, *Article Number*, Page Range.

ISBN 978-3-03928-486-3 (Pbk)
ISBN 978-3-03928-487-0 (PDF)

© 2020 by the authors. Articles in this book are Open Access and distributed under the Creative Commons Attribution (CC BY) license, which allows users to download, copy and build upon published articles, as long as the author and publisher are properly credited, which ensures maximum dissemination and a wider impact of our publications.

The book as a whole is distributed by MDPI under the terms and conditions of the Creative Commons license CC BY-NC-ND.

Contents

About the Special Issue Editors . . . . **vii**

Preface to "Metal Organic Frameworks" . . . . **ix**

Victoria F. Samanidou and Eleni A. Deliyanni
Metal Organic Frameworks: Synthesis and Application
Reprinted from: *Molecules* **2020**, *25*, 960, doi:10.3390/molecules25040960 . . . . **1**

Dimitrios A. Giannakoudakis and Teresa J. Bandosz
Building MOF Nanocomposites with Oxidized Graphitic Carbon Nitride Nanospheres: The Effect of Framework Geometry on the Structural Heterogeneity
Reprinted from: *Molecules* **2019**, *24*, 4529, doi:10.3390/molecules24244529 . . . . **4**

Gabriel González-Rodríguez, Iván Taima-Mancera, Ana B. Lago, Juan H. Ayala, Jorge Pasán and Verónica Pino
Mixed Functionalization of Organic Ligands in UiO-66: A Tool to Design Metal–Organic Frameworks for Tailored Microextraction
Reprinted from: *Molecules* **2019**, *24*, 3656, doi:10.3390/molecules24203656 . . . . **18**

Despoina Andriotou, Stavros A. Diamantis, Anna Zacharia, Grigorios Itskos, Nikos Panagiotou, Anastasios J. Tasiopoulos and Theodore Lazarides
Dual Emission in a Ligand and Metal Co-Doped Lanthanide-Organic Framework: Color Tuning and Temperature Dependent Luminescence
Reprinted from: *Molecules* **2020**, *25*, 523, doi:10.3390/molecules25030523 . . . . **32**

Xue-Xue Liang, Nan Wang, You-Le Qu, Li-Ye Yang, Yang-Guang Wang and Xiao-Kun Ouyang
Facile Preparation of Metal-Organic Framework (MIL-125)/Chitosan Beads for Adsorption of Pb(II) from Aqueous Solutions
Reprinted from: *Molecules* **2018**, *23*, 1524, doi:10.3390/molecules23071524 . . . . **46**

Mohammad S. Yazdanparast, Victor W. Day and Tendai Gadzikwa
Hydrogen-Bonding Linkers Yield a Large-Pore, Non-Catenated, Metal-Organic Framework with pcu Topology
Reprinted from: *Molecules* **2020**, *25*, 697, doi:10.3390/molecules25030697 . . . . **60**

Sofia C. Vardali, Natalia Manousi, Mariusz Barczak and Dimitrios A. Giannakoudakis
Novel Approaches Utilizing Metal-Organic Framework Composites for the Extraction of Organic Compounds and Metal Traces from Fish and Seafood
Reprinted from: *Molecules* **2020**, *25*, 513, doi:10.3390/molecules25030513 . . . . **68**

Dimitrios Giliopoulos, Alexandra Zamboulis, Dimitrios Giannakoudakis, Dimitrios Bikiaris and Konstantinos Triantafyllidis
Polymer/Metal Organic Framework (MOF) Nanocomposites for Biomedical Applications
Reprinted from: *Molecules* **2020**, *25*, 185, doi:10.3390/molecules25010185 . . . . **95**

Natalia Manousi, Dimitrios A. Giannakoudakis, Erwin Rosenberg and George A. Zachariadis
Extraction of Metal Ions with Metal–Organic Frameworks
Reprinted from: *Molecules* **2019**, *24*, 4605, doi:10.3390/molecules24244605 . . . . **123**

**Zoi-Christina Kampouraki, Dimitrios A. Giannakoudakis, Vaishakh Nair,
Ahmad Hosseini-Bandegharaei, Juan Carlos Colmenares and Eleni A. Deliyanni**
Metal Organic Frameworks as Desulfurization Adsorbents of DBT and 4,6-DMDBT from Fuels
Reprinted from: *Molecules* **2019**, *24*, 4525, doi:10.3390/molecules24244525 . . . . **144**

Natalia Manousi, George A. Zachariadis, Eleni A. Deliyanni and Victoria F. Samanidou
Applications of Metal-Organic Frameworks in Food Sample Preparation
Reprinted from: *Molecules* **2018**, *23*, 2896, doi:10.3390/molecules23112896 . . . . **166**

About the Special Issue Editors

Victoria Samanidou Interests: analytical chemistry; sample preparation; separations; HPLC; extraction techniques; development and optimization of methodology for sample preparation of various samples, e.g., food, biological fluids, etc., in terms of selective extraction of analytes; using modern sample pre-treatment techniques such as solid phase extraction, matrix solid phase dispersion, membranes, sonication, microwaves etc.; study of new chromatographic materials used in separation and sample preparation (polymeric sorbents, monoliths, carbon nanotubes, fused core particles, etc.) compared with conventional materials; application of HPLC in the analysis of different samples such as food, biological fluids, pharmaceuticals, environmental, forensics, etc.; application of ion chromatography in environmental pollution elimination.

Eleni Deliyanni Interests: materials chemistry; modification/impregnation of materials; synthesis and surface characterization of new adsorbent materials; carbonaceous materials/activated carbons/graphene oxide/graphene; graphene oxide based/polymer nanocomposite adsorbents; biomass conversion to activated carbon; adsorption/separation processes in environmental applications; activated carbons as adsorbents; advanced oxidation processes/catalytic oxidation; carbonaceous materials as metal-free catalysts; deep desulfurization of fuels.

Preface to "Metal Organic Frameworks"

The concept of metal–organic frameworks (MOFs) was first introduced in 1990 and they are nowadays among the most promising novel materials. MOEs belong to a new class of crystalline materials that consist of coordination bonds between metal clusters (e.g., metal carboxylate clusters and metal azolate clusters), metal atoms, or rod-shaped clusters, and multidentate organic linkers that contain oxygen or nitrogen donors (like carboxylates, azoles, nitriles, etc.), thus forming a three-dimensional structure. The properties of both metal ions and linkers determine the physical, structural, and morphological features of MOFs' networks (e.g., porosity, pore size, and pore surface). Additionally, the above-mentioned as well as the chemical features of the prepared frameworks can be controlled by a solvent system, pH, metal–ligand ratio, and temperature. Although MOFs were actually initially used in catalysis, for gas storage, separation, membranes, or electrochemical sensors, they were later introduced as solid phase extraction (SPE) sorbents. Initially, they were applied for polycyclic aromatic hydrocarbons (PAHs) in environmental water samples; subsequently, the range of applications was expanded to the field of analytical chemistry, both in chromatographic separation and sample preparation, with success, e.g., in SPE and solid phase microextraction (SPME).

Since then, the number of analytical applications implementing MOFs as sorbents in sample preparation approaches has increased, reinforcing that, at least theoretically, an infinite number of structures can be designed and synthesized, thus making tuneability one of the most unique characteristics of MOF materials. They have been designed in various shapes, such as columns, fibers, and films, so that they can be used to address more analytical challenges with improved analytical features. Going a step further, the design and synthesis of advantageous composites or the controllable incorporation of defects were revealed to be promising strategies that positively impact the desirable features and their stability and reusability. MOFs' exceptional properties attracted the interest of analytical chemists who have taken advantage of the unique structures and features, and have already introduced them into several sample pretreatment techniques, such as solid phase extraction, dispersive SPE, magnetic solid phase extraction, solid phase microextraction, stir bar adsorptive extraction, etc.

This Special Issue presents the recent developments in the synthesis and applications of MOFs. The outcomes are impressive as 10 manuscripts illustrate the impact of MOFs as useful tools in various fields like analytic methods, biofuels desulfurization, CO_2 capture and more. One communication report, four original research articles, and five comprehensive reviews are the contributions from research groups located in Greece, United States of America, Austria, Spain, Poland, Iran, India, and China. The Guest Editors wish to thank all authors for their contributions and hope that the readers will find all information provided in this Special Issue interesting and helpful.

Victoria Samanidou, Eleni Deliyanni
Special Issue Editors

Editorial

Metal Organic Frameworks: Synthesis and Application

Victoria F. Samanidou [1,*] and Eleni A. Deliyanni [2,*]

1. Laboratory of Analytical Chemistry, Department of Chemistry, Aristotle University of Thessaloniki; GR-54124 Thessaloniki, Greece
2. Laboratory of Chemical and Environmental Technology, Department of Chemistry, Aristotle University of Thessaloniki, GR-54124 Thessaloniki, Greece
* Correspondence: samanidu@chem.auth.gr (V.F.S.); lenadj@chem.auth.gr (E.A.D.); Tel.: +302310997698 (V.F.S.); +302310997808 (E.A.D.); Fax: +302310997719 (V.F.S.)

Received: 13 February 2020; Accepted: 14 February 2020; Published: 20 February 2020

The concept of metal–organic frameworks (MOFs) was first introduced in 1990; nowadays they are among the most promising novel materials. MOFs belong to a new class of crystalline materials that consist of coordination bonds between metal clusters (e.g., metal-carboxylate clusters and metal-azolate clusters), metal atoms, or rod-shaped clusters and multidentate organic linkers that contain oxygen or nitrogen donors (like carboxylates, azoles, nitriles, etc.); thus, a three-dimensional structure is formed [1].

The properties of both metal ions and linkers determine the physical, structural, and morphological features of MOF networks (e.g., porosity, pore size, and pore surface). Additionally, the aforementioned as well as the chemical features of the prepared frameworks can be controlled by the solvent system, pH, metal-ligand ratio, and temperature [1].

Although MOFs were initially used in catalysis, gas storage and separation, membranes, or electrochemical sensors, they were later introduced as SPE (Solid Phase Extraction) sorbents. Initially they were applied for PAHs (Polycyclic Aromatic Hydrocarbons) in environmental water samples, but subsequently, the range of applications was expanded to the field of analytical chemistry, both in chromatographic separation and sample preparation, with great success in, e.g., SPE and SPME (Solid Phase Micro-extraction). Since then, the number of analytical applications implementing MOFs as sorbents in sample preparation approaches has increased. This is reinforced by the fact that, at least theoretically, an infinite number of structures can be designed and synthesized, thus making tuneability one of the most unique characteristics of MOF materials. Moreover, they have been designed in various shapes, such as columns, fibers, and films, so that they can meet more analytical challenges with improved analytical features. Going a step further, the design and synthesis of advantageous composites or the controllable incorporation of defects has been shown to be a promising strategy with a positive impact on the desirable features and on stability/reusability [1].

The exceptional properties of MOFs have attracted the interest of analytical chemists who have taken advantage of their unique structures and features, and have already introduced them in several sample pretreatment techniques, such as solid phase extraction, dispersive SPE, magnetic solid phase extraction, solid phase microextraction, stir bar adsorptive extraction, etc. [1].

This Special Issue aims to present the recent developments in the synthesis and applications of MOFs.

The outcome is very impressive; ten manuscripts illustrate the impact of MOFs as useful tools in various fields like analytic methods, biofuels desulfurization, CO_2 capture, and more. Research groups located in Greece, United Stated of America, Austria, Spain, Poland, Iran, India, and China have contributed one communication report, four original research articles, and five comprehensive reviews [1–10].

Yazdanparast et al. present an unusual, noncatenated, large pore, pillared paddle-wheel MOF, providing an additional datapoint to support current postulation on the factors that may influence catenation in these frameworks. This information will be useful to MOF chemists who are interested in the well-defined multifunctionality of these materials.

In their research article, Andriotou et al. report on luminescence color tuning in a lanthanide metal-organic framework (LnMOF) ([La(bpdc)Cl(DMF)] (1); bpdc^{2-} = [1,1'-biphenyl]-4,4'-dicarboxylate, DMF = N,N-dimethylformamide) by introducing dual emission properties in a La^{3+} MOF scaffold through doping with the blue fluorescent 2,2'-diamino-[1,1'-biphenyl]-4,4'-dicarboxylate (dabpdc^{2-}) and the red emissive Eu^{3+}.

Giannakoudakis. and Bandosz in their research article, describe the building of MOF nanocomposites with oxidized graphitic carbon nitride nanospheres. A composite of the two most studied MOFs, i.e., copper-based Cu-BTC (HKUST-1) and zirconium-based Zr-BDC (UiO-66), with oxidized graphitic carbon nitride nanospheres was designed, synthesized, and characterized. The role of oxidized g-C_3N_4 during the synthesis of the composite was found to be different, depending on the geometry of the framework. In the case of the UiO-66-based composite, spherical particles were obtained after the growth of the framework around the oxidized and spherical g-C_3N_4 nanoparticles. For the HKUST-1-based composite, the growth of the octahedral framework units experienced geometrical constraints, resulting in more defects and the creation of mesoporosity. The formation of the composite upon the incorporation of the nanospheres led to differences in the amounts of the adsorbed CO_2.

Liang. et al. describe the facile preparation of a metal-organic framework (MIL-125)/chitosan beads for the adsorption of Pb(II) from aqueous solutions. In their research work, a novel composite of a titanium-based, metal-organic framework (MOF) with chitosan beads was synthesized following a template-free solvothermal approach under ambient conditions; the resulting composite presented a higher remediation capability compared to pure MOF.

González-Rodríguez et al. propose the mixed functionalization of organic ligands in UiO-66. Their study is intended to prepare and characterize UiO-66 derivatives incorporating different contents of nonfunctionalized and functionalized-organic ligands, including -NH_2 and -NO_2 groups, in the MOF structure through the mixed-linker approach. As a second goal, the paper evaluates the influence of such modifications on the resulting material when used as a sorbent in a D-µSPE method for different target analytes in water. The selected analytes presented a low to high size (to evaluate their influence when entering or not entering the pores of the MOF), while incorporating or not incorporating polar groups in their structures (to evaluate possible interactions between MOF pore functionalities and analyte groups).

Vardali et al. illustrated some novel approaches utilizing metal-organic framework composites for the extraction of organic compounds and metal traces from fish and seafood. The authors discuss the applications of MOFs and their composites/hybrids as potential media for the extraction, detection, or sensing of organic and inorganic pollutants from fish samples, prior to their determination using an instrumental technique. Emphasis is given to the extraction of antibiotics as well as metals from fish tissue, since they are considered significant contaminants in the marine environment.

In their review, Giliopoulos et al. examine the various types of polymer/MOF nanocomposites used in biomedical applications, and more specifically in drug delivery and imaging. They focus on the different approaches followed to produce the composites, and discuss their findings regarding the behavior of the composites in each application.

Manousi et al. provide a comprehensive review of the extraction of metal ions with MOFs. The authors discuss the applications of MOFs as potential sorbents for the extraction of metal ions prior to their determination from environmental, biological, and food samples. The application of subfamilies of MOFs, such as zeolitic imidazole frameworks (ZIFs) or covalent organic frameworks (COFs), is also discussed.

Kampouraki et al., describe the use of MOFs as desulfurization adsorbents of DBT and 4,6-DMDBT from fuels. In their review, applications of MOFs and their functionalized composites for adsorptive desulfurization of fuels are presented and discussed, as well as the main desulfurization mechanisms reported for the removal of thiophenic compounds by various frameworks. Prospective methods regarding the further improvement of the desulfurization capabilities of MOFs are also suggested.

Last but not least, Manousi et al. present applications of MOFs in food sample preparation. The authors identify applications of MOFs reported in the literature, including the use of metal-organic compounds and their derived carbons as absorbents in combination with dispersive sample preparation techniques, magnetic sample preparation techniques, in-tube sample preparation techniques, and online sample preparation techniques for the analysis of complex food samples, such as milk, tea and beverages, fruits and vegetables, meat, chicken, fish, etc. [1].

This special issue is accessible through the following link:

https://www.mdpi.com/journal/molecules/special_issues/MOFs

As guest editors for this Special Issue, we would like to thank all the authors and coauthors for their contributions, and all the reviewers for their time and effort in carefully evaluating the manuscripts, making recommendations that significantly improved the quality of original submissions. Last but not least, we would like to acknowledge the editorial office of the *Molecules* journal for their kind assistance in all stages of preparing this Special Issue.

We hope that readers will find the information provided in this Special Issue interesting and helpful.

Funding: This research received no external funding.

Conflicts of Interest: The authors declare no conflict of interest.

References

1. Manousi, N.; Zachariadis, G.; Deliyanni, E.; Samanidou, V. Applications of Metal-Organic Frameworks in Food Sample Preparation. *Molecules* **2018**, *23*, 2896. [CrossRef] [PubMed]
2. Yazdanparast, M.; Day, V.; Gadzikwa, T. Hydrogen-Bonding Linkers Yield a Large-Pore, Non-Catenated, Metal-Organic Framework with pcu Topology. *Molecules* **2020**, *25*, 697. [CrossRef] [PubMed]
3. Andriotou, D.; Diamantis, S.; Zacharia, A.; Itskos, G.; Panagiotou, N.; Tasiopoulos, A.; Lazarides, T. Dual Emission in a Ligand and Metal Co-Doped Lanthanide-Organic Framework: Color Tuning and Temperature Dependent Luminescence. *Molecules* **2020**, *25*, 523. [CrossRef] [PubMed]
4. Giannakoudakis, D.; Bandosz, T. Building MOF Nanocomposites with Oxidized Graphitic Carbon Nitride Nanospheres: The Effect of Framework Geometry on the Structural Heterogeneity. *Molecules* **2019**, *24*, 4529. [CrossRef] [PubMed]
5. Liang, X.; Wang, N.; Qu, Y.; Yang, L.; Wang, Y.; Ouyang, X. Facile Preparation of Metal-Organic Framework (MIL-125)/Chitosan Beads for Adsorption of Pb(II) from Aqueous Solutions. *Molecules* **2018**, *23*, 1524. [CrossRef] [PubMed]
6. González-Rodríguez, G.; Taima-Mancera, I.; Lago, A.; Ayala, J.; Pasán, J.; Pino, V. Mixed Functionalization of Organic Ligands in UiO-66: A Tool to Design Metal–Organic Frameworks for Tailored Microextraction. *Molecules* **2019**, *24*, 3656. [CrossRef] [PubMed]
7. Vardali, S.; Manousi, N.; Barczak, M.; Giannakoudakis, D. Novel Approaches Utilizing Metal-Organic Framework Composites for the Extraction of Organic Compounds and Metal Traces from Fish and Seafood. *Molecules* **2020**, *25*, 513. [CrossRef] [PubMed]
8. Giliopoulos, D.; Zamboulis, A.; Giannakoudakis, D.; Bikiaris, D.; Triantafyllidis, K. Polymer/Metal Organic Framework (MOF) Nanocomposites for Biomedical Applications. *Molecules* **2020**, *25*, 185. [CrossRef] [PubMed]
9. Manousi, N.; Giannakoudakis, D.; Rosenberg, E.; Zachariadis, G. Extraction of Metal Ions with Metal–Organic Frameworks. *Molecules* **2019**, *24*, 4605. [CrossRef] [PubMed]
10. Kampouraki, Z.; Giannakoudakis, D.; Nair, V.; Hosseini-Bandegharaei, A.; Colmenares, J.; Deliyanni, E. Metal Organic Frameworks as Desulfurization Adsorbents of DBT and 4,6-DMDBT from Fuels. *Molecules* **2019**, *24*, 4525. [CrossRef] [PubMed]

© 2020 by the authors. Licensee MDPI, Basel, Switzerland. This article is an open access article distributed under the terms and conditions of the Creative Commons Attribution (CC BY) license (http://creativecommons.org/licenses/by/4.0/).

Article

Building MOF Nanocomposites with Oxidized Graphitic Carbon Nitride Nanospheres: The Effect of Framework Geometry on the Structural Heterogeneity

Dimitrios A. Giannakoudakis [1,2] and Teresa J. Bandosz [1,*]

1. Department of Chemistry and Biochemistry, The City College of New York, New York, NY 10031, USA; DAGchem@gmail.com
2. Institute of Physical Chemistry, Polish Academy of Sciences, Kasprzaka 44/52, 01-224 Warsaw, Poland
* Correspondence: tbandosz@ccny.cuny.edu

Academic Editors: Victoria Samanidou, Eleni Deliyanni and Rafael Lucena
Received: 3 November 2019; Accepted: 10 December 2019; Published: 11 December 2019

Abstract: Composite of two MOFs, copper-based Cu-BTC (HKUST-1) and zirconium-based Zr-BDC (UiO-66), with oxidized graphitic carbon nitride nanospheres were synthesized. For comparison, pure MOFs were also obtained. The surface features were analyzed using x-ray diffraction (XRD), sorption of nitrogen, thermal analysis, and scanning electron microscopy (SEM). The incorporation of oxidized g-C_3N_4 to the Cu-BTC framework caused the formation of a heterogeneous material of a hierarchical pores structure, but a decreased surface area when compared to that of the parent MOF. In the case of UiO-66, functionalized nanospheres were acting as seeds around which the crystals grew. Even though the MOF phases were detected in both materials, the porosity analysis indicated that in the case of Cu-BTC, a collapsed MOF/nonporous and amorphous matter was also present and the MOF phase was more defectous than that in the case of UiO-66. The results suggested different roles of oxidized g-C_3N_4 during the composite synthesis, depending on the MOF geometry. While spherical units of UiO-66 grew undisturbed around oxidized and spherical g-C_3N_4, octahedral Cu-BTC units experienced geometrical constraints, leading to more defects, a disturbed growth of the MOF phase, and to the formation of mesopores at the contacts between the spheres and MOF units. The differences in the amounts of CO_2 adsorbed between the MOFs and the composites confirm the proposed role of oxidized g-C_3N_4 in the composite formation.

Keywords: metal organic framework composites; oxidized graphitic carbon nitride nanoparticles; porosity; structural heterogeneity

---

1. Introduction

Highly porous metal–organic frameworks (MOFs) are synthesized by the self-assembly of metal ions or clusters of them (as coordination centers) with polyatomic organic bridging linkages. In this process, 3D microporous structures are formed [1–3]. The diversity of the metal centers and organic ligands leads to materials of particular crystallographic structure, texture, and chemistry. Due to these properties, MOFs have been tested for various applications such as gas separation/storage [4–9], purification [10–12], sensing [13–16], electrodes for batteries [17,18], microextraction [19,20], detoxification of chemical warfare agents [21–25], and heterogeneous catalysis [26–28].

Even though MOFs can be considered as perfect porous materials of well-described geometry, this "perfection" has been recently found as limiting their performance, especially in separation and catalysis. In many of these applications, the hierarchical pore structure is needed and thus the homogeneity of the MOFs' pore system, mainly related to micropores of specific sizes, can be disadvantageous for mass transfer processes. Moreover, uniformed chemistry, although advantageous for some applications, might limit the number of specific interactions/adsorption or catalytic centers. Therefore, the efforts have been intensified to introduce defects to the MOF structure targeting specific applications. Examples include mixed linkers [28–30], HCl treatment [31,32], variations in the synthesis conditions [33], the addition of molecular guests [34–38] or the incorporation of modified linkers [39,40]. These processes result in crystal imperfection, partial ligand replacement, or in nonbridging ligands, affecting the porosity, and the population, dispersion, and availability of active centers.

The composites of MOFs with graphite oxide (GO) showed an increased pore volume, conductivity, and chemical heterogeneity [41–43]. This trend was an outcome of the reaction of the copper centers of Cu-BTC and the O-containing (epoxy, carboxylic, hydroxyl, and sulfonic) or N-containing functional groups of the 2-D GO phase [42–44]. The oxygen groups of GO were suggested to act either as equatorial or axial linkers, replacing BTC or water molecules, respectively.

Since for building MOF-based composites, the geometry and morphology of the modifier is important, graphitic carbon nitride, g-C_3N_4, has also been used for this purpose. In its unoxidized form, it is an n-type semiconductor with a tunable band gap near 2.7 eV. g-C_3N_4 has a flake-like structure similar to that of graphite with mainly carbon and nitrogen organized in triazine and tri-s-triazine (or s-heptazine) units [45]. g-C_3N_4 was used to form composites with MIL-88A [46] to efficiently separate the photoinduced charge carriers. For its composites with Ti-based MOF [47] (MIL-125(Ti)), an enhanced photo-degradation of Rhodamine B was reported. For the synthesis process leading to true composites and not to physical mixtures, the interactions of a MOF phase and modifier functional groups are important. Thus, owing to these interactions, the composites of Cu-BTC and oxidized g-C_3N_4 had hierarchical porosity and exhibited photoactive properties [23].

Even though structural or chemical defects were not the focus of the synthesis procedure at the time of the introduction of MOF/other phase composites, the published results showed some distortion in the crystal structure, along with an increase in the porosity and in the population of metal centers [40,48]. Therefore, building the MOF composites with another phase can also be considered as a materials' design strategy for introducing some defects to MOF crystals. Since these composites deserve another look at the origin of their surface activity, the objective of this paper was to present the comparison of the surface properties of the composites of two popular MOFs, HKUST-1 or Cu-BTC and UiO-66 with oxidized graphic carbon nitride nanospheres, with emphases on the formation of defects or/and new, physical/textural, optical, and chemical features. Since in both cases the same modifier is used, in the comparison presented, we focus on the geometry of MOF and its effects on the final properties of the composites.

2. Results and Discussion

The synthesized composites of oxidized g-C_3N_4 with Cu-BTC and UiO-66 are referred to as CuBTC-C and UiO66-C, respectively. They contain ~25% and ~10% of the oxidized g-C_3N_4 (gCNox) phase, respectively. In the evaluation of the outcomes of the synthesis of these materials, the analysis of the x-ray diffraction (XRD) patterns is important to assess the MOF structure features, formed in the presence of another phase. XRD patterns of the composites and their parent MOFs are presented in Figure 1. The patterns of Cu-BTC and UiO-66 follow those reported in the literature [49–51]. The preserved MOF structure was found in both composites. While in the case of CuBTC-C, the diffraction peaks were of a lower intensity than those for the parent MOF, the trend was the opposite in the case of UiO66-C. This suggests a different role of the modifier in the crystallization processes. The diffractogram of CuBTC-C indicates that the spherical nanoparticles of oxidized g-C_3N_4 with sizes of 10–50 nm [52] led to variations in the crystallization process, which caused minor changes in the lattice structure and morphology. The x-ray diffraction pattern of gCNox revealed two peaks at 27.6° and 13.5°, related to interplanar stacked graphitic layers [53]. For CuBTC-C, a broad and low intensity peak with a maximum at 26.7° was visible. The peak at 13.5° was not detected due to its overlap with an intense reflection of the framework. In the case of UiO66-C, where only 10% of the modifier was added, the absence of the peaks could be due to either the high dispersion of oxidized g-C_3N_4 or its small content.

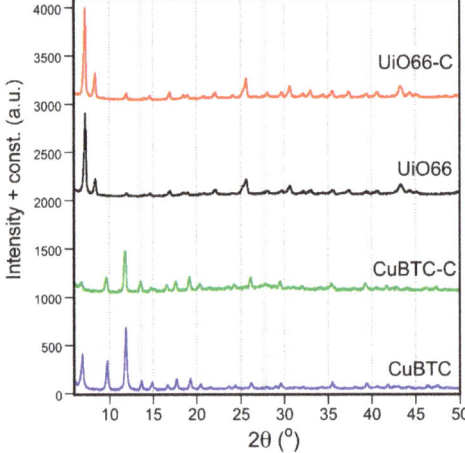

Figure 1. Comparison of x-ray diffraction patterns for MOFs and their composites.

The morphology of the MOFs and their composites is compared in Figure 2. The CuBTC and CuBTC-C had octahedral shaped crystals, typical of this particular MOF. However, the crystals of the composite showed the visible effect of distortion demonstrated in their blunter edges and rough surfaces. That roughness was caused by the spherical nanoparticles, likely oxidized g-C_3N_4 [23], visible also on the crystals' surfaces. In the case of UiO-66, the aggregates of semi-spherical particles with sizes between 90 to 190 nm were visible (Figure 2). For its composite, the aggregates were slightly smaller, and knowing that the sizes of oxidized g-C_3N_4 nanospheres are between 10–50 nm [23,52], it is not possible to determine the chemical homogeneity level of the material based only on the SEM images.

Figure 2. SEM images of CuBTC (**a**), CuBTC-C (**b**), UiO66 (**c**), and UiO66-C (**d**).

Since separation and catalysis are our target applications, the porosity of the synthesized materials was evaluated in detail from measured nitrogen adsorption isotherms (Figure 3a). The differences in the nitrogen uptake and in the shapes of the isotherms for the composites in comparison to those for pure MOFs are related to the alterations in the porous structure, upon the formation of the composites, especially for CuBTC-C. For this sample, the amount of nitrogen adsorbed decreased almost twice in comparison with that on CuBTC and the isotherm suggests the existence of mesopores. On the other hand, for UiO66-C, only small decreases in the amount adsorbed was seen in comparison to that on UiO66.

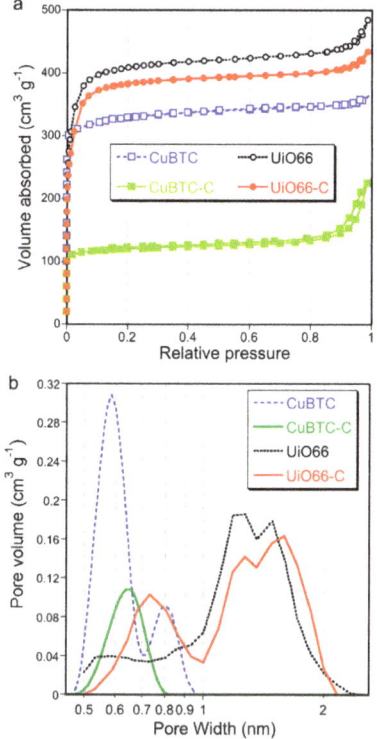

Figure 3. Nitrogen adsorption isotherms (**a**) and pore size distributions (**b**).

The pore size distributions (PSDs) were calculated from the isotherms using Non-Linear Density Functional Theory (NLDFT). Even though a specific kernel for this kind of material does not exist, the comparison of the results obtained for the same group of materials was considered as bringing meaningful information on the trend of textural alterations. The results suggest a more homogeneous distribution of micropores in CuBTC-C than that in CuBTC. The former sample also showed the presence of large pores with sizes between 5–50 nm (predominant 50 nm). The agreement of these pore sizes with the sizes of the oxidized g-C_3N_4 nanospheres suggests that these pores are a consequence of the incorporation of these nanoparticles inside the framework's matrix. For UiO66-C, the formation of more pores with sizes between 0.7–1 nm (increase in their ratio to total pore volume) and the disappearance of small mesopores were the only visible changes in the PSD (Figure 3b).

The comparison of the pore volumes in the range of ultramicro-, supermicro-, and meso-pores for our samples is presented in Figure 4a. Figure 4b collects the percentages of the volumes in each range of the pore sizes per the total pores volume. In the case of CuBTC, the addition of the modifier led to a 50% decrease in the total pore volume. The volumes of the ultramicropores (<0.7 nm) and of the supermicropores (0.7–2 nm) decreased around 60% and 77%, respectively. That marked decrease in the volume of the supermicropores suggests that the nanospheres not only played a significant role in acting as linkers, but they also affected the crystallization/formation of the MOF phase and led to the formation of some amorphous or/and nonporous phases in the composite. Another plausible explanation of the decreased microporosity can be the blockage of the entrance of these pores by the gCNox nanoparticles. On the other hand, the volume of the mesopores in CuBTC-C increased three times when compared to that in CuBTC. The complex role of gCNox in the composite formation was also reflected in the ratio of ultramicro- to supermicro-pores (Figure 4b), which decreased from 0.65 for pure MOF, to 0.36 for the composite. For the UiO66 composite, the additive affected the structural features to a smaller extent and in a different way than in the case of Cu-BTC. The volumes of the ultramicro- and supermicro-pores decreased by 13 and 4%, respectively. The distribution of the PSDs indicated that the addition of nanospheres led to the formation of pores in the range of 0.6 to 0.9 nm. This, along with the same morphology of the composite as that of UiO66 (as seen on SEM images in Figure 2c,d), suggests that the nanospheres acted as nucleation centers, and the new pores were formed at the interface of the nanospheres and the MOF units. It is also interesting that the volume of the mesopores slightly decreased for this composite.

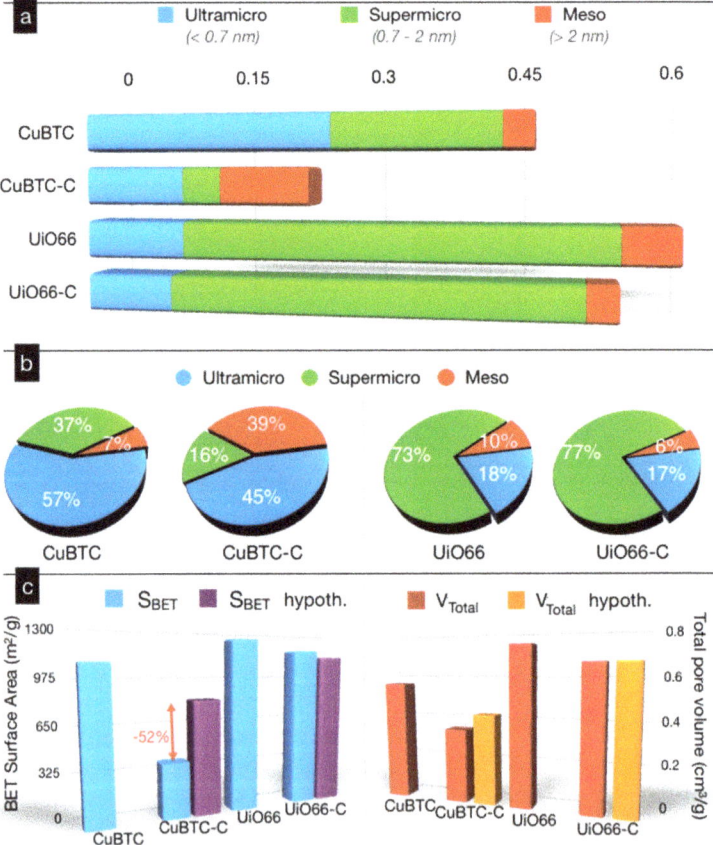

Figure 4. The comparison of the volumes of ultramicro-, supermicro-, and meso-pores (**a**), the percentages of each size range of pores (**b**), and a comparison of the measured and hypothetical (assuming physical mixtures) surface areas (S_{BET}) and total pore volumes (V_{Total}) (**c**).

The extent of the effects of the same modifier addition on the alteration of the pore structure was also analyzed by comparing the measured surface areas and total pore volumes to those calculated for the hypothetical physical mixture (taking into consideration the contents of both phases and their specific contributions to porosity) (Figure 4c). For CuBTC-C, these parameters decreased 52% when compared to the physical mixture, indicating a marked effect of 25 wt.% oxidized g-C_3N_4 on the final porosity. Oxidized g-C_3N_4 is basically not very porous (surface area of 84 m^2/g and the total pore volume of 0.482 cm^3/g [52]) and its addition can contribute to the so called mass dilution effect in the physical mixture. The greater decrease of more than 25% supports a nonporous phase precipitation during composite synthesis and/or blocking of some microporosity of the MOF units by gCNox entities. On the other hand, the surface area of UiO66-C was 4% higher than that of the hypothetical physical mixture due to the formation of new pores, as discussed above.

Thermal analysis experiments were performed in order to evaluate how the changes in the porous structure and chemistry affected the thermal stability of the composites. The thermogravimetric (TG) and derivative thermogravimetric (DTG) curves under a helium atmosphere are collected in Figure 5. It should be mentioned here that the weight loss of gCNox occurs continuously/gradually from room temperature up to complete combustion at 720 °C [23,52]. The thermal decomposition patterns of UiO-66 and UiO66-C are almost identical, suggesting limited chemical interactions of the MOF matrix with the nanospheres. The decomposition of the zirconium-based frameworks is visible as a peak at 520 °C revealed on the DTG curves for both samples. For the composite, the total weight loss was larger than that for the pure MOF due to the decomposition of the gCNox phase. The addition of the gCNox phase also led to a decrease in the affinity to retain water/decrease in hydrophilicity when compared to UiO66. In the case of CuBTC-C, the weight loss pattern revealed more pronounced differences in comparison to that for CuBTC, suggesting chemical heterogeneity and the involvement of nanospheres as linkers [54]. This is supported by the weight loss in the range from 160 to 260 °C, revealed only for the composite. The decomposition of CuBTC occurred between 310 and 370 °C and is seen as a peak on the DTG curve with a maximum at 340 °C. For the composite, the decomposition of the MOF phase started at a slightly higher temperature.

Since g-C_3N_4 is photoactive, its effect on the optical features of the composites was also evaluated. Defuse reflectance UV–Vis–IR spectra are collected in Figure 6. The coordination of the BTC ligands with the copper centers can occur in two planar symmetric bonding directions and in an axial direction [23,55]. For CuBTC-C, the latter coordination did not take place since its absorption spectrum did not show the characteristic absorption in the range from 450 to 530 nm [23]. The lack of this feature supports that the nanospheres acted as linkers and introduced a distortion of the ideal octahedral square grid due to π–π interactions with the BTC units [55]. Some alteration of the optical features was also observed in the case of UiO66-C. The broad absorption in the lower range of the visible range of light, revealed for UiO66, disappeared for the composite. For UiO66, absorption occurs in the ultraviolent range, up to 315 nm (~4 eV). Taddei and co-workers reported the band gap of this MOF as 4.1 eV (302 nm) [28] and showed that the defect engineering of UIO-66 based on modulated synthesis or post-synthetic linker exchange led to a decrease in the optical band gap. In the case of UiO66-C, the light absorption starting at 400 nm (3.1 eV) supports the decrease in the band gap compared to the pure UiO66.

The CO_2 adsorption isotherms measured on our materials are presented in Figure 7a. The comparison of the amounts adsorbed at 1 atm and at 25 °C (expressed as mg/g) is included in Figure 7b. In the case of UiO-66-C, a 13% increase in the amount of CO_2 adsorbed compared to that on MOF is linked to the formation of the modifier/MOF units' interface providing small pores where CO_2 could be adsorbed. UiO-66 is not expected to interact specifically with carbon dioxide molecules and Cao et al. presented a similar CO_2 adsorption capability for UiO-66 [56] without the loss of adsorption even after five cycles. The CO_2 adsorption results revealed an opposite trend in the case of copper frameworks, since the composite showed an 8% smaller uptake. Considering that the composite consisting of 25% gCNox adsorbed a limited amount of CO_2, the addition of gCNox is beneficial for CO_2 adsorption in the MOF phase. It is linked to the high level of defects in the latter and thus there is a higher availability of open copper centers for interaction with CO_2 molecules [40,49]. On CuBTC-C and UiO66-C, 16% and 19% more CO_2, respectively, is adsorbed than on the hypothetical mixtures of the components. The mechanisms of the CO_2 adsorption on both MOFs support that in the case of CuBTC-C, mainly chemical/structural defects are responsible for the observed trend while those in UiO66-C are due to the development of small pores on the modifier/MOF unit interface. The comparison of the quantities of CO_2 adsorbed on the materials tested are presented in Table 1. When the amount adsorbed is recalculated per units amount of the MOF phase, the amount adsorbed in the composites was about 25% higher than those on pure MOF. This effect is especially visible when the amount adsorbed per units of total pore volume of the adsorbent is compared. In such cases, CuBTC-C adsorbs 78% more CO_2 than CuBTC.

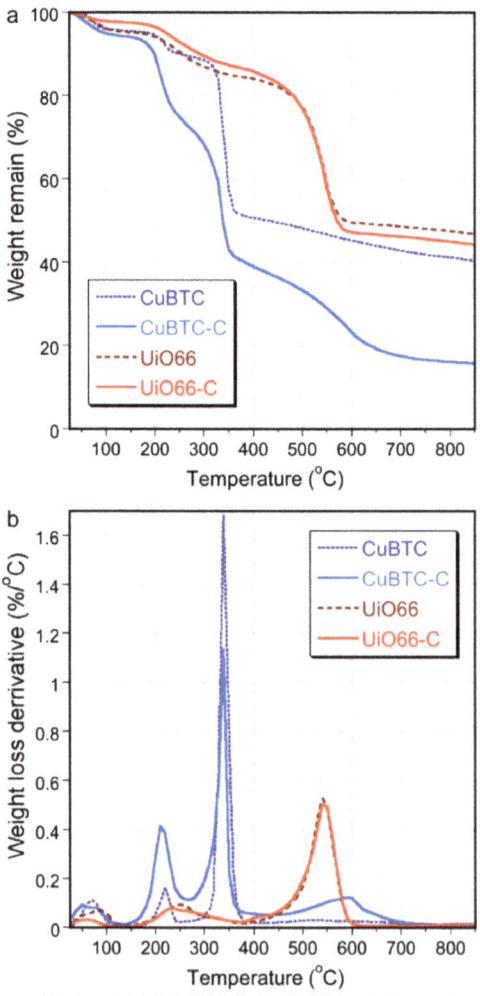

Figure 5. TG (a) and DTG (b) curves for the pure MOFs and their composites (measured in helium).

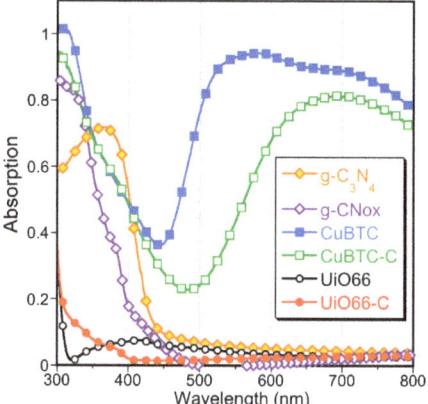

Figure 6. UV–Vis–NIR absorption spectra of the materials.

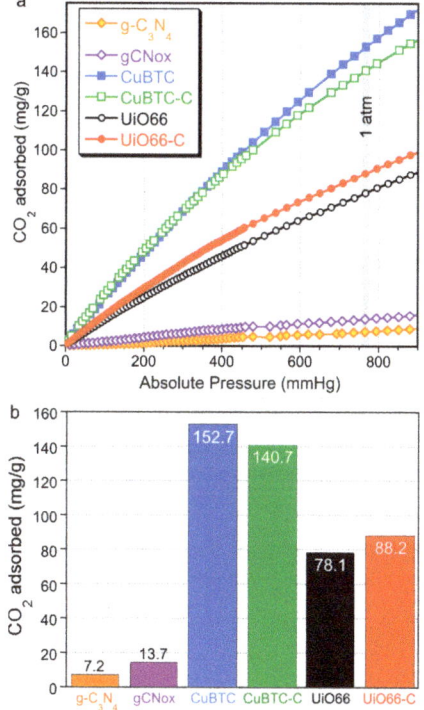

Figure 7. The CO_2 adsorption isotherms (**a**) and the mg of CO_2 adsorbed per gram of the materials at 1 atm (**b**).

Table 1. Comparison of quantities of CO_2 adsorbed on the metal organic frameworks and their composites.

Quantity Adsorbed	CuBTC	CuBTC-C	UiO66	UiO66-C
mg/g (as in bars Figure)	152.7	140.7 (−8%)	78.1	88.2 (+13%)
mg/g of MOF phase	152.7	187.6 (+23%)	78.1	98.0 (+25%)
mg/cm^3 of total pore volume	330	588 (+78%)	130	162 (+25%)

The marked differences in the surface heterogeneity levels between the two composites with the same modifier but with different MOF (which also naturally must include the defects in the MOF structure in the broad sense of this word) are likely to be caused by the differences in the MOF structure geometry. While Cu-BTC is considered as having a simple cubic geometry, UiO-66 is more complex, both in its chemistry and the crystal structure. These differences might lead to the distinct levels of compatibility with the geometry and chemistry of the spherical modifier. It has been previously found that forming the composites of enhanced porosity such as those of MOF and 2-D graphite oxide (GO), besides the presence of functional groups that work as linkers [42,43,57,58], some compatibility of the MOF geometry and that of a modifier is required [57]. This is the case of CuBTC, whose units could align parallel to the flat surface of GO, leading to a significant increase in the porosity. The opposite effect was reported for MIL-125 (Ti-benzenedicarboxylate) [47], whose geometry prevented the porous composite formation [57]. Following this line of reasoning, the opposite effects are expected in the case of spherical modifiers. Thus, in the case of UiO-66, the nanospheres are considered as seeds around which, with the involvement of their functional groups, MOF crystals grow. This could explain a lack of clear distinction of oxidized g-C_3N_4 nanospheres in the SEM images and the small increase in the volume of ultramicropores of specific sizes. These pores likely represent the MOF/modifier interface. In the case of Cu-BTC, 25% of the geometrically incompatible spherical modifier not only did not contribute efficiently to the growth of the interface porosity/defects, but probably disturbed the yield of porous Cu-BTC units. The gCNox presence in the composite brought the mesoporosity formed between the units of MOF and modifier, but decreased the microporosity by hindering the MOF growing process. The visualization of these effects on the structure of the composites is presented in Figure 8.

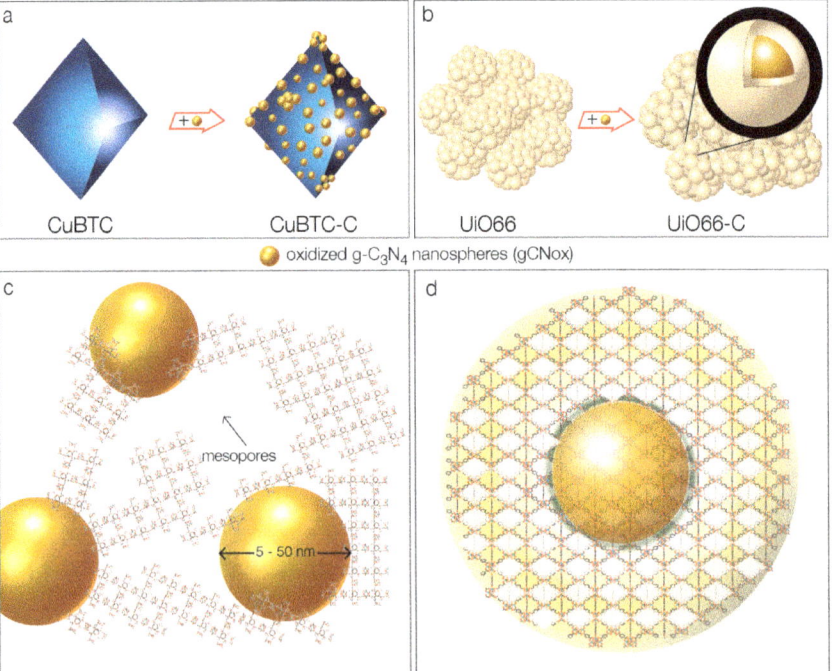

Figure 8. Visualization of the composites' formation processes for CuBTC (**a,c**) and UiO66 (**b,d**).

3. Conclusions

The differences in the surface features of the composites of two distinctive MOFs, Cu-BTC (or HKUST-1), and UiO-66 (Zr-BDC), with the same modifier, the oxidized g-C_3N_4 nanospheres, indicate the importance of the geometrical compatibility between both composite constituents for the full utilization/development of interlayer space. Since the functional groups of the modifier are expected to work as linkers for the MOF units, those units have to be able to find the anchoring points/groups that will not be an obstacle to crystal growth, and this is apparently the case of the composite with the UiO-66. In the case of CuBTC, its crystals could not grow undisturbed on the spherical surface of the modifiers and this led to a significant obstruction in the MOF formation process. Nevertheless, some MOF units were formed, and they coexisted with the spheres of the modifier, resulting in the development of mesoporosity and hierarchical pore structure beneficial for mass transfer process. That disturbance in the MOF formation process led to the availability of more open metal sites, increasing CO_2 adsorption on the composite per both unit mass and unit volume in the final materials.

Even though the detailed description of the defects formed in the composites is beyond the scope of this paper, we have shown that the composite formation by using nanoparticles as MOF indeed introduced structural and thus chemical surface heterogeneity that could enrich the application of these kinds of materials. We foresee that this approach can lead to tuning the structural, morphological, physico-chemical, or photochemical properties of the frameworks, bringing simultaneously new features of the unique MOF–modifier interfaces.

4. Experimental

4.1. Materials

The Hummers method was followed for the synthesis of oxidized graphitic carbon nitride nanoparticles (gCNox), starting with graphitic carbon nitride (g-C_3N_4) as the precursor [52,59]. The latter was obtained by the thermal treatment in air of dicyandiamide (Sigma-Aldrich) at 550 °C for 4 h in a horizontal furnace [60,61]. Details regarding the synthetic process of gCNox can be seen elsewhere [52].

The Cu-based MOFs were obtained following the synthetic protocol reported by Millward et al. [62]. For the composite, the targeted presentence of gCNox at the final material's mass was 25 wt%. For the homogeneous dispersion/mixing, 5 min of mechanical stirring (600 rpm) and 30 min of sonication were performed after the addition of gCNox in the Cu-BTC precursor solutions. The remaining steps for the synthesis of the frameworks can be seen elsewhere [23,42]. The pure MOF is referred to as CuBTC and its composite with gCNox as CuBTC-C.

The Zr-based MOFs were obtained following a scaled-up synthetic process reported by Farha and Hupp with some minimal alterations [51]. In a glass reaction vessel were placed 30 mL of dimethylformamide (DMF), 6 mL of concentrated HCl, and 750 mg of $ZrCl_4$. For complete dissolvement, 20 min of sonication was performed in an ultrasonication bath. The linker (terephthalic acid (BDC), 738 g) was dispersed in 20 mL DMF, and after 5 min of sonication, was inserted in the glass reaction vessel. The latter was sonicated for 20 min and afterward sealed hermetically and placed for 16 h in a furnace at 80 °C. After filtration and washing with DMF and ethanol, the received white powder was dried in a vacuum oven for 12 h (135 °C and 660 Torr) and the yield was found 83.2%. The composite was synthesized in the same way by adding 90 mg of gCNox (targeting a 10 wt% considering the yield) in together with the $ZrCl_4$. The pure MOF is referred to as UiO66, while the composite is UiO66-C. The obtained dried powders were activated in high vacuum (10^{-4} Torr) at 150 °C (using ASAP 2020, Micromeritics) and were kept in hermetically closed vials prior the use [63].

4.2. Methods

The x-ray diffraction patterns were collected from 6 to 50 °C 2θ on a Philips Pert x-ray diffractometer (CuK$_\alpha$ radiation at 40 mA and 40 kV). A Zeiss Supra 55 VP microscope, equipped with a backscatter electron detector (acceleration voltage of 5 keV), was used to collect the SEM images. Nitrogen adsorption/desorption isotherms were measured at −196 °C on an ASAP 2020 (Micromeritic). The samples were outgassed at 120 °C for 16 h. From the nitrogen isotherms, the specific surface areas (S$_{BET}$) were calculated using the Brunauer–Emmet–Teller method. The total pore volume (V$_{Total}$) was evaluated based on the amount of nitrogen adsorbed at a relative pressure of ~0.99. The Non-Local Density Functional Theory (NLDFT) method was applied to calculate the pore size distributions (PSD) and the volume of ultramicropores (<0.7 nm), supermicropores (0.7–2 nm), and mesopores (>2 nm) [64–66]. Using the same DFT kernel for all samples allowed us to establish the trend in the PSDs. An SDT Q600 (TA instruments) thermal analyzer was used to measure the thermogravimetric (TG) curves from which derivative thermogravimetric (DTG) curves were obtained. The experiments were run in helium from room temperature to 1000 °C at a heating rate of 10 °C/min. Defuse reflectance (DR) UV–Vis–NIR spectroscopy was performed by using a spectrophotometer (Jasco V-570) equipped with an integrating sphere using Spectralon [poly(tetra-fluoroethylene)] as the baseline [67,68]. CO_2 adsorption isotherms were measured using an ASAP 2020 (Micromeritics) under low pressure (0–900 mmHg). The experiments were performed at a constant temperature of 25 °C by immersing the tube inside a water bath in which water was circulated.

Author Contributions: D.A.G. contributed to conceptualization, sample syntheses, experimental analyses, data interpretation, and writing the manuscript. T.J.B. established and contributed to conceptualization, data analysis and interpretation, and writing the manuscript.

Funding: This research received no external funding.

Conflicts of Interest: The authors declare no conflict of interest.

References

1. Furukawa, H.; Cordova, K.E.; O'Keeffe, M.; Yaghi, O.M. The Chemistry and Applications of Metal-Organic Frameworks. *Science* **2010**, *9*, 1230444. [CrossRef]
2. Jiao, L.; Seow, J.Y.R.; Skinner, W.S.; Wang, Z.U.; Jiang, H.L. Metal–organic frameworks: Structures and functional applications. *Mater. Today* **2019**, *27*, 43–68. [CrossRef]
3. Zhou, H.C.; Long, J.R.; Yaghi, O.M. Introduction to metal-organic frameworks. *Chem. Rev.* **2012**, *112*, 673–674. [CrossRef] [PubMed]
4. Oh, H.; Savchenko, I.; Mavrandonakis, A.; Heine, T.; Hirscher, M. Highly Effective Hydrogen Isotope Separation in Nanoporous Metal–Organic Frameworks with Open Metal Sites: Direct Measurement and Theoretical Analysis. *ACS Nano* **2014**, *8*, 761–770. [CrossRef] [PubMed]
5. Manousi, N.; Zachariadis, G.A.; Deliyanni, E.A.; Samanidou, V.F. Applications of metal-organic frameworks in food sample preparation. *Molecules* **2018**, *23*, 2896. [CrossRef] [PubMed]
6. Krause, S.; Bon, V.; Senkovska, I.; Stoeck, U.; Wallacher, D.; Többens, D.M.; Zander, S.; Pillai, R.S.; Maurin, G.; Coudert, F.-X.; et al. A pressure-amplifying framework material with negative gas adsorption transitions. *Nature* **2016**, *532*, 348–352. [CrossRef]
7. Senkovska, I.; Kaskel, S. High pressure methane adsorption in the metal-organic frameworks Cu3(btc)2, Zn2(bdc)2dabco, and Cr3F(H2O)2O(bdc)3. *Microporous Mesoporous Mater.* **2008**, *112*, 108–115. [CrossRef]
8. Van Assche, T.R.C.; Duerinck, T.; Van Der Perre, S.; Baron, G.V.; Denayer, J.F.M. Prediction of molecular separation of polar-apolar mixtures on heterogeneous metal-organic frameworks: HKUST-1. *Langmuir* **2014**, *30*, 7878–7883. [CrossRef]
9. Spanopoulos, I.; Tsangarakis, C.; Klontzas, E.; Tylianakis, E.; Froudakis, G.; Adil, K.; Belmabkhout, Y.; Eddaoudi, M.; Trikalitis, P.N. Reticular Synthesis of HKUST-like tbo-MOFs with Enhanced CH4 Storage. *J. Am. Chem. Soc.* **2016**, *138*, 1568–1574. [CrossRef]

10. Liang, X.X.; Wang, N.; Qu, Y.L.; Yang, L.Y.; Wang, Y.G.; Ouyang, X.K. Facile preparation of metal-organic framework (MIL-125)/chitosan beads for adsorption of Pb(II) from aqueous solutions. *Molecules* **2018**, *23*, 1524. [CrossRef]
11. Barea, E.; Montoro, C.; Navarro, J.A.R. Toxic gas removal—Metal-organic frameworks for the capture and degradation of toxic gases and vapours. *Chem. Soc. Rev.* **2014**, *43*, 5419–5430. [CrossRef] [PubMed]
12. Vellingiri, K.; Deng, Y.X.; Kim, K.H.; Jiang, J.J.; Kim, T.; Shang, J.; Ahn, W.S.; Kukkar, D.; Boukhvalov, D.W. Amine-Functionalized Metal-Organic Frameworks and Covalent Organic Polymers as Potential Sorbents for Removal of Formaldehyde in Aqueous Phase: Experimental Versus Theoretical Study. *ACS Appl. Mater. Interfaces* **2019**, *11*, 1426–1439. [CrossRef] [PubMed]
13. Travlou, N.A.; Singh, K.; Rodríguez-Castellón, E.; Bandosz, T.J. Cu–BTC MOF–graphene-based hybrid materials as low concentration ammonia sensors. *J. Mater. Chem. A* **2015**, *3*, 11417–11429. [CrossRef]
14. Raza, W.; Kukkar, D.; Saulat, H.; Raza, N.; Azam, M.; Mehmood, A.; Kim, K.-H. Metal-organic frameworks as an emerging tool for sensing various targets in aqueous and biological media. *Trends Anal. Chem.* **2019**, *120*, 115654. [CrossRef]
15. Vikrant, K.; Tsang, D.C.W.; Raza, N.; Giri, B.S.; Kukkar, D.; Kim, K.H. Potential Utility of Metal-Organic Framework-Based Platform for Sensing Pesticides. *ACS Appl. Mater. Interfaces* **2018**, *10*, 8797–8817. [CrossRef]
16. Drache, F.; Bon, V.; Senkovska, I.; Adam, M.; Eychmuller, A.; Kaskel, S. Vapochromic Luminescence of a Zirconium-Based Metal-Organic Framework for Sensing Applications. *Eur. J. Inorg. Chem.* **2016**, *2016*, 4483–4489. [CrossRef]
17. Shrivastav, V.; Sundriyal, S.; Goel, P.; Kaur, H.; Tuteja, S.K.; Vikrant, K.; Kim, K.H.; Tiwari, U.K.; Deep, A. Metal-organic frameworks (MOFs) and their composites as electrodes for lithium battery applications: Novel means for alternative energy storage. *Coord. Chem. Rev.* **2019**, *393*, 48–78. [CrossRef]
18. Zhang, H.; Nai, J.; Yu, L.; Lou, X.W. (David) Metal-Organic-Framework-Based Materials as Platforms for Renewable Energy and Environmental Applications. *Joule* **2017**, *1*, 77–107. [CrossRef]
19. González-Rodríguez, G.; Taima-Mancera, I.; Lago, A.B.; Ayala, J.H.; Pasán, J.; Pino, V. Mixed functionalization of organic ligands in UiO-66: A tool to design metal-organic frameworks for tailored microextraction. *Molecules* **2019**, *24*, 3656. [CrossRef]
20. Hashemi, B.; Zohrabi, P.; Raza, N.; Kim, K.H. Metal-organic frameworks as advanced sorbents for the extraction and determination of pollutants from environmental, biological, and food media. *Trends Anal. Chem.* **2017**, *97*, 65–82. [CrossRef]
21. Giannakoudakis, D.A.; Bandosz, T.J. *Detoxification of Chemical Warfare Agents*, 1st ed.; Springer International Publishing: Cham, Switzerland, 2018; ISBN 978-3-319-70759-4.
22. Montoro, C.; Linares, F.; Quartapelle Procopio, E.; Senkovska, I.; Kaskel, S.; Galli, S.; Masciocchi, N.; Barea, E.; Navarro, J.A.R. Capture of nerve agents and mustard gas analogues by hydrophobic robust MOF-5 type metal-organic frameworks. *J. Am. Chem. Soc.* **2011**, *133*, 11888–11891. [CrossRef]
23. Giannakoudakis, D.A.; Travlou, N.A.; Secor, J.; Bandosz, T.J. Oxidized g-C_3N_4 Nanospheres as Catalytically Photoactive Linkers in MOF/g-C_3N_4 Composite of Hierarchical Pore Structure. *Small* **2017**, *13*, 1601758. [CrossRef]
24. Mondloch, J.E.; Katz, M.J.; Isley, W.C., III; Ghosh, P.; Liao, P.; Bury, W.; Wagner, G.W.; Hall, M.G.; DeCoste, J.B.; Peterson, G.W.; et al. Destruction of chemical warfare agents using metal–organic frameworks. *Nat. Mater.* **2015**, *14*, 512–516. [CrossRef]
25. Zhao, J.; Lee, D.T.; Yaga, R.W.; Hall, M.G.; Barton, H.F.; Woodward, I.R.; Oldham, C.J.; Walls, H.J.; Peterson, G.W.; Parsons, G.N. Ultra-Fast Degradation of Chemical Warfare Agents Using MOF-Nanofiber Kebabs. *Angew. Chemie Int. Ed.* **2016**, *55*, 13224–13228. [CrossRef]
26. Wang, C.C.; Li, J.R.; Lv, X.L.; Zhang, Y.Q.; Guo, G.S. Photocatalytic organic pollutants degradation in metal-organic frameworks. *Energy Environ. Sci.* **2014**, *7*, 2831–2867. [CrossRef]
27. Liu, J.; Chen, L.; Cui, H.; Zhang, J.; Zhang, L.; Su, C.Y. Applications of metal–organic frameworks in heterogeneous supramolecular catalysis. *Chem. Soc. Rev.* **2014**, *43*, 6011–6061. [CrossRef]
28. Taddei, M.; Schukraft, G.M.; Warwick, M.E.A.; Tiana, D.; McPherson, M.J.; Jones, D.R.; Petit, C. Band gap modulation in zirconium-based metal-organic frameworks by defect engineering. *J. Mater. Chem. A* **2019**, *7*, 23781–23786. [CrossRef]
29. Fang, Z.; Bueken, B.; De Vos, D.E.; Fischer, R.A. Defect-Engineered Metal-Organic Frameworks. *Angew. Chemie Int. Ed.* **2015**, *54*, 7234–7254. [CrossRef]

30. Ethiraj, J.; Albanese, E.; Civalleri, B.; Vitillo, J.G.; Bonino, F.; Chavan, S.; Shearer, G.C.; Lillerud, K.P.; Bordiga, S. Carbon dioxide adsorption in amine-functionalized mixed-ligand metal-organic frameworks of UiO-66 topology. *ChemSusChem* **2014**, *7*, 3382–3388. [CrossRef]
31. Van de Voorde, B.; Stassen, I.; Bueken, B.; Vermoortele, F.; De Vos, D.; Ameloot, R.; Tan, J.-C.; Bennett, T.D. Improving the mechanical stability of zirconium-based metal–organic frameworks by incorporation of acidic modulators. *J. Mater. Chem. A* **2015**, *3*, 1737–1742. [CrossRef]
32. Peterson, G.W.; Destefano, M.R.; Garibay, S.J.; Ploskonka, A.; McEntee, M.; Hall, M.; Karwacki, C.J.; Hupp, J.T.; Farha, O.K. Optimizing Toxic Chemical Removal through Defect-Induced UiO-66-NH$_2$ Metal–Organic Framework. *Chem. A Eur. J.* **2017**, *23*, 15913–15916. [CrossRef]
33. DeStefano, M.R.; Islamoglu, T.; Garibay, S.J.; Hupp, J.T.; Farha, O.K. Room-Temperature Synthesis of UiO-66 and Thermal Modulation of Densities of Defect Sites. *Chem. Mater.* **2017**, *29*, 1357–1361. [CrossRef]
34. Gil-San-Millan, R.; López-Maya, E.; Hall, M.; Padial, N.M.; Peterson, G.W.; DeCoste, J.B.; Rodríguez-Albelo, L.M.; Oltra, J.E.; Barea, E.; Navarro, J.A.R. Chemical Warfare Agents Detoxification Properties of Zirconium Metal-Organic Frameworks by Synergistic Incorporation of Nucleophilic and Basic Sites. *ACS Appl. Mater. Interfaces* **2017**, *9*, 23967–23973. [CrossRef]
35. Huang, Y.; Qin, W.; Li, Z.; Li, Y. Enhanced stability and CO$_2$ affinity of a UiO-66 type metal-organic framework decorated with dimethyl groups. *Dalt. Trans.* **2012**, *41*, 9283–9285. [CrossRef]
36. Cmarik, G.E.; Kim, M.; Cohen, S.M.; Walton, K.S. Tuning the adsorption properties of uio-66 via ligand functionalization. *Langmuir* **2012**, *28*, 15606–15613. [CrossRef]
37. Garibay, S.J.; Cohen, S.M. Isoreticular synthesis and modification of frameworks with the UiO-66 topology. *Chem. Commun.* **2010**, *46*, 7700–7702. [CrossRef]
38. Deng, H.; Doonan, C.J.; Furukawa, H.; Ferreira, R.B.; Towne, J.; Knobler, C.B.; Wang, B.; Yaghi, O.M. Multiple functional groups of varying ratios in metal-organic frameworks. *Science* **2010**, *327*, 846–850. [CrossRef]
39. Wu, H.; Chua, Y.S.; Krungleviciute, V.; Tyagi, M.; Chen, P.; Yildirim, T.; Zhou, W. Unusual and highly tunable missing-linker defects in zirconium metal-organic framework UiO-66 and their important effects on gas adsorption. *J. Am. Chem. Soc.* **2013**, *135*, 10525–10532. [CrossRef]
40. Zhao, Y.; Seredych, M.; Zhong, Q.; Bandosz, T.J. Aminated graphite oxides and their composites with copper-based metal-organic framework: In search for efficient media for CO$_2$ sequestration. *RSC Adv.* **2013**, *3*, 9932–9941. [CrossRef]
41. Petit, C.; Bandosz, T.J. Engineering the surface of a new class of adsorbents: Metal-organic framework/graphite oxide composites. *J. Colloid Interface Sci.* **2015**, *447*, 139–151. [CrossRef]
42. Petit, C.; Burress, J.; Bandosz, T.J. The synthesis and characterization of copper-based metal-organic framework/graphite oxide composites. *Carbon* **2011**, *49*, 563–572. [CrossRef]
43. Petit, C.; Bandosz, T.J. MOF-graphite oxide composites: Combining the uniqueness of graphene layers and metal-organic frameworks. *Adv. Mater.* **2009**, *21*, 4753–4757. [CrossRef]
44. Giannakoudakis, D.A.; Bandosz, T.J. Graphite Oxide Nanocomposites for Air Stream Desulfurization. In *Composite Nanoadsorbents*; Elsevier: Amsterdam, The Netherlands, 2019; pp. 1–24. ISBN 9780128141328.
45. Dong, G.; Zhang, Y.; Pan, Q.; Qiu, J. A fantastic graphitic carbon nitride (g-C3N4) material: Electronic structure, photocatalytic and photoelectronic properties. *J. Photochem. Photobiol. C Photochem. Rev.* **2014**, *20*, 33–50. [CrossRef]
46. Hong, J.; Chen, C.; Bedoya, F.E.; Kelsall, G.H.; O'Hare, D.; Petit, C. Carbon nitride nanosheet/metal–organic framework nanocomposites with synergistic photocatalytic activities. *Catal. Sci. Technol.* **2016**, *6*, 5042–5051. [CrossRef]
47. Wang, H.; Yuan, X.; Wu, Y.; Zeng, G.; Chen, X.; Leng, L.; Li, H. Synthesis and applications of novel graphitic carbon nitride/metal-organic frameworks mesoporous photocatalyst for dyes removal. *Appl. Catal. B Environ.* **2015**, *174–175*, 445–454. [CrossRef]
48. Zhao, Y.; Seredych, M.; Jagiello, J.; Zhong, Q.; Bandosz, T.J. Insight into the mechanism of CO$_2$ adsorption on Cu-BTC and its composites with graphite oxide or aminated graphite oxide. *Chem. Eng. J.* **2014**, *239*, 399–407. [CrossRef]
49. Zhao, J.; Nunn, W.T.; Lemaire, P.C.; Lin, Y.; Dickey, M.D.; Oldham, C.J.; Walls, H.J.; Peterson, G.W.; Losego, M.D.; Parsons, G.N. Facile Conversion of Hydroxy Double Salts to Metal-Organic Frameworks Using Metal Oxide Particles and Atomic Layer Deposition Thin-Film Templates. *J. Am. Chem. Soc.* **2015**, *137*, 13756–13759. [CrossRef]

50. Klimakow, M.; Klobes, P.; Rademann, K.; Emmerling, F. Characterization of mechanochemically synthesized MOFs. *Microporous Mesoporous Mater.* **2012**, *154*, 113–118. [CrossRef]
51. Katz, M.J.; Brown, Z.J.; Colón, Y.J.; Siu, P.W.; Scheidt, K.A.; Snurr, R.Q.; Hupp, J.T.; Farha, O.K. A facile synthesis of UiO-66, UiO-67 and their derivatives. *Chem. Commun.* **2013**, *49*, 9449. [CrossRef]
52. Giannakoudakis, D.A.; Seredych, M.; Rodríguez-Castellón, E.; Bandosz, T.J. Mesoporous Graphitic Carbon Nitride-Based Nanospheres as Visible-Light Active Chemical Warfare Agents Decontaminant. *ChemNanoMat* **2016**, *2*, 268–272. [CrossRef]
53. Groenewolt, M.; Antonietti, M. Synthesis of g-C3N4 nanoparticles in mesoporous silica host matrices. *Adv. Mater.* **2005**, *17*, 1789–1792. [CrossRef]
54. Petit, C.; Mendoza, B.; Bandosz, T.J. Reactive adsorption of ammonia on Cu-based MOF/graphene composites. *Langmuir* **2010**, *26*, 15302–15309. [CrossRef]
55. Maiti, S.; Pramanik, A.; Manju, U.; Mahanty, S. Reversible Lithium Storage in Manganese 1,3,5-Benzenetricarboxylate Metal–Organic Framework with High Capacity and Rate Performance. *ACS Appl. Mater. Interfaces* **2015**, *7*, 16357–16363. [CrossRef]
56. Cao, Y.; Zhao, Y.; Lv, Z.; Song, F.; Zhong, Q. Preparation and enhanced CO_2 adsorption capacity of UiO-66/graphene oxide composites. *J. Ind. Eng. Chem.* **2015**, *27*, 102–107. [CrossRef]
57. Petit, C.; Bandosz, T.J. Synthesis, characterization, and ammonia adsorption properties of mesoporous metal-organic framework (MIL(Fe))-graphite oxide composites: Exploring the limits of materials fabrication. *Adv. Funct. Mater.* **2011**, *21*, 2108–2117. [CrossRef]
58. Petit, C.; Bandosz, T.J. Exploring the coordination chemistry of MOF–graphite oxide composites and their applications as adsorbents. *Dalt. Trans.* **2012**, *41*, 4027. [CrossRef]
59. Hummers, W.S.; Offeman, R.E. Preparation of Graphitic Oxide. *J. Am. Chem. Soc.* **1958**, *80*, 1339. [CrossRef]
60. Seredych, M.; Łoś, S.; Giannakoudakis, D.A.; Rodríguez-Castellón, E.; Bandosz, T.J. Photoactivity of g-C3N4/S-Doped Porous Carbon Composite: Synergistic Effect of Composite Formation. *ChemSusChem* **2016**, *9*, 795–799. [CrossRef]
61. Zhang, J.; Chen, X.; Takanabe, K.; Maeda, K.; Domen, K.; Epping, J.D.; Fu, X.; Antonietti, M.; Wang, X. Synthesis of a carbon nitride structure for visible-light catalysis by copolymerization. *Angew. Chemie Int. Ed.* **2010**, *49*, 441–444. [CrossRef]
62. Millward, A.R.; Yaghi, O.M. Metal-organic frameworks with exceptionally high capacity for storage of carbon dioxide at room temperature. *J. Am. Chem. Soc.* **2005**, *127*, 17998–17999. [CrossRef]
63. Asha, P.; Sinha, M.; Mandal, S. Effective removal of chemical warfare agent simulants using water stable metal–organic frameworks: Mechanistic study and structure–property correlation. *RSC Adv.* **2017**, *7*, 6691–6696. [CrossRef]
64. Zdravkov, B.D.; Čermák, J.J.; Šefara, M.; Janků, J. Pore classification in the characterization of porous materials: A perspective. *Cent. Eur. J. Chem.* **2007**, *5*, 385–395.
65. Jagiello, J.; Olivier, J.P. Carbon slit pore model incorporating surface energetical heterogeneity and geometrical corrugation. *Adsorption* **2013**, *19*, 777–783. [CrossRef]
66. Bashkova, S.; Bandosz, T.J. The effects of urea modification and heat treatment on the process of NO_2 removal by wood-based activated carbon. *J. Colloid Interface Sci.* **2009**, *333*, 97–103. [CrossRef] [PubMed]
67. Colmenares, J.C.; Lisowski, P.; Łomot, D.; Chernyayeva, O.; Lisovytskiy, D. Sonophotodeposition of Bimetallic Photocatalysts Pd-Au/TiO_2: Application to Selective Oxidation of Methanol to Methyl Formate. *ChemSusChem* **2015**, *8*, 1676–1685. [CrossRef] [PubMed]
68. Ouyang, W.; Kuna, E.; Yepez, A.; Balu, A.; Romero, A.; Colmenares, J.; Luque, R. Mechanochemical Synthesis of TiO_2 Nanocomposites as Photocatalysts for Benzyl Alcohol Photo-Oxidation. *Nanomaterials* **2016**, *6*, 93. [CrossRef]

© 2019 by the authors. Licensee MDPI, Basel, Switzerland. This article is an open access article distributed under the terms and conditions of the Creative Commons Attribution (CC BY) license (http://creativecommons.org/licenses/by/4.0/).

Article

Mixed Functionalization of Organic Ligands in UiO-66: A Tool to Design Metal–Organic Frameworks for Tailored Microextraction

Gabriel González-Rodríguez [1], Iván Taima-Mancera [1], Ana B. Lago [2], Juan H. Ayala [1], Jorge Pasán [2,*] and Verónica Pino [1,3,*]

1. Departamento de Química, Unidad Departamental de Química Analítica, Universidad de La Laguna (ULL), Tenerife, 38206 La Laguna, Spain; alu0100995312@ull.edu.es (G.G.-R.); ivan.taima.13@ull.edu.es (I.T.-M.); jayala@ull.edu.es (J.H.A.)
2. Laboratorio de Rayos X y Materiales Moleculares (MATMOL), Departamento de Física, Universidad de La Laguna (ULL), Tenerife, 38206 La Laguna, Spain; alagobla@ull.edu.es
3. University Institute of Tropical Diseases and Public Health, Universidad de La Laguna (ULL), Tenerife, 38206 La Laguna, Spain
* Correspondence: veropino@ull.edu.es (V.P.); jpasang@ull.edu.es (J.P.); Tel.: +34-922-318990 (V.P.)

Received: 12 September 2019; Accepted: 3 October 2019; Published: 10 October 2019

Abstract: The mixed-ligand strategy was selected as an approach to tailor a metal–organic framework (MOF) with microextraction purposes. The strategy led to the synthesis of up to twelve UiO-66-based MOFs with different amounts of functionalized terephthalate ligands (H-bdc), including nitro (-NO_2) and amino (-NH_2) groups (NO_2-bdc and NH_2-bdc, respectively). Increases of 25% in ligands were used in each case, and different pore environments were thus obtained in the resulting crystals. Characterization of MOFs includes powder X-ray diffraction, infrared spectroscopy, and elemental analysis. The obtained MOFs with different degrees and natures of functionalization were tested as sorbents in a dispersive miniaturized solid-phase extraction (D-µSPE) method in combination with high-performance liquid chromatography (HPLC) and diode array detection (DAD), to evaluate the influence of mixed functionalization of the MOF on the analytical performance of the entire microextraction method. Eight organic pollutants of different natures were studied, using a concentration level of 5 µg·L^{-1} to mimic contaminated waters. Target pollutants included carbamazepine, 4-cumylphenol, benzophenone-3, 4-tert-octylphenol, 4-octylphenol, chrysene, indeno(1,2,3-cd)pyrene, and triclosan, as representatives of drugs, phenols, polycyclic aromatic hydrocarbons, and disinfectants. Structurally, they differ in size and some of them present polar groups able to form H-bond interactions, either as donors (-NH_2) or acceptors (-NO_2), permitting us to evaluate possible interactions between MOF pore functionalities and analytes' groups. As a result, extraction efficiencies can reach values of up to 60%, despite employing a microextraction approach, with four main trends of behavior being observed, depending on the analyte and the MOF.

Keywords: metal–organic frameworks; dispersive miniaturized solid-phase extraction; mixed functionalization; interactions MOF–analyte; UiO-66

1. Introduction

Metal–organic frameworks (MOFs) are having enormous success as novel sorbent materials in analytical solid-phase extraction (SPE) approaches, particularly when performing in dispersive and miniaturized modes (D-µSPE) [1–5]. The synergies of MOFs' features, such as their impressive surface area, synthetic tuneability, and chemical stability [6,7], and those of D-µSPE, such as method simplicity and a high efficiency [8,9], are among the reasons justifying the high number of recent studies in the field.

A step forward in ensuring the true expansion of MOFs as competitive materials for D-µSPE requires not only the assurance of a better performance than that resulting from commercial materials (exhaustive comparison and inter-laboratory validation) [5,10], but also deep evaluation of the main factors of MOFs justifying the improved analytical performance for target compounds [11,12]. Gaining an understanding of the process can serve as the basis of proper MOF design.

A number of studies using MOFs in D-µSPE have pointed out the complexity of the systems, indicating the pore environment, pore size, and pore aperture widths of the MOF as the most influential factors, together with a clear influence of the metal nature (particularly the presence of unsaturated metal sites) [11]. Lirio et al. also pointed out the influence of the metal, particularly the radius of the metal [13]. Taima-Mancera et al. showed the positive effects of incorporating polar functionalities in the organic ligand of the MOFs used in D-µSPE when intending to extract polar analytes of a small size [12], with this idea having been further expanded by Boontongto et al. to other application studies for polar analytes [14].

Computational and modeling studies are also powerful tools for evaluating the adequacy of MOFs for different applications, but have hardly been tested for MOFs in D-µSPE. Therefore, the majority of studies have been devoted to the evaluation of gas storage applications of MOFs or in catalysis studies [15–17]. However, it is fair to mention the studies of Gao et al. [18], which computationally selected the MOF MIL-53(Al) as adequate material for a D-µSPE method to determine a group of estrogens and glucocorticoids, and proved it with experimental studies.

In spite of the abovementioned studies that have tried to provide insights linked to the nature of the MOFs to improve their performance for target analytes in D-µSPE, most MOFs in reported applications are archetypical MOFs. This is particularly true for those that are currently commercialized, such as MIL-53(Al) [11] HKUST-1 [19], and MIL-100(Fe) [20].

Currently, Zr-based MOFs are widely studied in a number of fields (not exclusively in D-µSPE) because of their high chemical stability [21,22]. Among Zr-based MOFs, UiO-66(Zr) and UiO-66-type MOFs have been studied the most, given their superior chemical and hydrothermal stability, together with their simple (and mild) preparation [23] and green aspects [24]. Therefore, they have appeared as sorbents in a number of recent D-µSPE studies [12,25–27].

Regarding the modifications of UiO-66 to obtain a number of derivatives, it is interesting to mention the mixed-linker approach, which consists of incorporating two or more linkers with similar sizes, but different functional groups [28]. In this way, the resulting framework will present the properties modulated by the relative amounts of functional groups incorporated [29]. Cohen et al. were the first to incorporate -Br and -NH$_2$ functionalities into the organic ligand (1,4-benzenedicarboxylic acid) of UiO-66 [30], and obtained a more thermally-stable derivative than the neat UiO-66. The incorporation of different contents of -NH$_2$ functionalization into UiO-66 has also been proposed, not only resulting in a thermally-stable superior material, but also permitting the porosity to be tuned by varying the ratios of non-functionalized ligand *versus* -NH$_2$-functionalized ligand [31].

Given these considerations, the current study intends to prepare and characterize UiO-66 derivatives incorporating different contents of non-functionalized and functionalized-organic ligand—including -NH$_2$ and -NO$_2$ groups—in the MOF structure through the mixed-linker approach. As a second goal, it pursues an evaluation of the influence of such modifications in the resulting material when used as a sorbent in a D-µSPE method for different target analytes in water. The selected analytes present a low to high size (to evaluate their influence when entering or when not entering the MOFs' pores), while incorporating or not incorporating polar groups in their structures (to evaluate possible interactions between MOF pore functionalities and analytes' groups).

2. Experiment

2.1. Chemicals, Reagents, and Materials

Six out of the eight target analytes were obtained as solid products from Sigma-Aldrich (Steinheim, Germany): carbamazepine (Cbz, 99.0%), 4-cumylphenol (CuP, 99%), 4-tert-octylphenol (t-OP, 97%), 4-octylphenol (OP, 99%), benzophenone-3 (BP-3, 99.5%), and chrysene (Chy, 98%). The remaining two target analytes, indeno(1,2,3-cd)pyrene (Ind) and triclosan (Tr), were purchased separately as standard solutions, with a concentration of 10 mg·L^{-1} in acetonitrile (ACN), by Dr. Ehrenstorfer GmbH (Augsburg, Germany). Chemical structures of the studied analytes are included in Table 1. A standard solution containing all eight target compounds was prepared in ACN Chromasolv™ liquid chromatography (LC) grade, purchased from Honeywell Fluka™ (Seelze, Germany), at a concentration of 5 mg·L^{-1}, and stored at 4 °C. Aqueous working standard solutions at 5 µg·L^{-1} containing all compounds were utilized in the D-µSPE method.

Table 1. Several physicochemical properties of the analytes studied (SciFinder® 2019).

Analyte (Abbreviation)	Structure	Molecular Formula Molecular Weight (g·mol^{-1})	Molar Volume [1] (Å^3·molecule^{-1})	pK$_a$	Vapor Pressure at 25 °C (N·m^{-2})	Log K$_{ow}$ [2]
Carbamazepine (Cbz)		C$_{15}$H$_{12}$N$_2$O 236.27	310	13.9	7.71 × 10^{-5}	1.90
4-Cumylphenol (CuP)		C$_{15}$H$_{16}$O 212.29	334	10.6	6.64 × 10^{-3}	4.24
Benzophenone-3 (BP-3)		C$_{14}$H$_{12}$O$_3$ 228.24	315	7.6	7.01 × 10^{-4}	4.00
Triclosan (Tr)		C$_{12}$H$_7$Cl$_3$O$_2$ 289.54	323	7.8	4.35 × 10^{-3}	5.34
4-tert-Octylphenol (t-OP)		C$_{14}$H$_{22}$O 206.32	366	10.2	2.64 × 10^{-1}	5.18
4-Octylphenol (OP)		C$_{14}$H$_{22}$O 206.32	365	10.2	3.33 × 10^{-2}	5.63
Chrysene (Chy)		C$_{18}$H$_{12}$ 228.29	318	-	1.13 × 10^{-5}	5.73
Indeno[1,2,3-cd]pyrene (Ind)		C$_{22}$H$_{12}$ 276.33	333	-	2.08 × 10^{-7}	6.65

[1] 20 °C and 1.01·10^5 N·m^{-2}; [2] octanol/water partition coefficient.

Reagents included in the synthesis of the UiO-66 MOF and its functionalized derivatives were ZrCl$_4$ (98%), HCl (37%, v/v), 1,4-benzenedicarboxylic acid (H-bdc, 98%), 2-amino-1,4-benzenedicarboxylic acid (NH$_2$-bdc, 99%), and 2-nitro-1,4-dicarboxylic acid (NO$_2$-bdc, ≥99%), purchased from Sigma-Aldrich. Dimethylformamide (DMF, ≥99.5%) was acquired from Merck KGaA (Darmstadt, Germany), and methanol (≥99.8%) was purchased from PanReac AppliChem (Barcelona, Spain).

The synthesis of MOFs required Teflon (PTFE®) solvothermal reactors of a 45 mL capacity and stainless-steel autoclaves, all from Parr Instrument Company (Moline, IL, USA).

Ultrapure water (Milli-Q, ultrapure grade) was obtained through the purification system A10 MilliPore (Watford, UK). High-performance liquid chromatography (HPLC) mobile phases were prepared with ultrapure water and ACN Chromasolv™ LC-MS grade. Both phases were filtered with 0.45 µm Durapore® membrane filters of Sigma-Aldrich.

Additionally, 0.2 µm polyvinylidene fluoride (PVDF) syringe filters Whatman™, purchased from GE Healthcare (Buckinghamshire, UK), were used when filtrating desorption solvents after application of the D-µSPE method. The microextraction method also required glass centrifuge tubes of a 28 mL capacity from Pyrex® (Corning Inc., Staffordshire, UK), with a size of 10 × 2.6 cm.

2.2. Instrumentation

In the D-µSPE procedure, a vortexer from Reax-Control Heidolph™ GmbH (Schwabach, Germany) and the centrifuge model 5720 Eppendorf™ (Hamburg, Germany) were utilized.

The HPLC model 1260 Infinity was purchased from Agilent Technologies (Santa Clara, CA, USA). A Rheodyne injection valve with an injection loop of 20 µL, supplied by Supelco (Bellefonte, PA, USA), was included in the system. The chromatographic separation used an ACE Ultra Core 5 SuperC18 (5 µm, 150 × 4.6 mm) analytical column, obtained from Symta (Madrid, Spain), with the safeguard column Pelliguard LC-18 purchased from Supelco. Ultrapure water and ACN were employed as mobile phases using a linear gradient at a constant flow rate of 1 mL·min^{-1}. The chromatographic method started at 50% (v/v) of ACN for 5 min and then increased up to 80% (v/v) for 2 min and up to 83% (v/v) in the next 2.5 min, before finally reaching 100% (v/v) of ACN in the next 3.5 min.

The detection of analytes was achieved with a diode array detection (DAD) 1260 Infinity model, purchased from Agilent Technologies. The quantification wavelengths of the DAD were set at 254 nm for Ind; 270 nm for Chy; 280 nm for CuP, t-OP, and OP; and 289 nm for Cbz, BP-3, and Tr.

The Universal UF30 oven, supplied by Memmert (Schwabach, Germany), was used in MOF synthesis.

All the MOFs were characterized by powder X-ray diffraction using a PANalytical Empyrean diffractometer (Eindhoven, The Netherlands) with Cu Kα radiation (λ = 1.5418 Å) and operating with Bragg–Brentano geometry. Measurements were carried out at room temperature in the range from 5.01° to 80.00° (0.02° steps), with a total exposure time of 12 min.

A Gemini V2365 model, supplied by Micromeritics (Norcross, GA, US), was used to measure the nitrogen adsorption isotherms with a surface area analyzer at 77 K in the range $0.02 \leq P/P_0 \leq 1.00$. The Brunauer, Emmett and Teller (BET) method was used to calculate the surface area.

An infrared spectroscopy instrument with Fourier transformed (FT-IR) model IFS 66/S from Bruker (MA, US) was used.

Elemental analyses (C, H, N) were carried out with the elemental analyzer CNHS Flash EA 1112 from Thermo Fisher Scientific (Massachusetts, MA, USA).

2.3. Procedures

2.3.1. Synthesis of MOFs

The synthesis of UiO-66 and its functionalized variants (Scheme 1) followed the procedure reported by Taima-Mancera et al. [12,32]. Briefly, UiO-66 required 233 mg of ZrCl$_4$ (1 mmol) and 246 mg of H-bdc (1.5 mmol). These reagents were dissolved in 15 mL of DMF, with 1 mL of concentrated HCl as a synthetic modulator. The resulting solution was heated in a solvothermal reactor at 150 °C for 24 h. Once cooled at room temperature, the obtained solid was filtered, washed twice with DMF (24 h each), filtered again, and then washed with methanol (24 h). Finally, the product was heated at 150 °C for one day in order to activate the MOF.

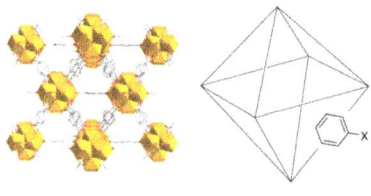

X = H, NH$_2$ or NO$_2$

Scheme 1. Schematic structure of UiO-66 and its derivatives.

UiO-66-functionalized variants were prepared analogously, replacing H-bdc with the equivalent molar amounts of NH_2-bdc or NO_2-bdc, depending on the specific MOF under preparation. In this sense, different ratios of NH_2-bdc, NO_2-bdc, and H-bdc were included in the synthetic approach, with the purpose of obtaining MOFs with different amounts of functionalities. The final set includes the preparation of up to 12 derivatives of the UiO-66 MOF, with the specific contents included in Figure 1.

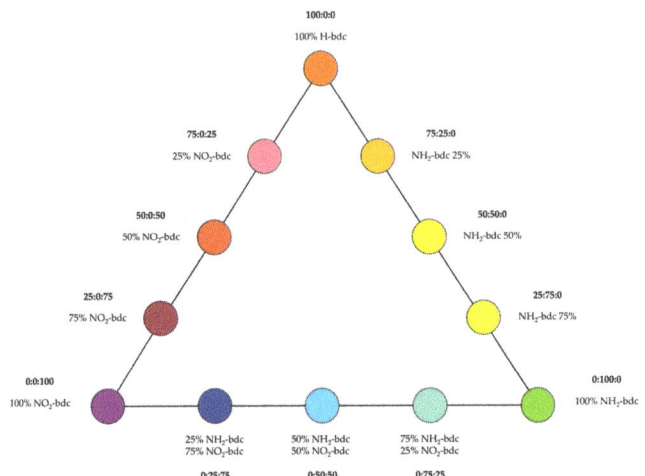

Figure 1. UiO-66-based metal–organic frameworks (MOFs) prepared with different contents of functional groups, labeled in the triangular diagram as percentages of functionalized terephthalate ligands.

2.3.2. Dispersive Miniaturized Solid-Phase Extraction (D-μSPE) Method

The extraction procedure followed our previous studies on the use of the MOF UiO-66 for determining endocrine disrupting chemicals using D-μSPE-HPLC-DAD [12], but minimizing the initial content of MOF. Therefore, the current study required the use of a lower amount of MOF and employed 10 mg rather than 20 mg. In summary, 10 mg of the UiO-66-based MOF were used when analyzing 20 mL of an aqueous standard containing target analytes. The microextraction took place in Pyrex® tubes subjected to 3 min of vortex, to increase the strength of the sorbent (MOF)–analyte interactions. Afterwards, phases were separated by centrifugation (2504× g for 5 min), followed by separation of the supernatant with a Pasteur pipette. Desorption took place using 500 μL of ACN under 5 min of vortex agitation, followed by 5 min of centrifugation (2504× g). Finally, the desorption solution was filtered through 0.2 μm PVDF syringe filters before HPLC injection. The entire procedure is schematized in Figure S1 of the ESM.

3. Results and Discussion

3.1. Characterization of the UiO-66-Based MOFs Obtained with the Mixed-Linker Approach

Complete characterization by powder X-ray diffraction, nitrogen adsorption isotherms, elemental analysis, and infrared spectroscopy took place for the synthesized and activated UiO-66-based MOFs (Figure 1). Powder X-ray diffraction patterns were obtained in order to verify the crystalline structure of all MOFs through a comparison with the simulated one for the UiO-66 [33]. Furthermore, N_2 adsorption isotherms were used for the calculation of Brunauer–Emmett–Teller (BET) surface areas. Likewise, infrared spectra were utilized to identify nitro- and amino-functional groups in the MOFs, and the elemental analysis was carried out to evaluate whether the degree of functionalization was correctly introduced in the resulting MOFs.

All powder X-ray diffraction patterns are included in Figures S2–S4 of the ESM. It is important to highlight that all obtained UiO-66-based MOFs using the current approach were crystalline and topologically identical.

The BET surface areas were calculated from the nitrogen adsorption isotherms for all MOFs, and the obtained data is shown in Table S1 of the ESM. The trend shows a decrease in the surface area with the increased degree of functionalization. For example, UiO-66 showed a BET surface area value of 1175 m^2 g^{-1}, whereas the increasing content of the amino group as functionalization (from 25% to 100%) showed decreasing values, down to 678 m^2 g^{-1}. For the nitro group as UiO-66 functionalization, values also decreased from 717 m^2 g^{-1} at 25% to 604 m^2 g^{-1} at 100%. If considering the amino/nitro group mixed functionalization in the UiO-66-based MOFs, the values range from the 100% nitro group to 100% amino group.

Figure S5 of the ESM includes the infrared spectra for the NH$_2$-bdc:NO$_2$-bdc series, whereas Figure 2 shows a zoom from 500 to 1700 cm^{-1} for such series. The FT-IR spectra of the MOFs display features corresponding to the bdc ligand and to the amino or nitro groups present in each MOF. It can be clearly observed that the intensity of the IR band at 1257 cm^{-1} (attributed to the symmetric in-plane bending or deformation mode of the -NH$_2$ group) increases with the content of NH$_2$-bdc, thus supporting the proper inclusion of the amino functionality [34] and its increasing content in the series. Moreover, a broad band at 3367 cm^{-1} is observed for amino derivatives (N–H stretching modes). This band is more defined when the content of amino groups increases, as occurred for the 25:75:0 and 0:100:0 MOFs in Figure S6 of the ESM. Furthermore, the intensity of the band associated with the nitro functionalization at 1546 cm^{-1} (N–O stretching modes) [34] rises when increasing the amount of nitro groups in the MOF.

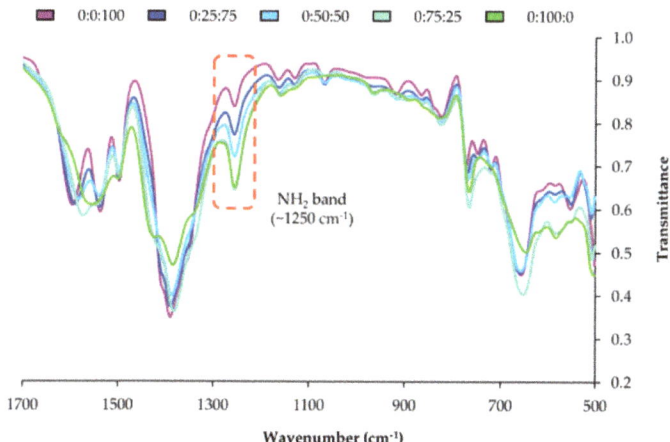

Figure 2. Zoom in the infrared spectra of the NO$_2$-bdc/NH$_2$-bdc series, from 500 to 1700 cm^{-1}. The code H-bdc: NH$_2$-bdc:NO$_2$-bdc is included for each MOF.

The formulae proposed for the different functionalized compounds are supported by the CHN elemental analysis. Table S2 of the ESM presents elemental analysis data, where the match between the experimental data and the calculated data can be observed. A higher degree of functionalization with both NH$_2$-bdc and NO$_2$-bdc implies an increase in the nitrogen content.

3.2. Analytical Performance of the D-μSPE-HPLC-DAD Method When Using Derivatives of UiO-66

All MOFs were used as sorbents in the D-μSPE-HPLC-DAD method following the conditions described in Section 2.3.2, with experiments carried out in triplicate. The target compounds included eight endocrine disrupting chemicals, specifically four small-sized analytes with polar functional groups in their structures (Cbz, CuP, Tr, and BP-3) and four heavier compounds (t-OP, OP, Chy, and Ind), two of which had no polar group in their structures. These common pollutants in water were selected in the current study to cover a wide range of possible MOF–analyte interactions. The main purpose was to evaluate the effects that mixed functionalization in the MOFs exerted on extraction efficiencies for these analytes in water. To extract proper conclusions, it is important to take into account the following considerations: (i) the amino group in the MOF can act as a hydrogen bond donor and it is an electron donating group towards the terephthalate ring, but the nitro group is a hydrogen bond acceptor and an electron withdrawal group; (ii) the pore window of the MOF reduces when the degree of functionalization increases; and (iii) the functionalization groups are mainly located in the pores, but they are also present in the external surface of the MOF crystallites.

If taking the extraction efficiency (of the microextraction method) as the main feature to evaluate the influence of the MOFs' nature in the method, four main trends can be observed in this study, as summarized in Figure 3. The extraction efficiency (E_R) was calculated as the ratio of the real enrichment factor of the microextraction procedure (calculated in the analytical method) and the theoretical maximum enrichment factor (40) [12], and expressed as %.

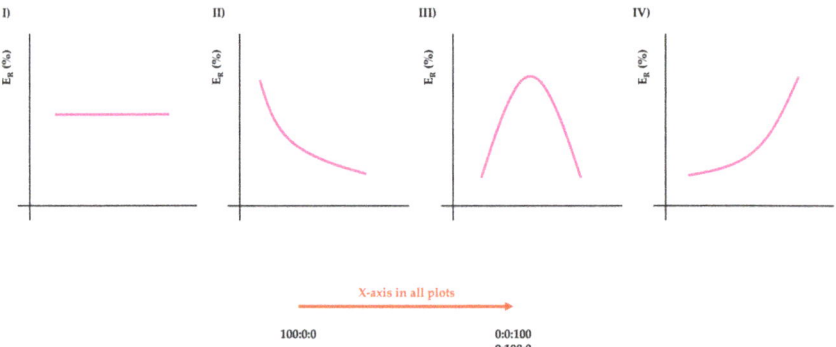

Figure 3. Types of general trends observed when extracting the target analytes by dispersive miniaturized solid-phase extraction (D-µSPE)-high-performance liquid chromatography (HPLC)-diode array detection (DAD) using different UiO-66-based MOFs as sorbents, as a function of the extraction efficiency (E_R in %). X-axis goes generically from the neat UiO-66 (H-bdc), 100:0:0, to increasing amounts of functionalization, to reach 0:0:100 (NO_2-bdc) or 0:100:0 (NH_2-bdc).

In trend I, there is no variation in the extraction efficiency, independently of the degree and type of functionalization. This means that the analyte is not interacting with the functional groups introduced, and it is not affected by the decrease of the pore window of the MOFs.

Trend II shows a decrease in the extraction efficiency with an increase in the functionalization. This situation is attributable to analytes with a critical size, highly affected by the decrease in the MOF's pore window, or to analytes able to establish strong interactions with the functional groups, thus precluding a good desorption process from the MOF once trapped (Figure S1 of the ESM).

Trend III shows a maximum at intermediate degrees of functionalization, with extraction efficiencies achieved with the neat MOFs lower than those achieved for the mixed ones. In this case, the analytes benefit from both types of functionalization, for example, larger window aperture due to bare terephthalic and hydrogen bond interactions caused by amino-modified ligands.

Trend IV is indeed the opposite to trend II: the functionalization increases the extraction efficiency of the analyte by the MOF. In this case, the window aperture is not a problem for the analyte and the interactions established with the functional groups increase the adsorption capability of the material.

3.2.1. Analytical Performance When Using the H-UiO-66 to NO$_2$-UiO-66 Series

Figure S7 of the ESM shows a comparison of the extraction efficiency for all the analytes when the amount of nitro functionalization in terephthalate ligands of the MOF increases. In general, the extraction efficiencies are better when using an intermediate degree of functionalization, indicating that the mixed-ligand approach produces small, but significant, improvements with respect to the bare UiO-66 MOF. Trend III is particularly clear in the case of carbamazepine, triclosan, and benzophenone-3, which are small-size analytes able to penetrate into the pores, and their polar groups (amide in Cbz, hydroxyl in Tr and BP-3) are able to establish hydrogen bond interactions with the nitro group of the MOF.

Chy, Ind, and OP also present better extraction efficiencies when using mixed-ligand UiO-66 MOFs, although the variation is less pronounced than for the smaller analytes. The size of these analytes is at the limit of the window aperture of the UiO-66 pores, and most likely, their adsorption occurs on the external surface of the crystallites. The addition of functional groups to the terephthalic ligands reduces the window aperture, and therefore, the small increase in the extraction efficiency may come from interactions with the nitro groups on the external surface of the UiO-66 crystallites.

3.2.2. Analytical Performance When Using the H-UiO-66 to NH$_2$-UiO-66 Series

The extraction efficiency for all the analytes when increasing the number of amino groups in the UiO-66 MOF is depicted in Figure S8 of the ESM. In this case, there are a variety of trends. For Cbz, there is an increase (trend IV) in the extraction efficiency with the increase in amino groups, indicating that the capability of the MOF to establish hydrogen bonds may be critical for this analyte. The amide–amino interaction is most likely responsible for this better E_R. A similar trend (IV) is observed for BP-3, where the acceptor groups of the benzophenone-3 can establish H-bonds with the donor amino groups of the MOF.

Furthermore, CuP and Tr show a continuous decrease in E_R values when the amount of amino groups increases (trend II). It seems that the presence of functional groups does not favor the extraction recovery. If we consider that the two analytes have similar structural features (two benzene rings and hydroxyl groups), this trend is most likely due to the limitation in the pore window size and the presence of polar groups in the pore walls.

As occurs for the nitro series, Chy and Ind experience better recoveries with mixed MOFs, with 50:50:0 functionalization being the one that produces the best results. In the case of the OP and t-OP analytes, they show different trends; for t-OP, the E_R drops after 50% of amino-terephthalic incorporation, and for OP, the E_R rises somewhat when the number of amino groups increases.

3.2.3. Analytical Performance When Using the NH$_2$-bdc to NO$_2$-bdc Series

The trends for the extraction efficiencies when a mixture of nitro- and amino-terephthalic ligands are used in the synthesis of UiO-66 are shown in Figure S9 of the ESM. In this case, it seems that the mixture clearly favors the recovery of the analytes, since in all the cases, except for OP, the E_R is better when using mixed-ligand MOFs. The reasons may not be the same for all the analytes, but the introduction of some amino groups (less bulky, H-bond donor) produces a better environment for analyte adsorption. In some cases, although a significant enhancement of the E_R is produced (BP-3, Chy, and Ind) with 0:25:75, the increase of amino groups content does not imply a better E_R.

3.3. Study of the Analytical Performance of the D-μSPE-HPLC-DAD Method Focusing on the Analyte's Structure and Possible MOF–Analyte Interactions

The individual analysis carried out by series (Section 3.2) gives a limited vision of the influence of functionalization in UiO-66. If intending the extraction of a specific target analyte by D-μSPE-HPLC-DAD, it is important to have an overview to obtain semi-quantitative conclusions related to which functionalization in the MOF is best for an individual analyte in the method.

Figures 4–6 include the extraction efficiency achieved with each of the twelve MOFs tested as sorbents using D-µSPE-HPLC-DAD, grouped in three different plots as a function of structural similarities of analytes. This gives a general overview by the analyte's nature when using all MOFs. It is important to note that all studies were accomplished using the eight analytes present all together in the aqueous standard subjected to the entire method, and thus, effects in the E_R values coming from analyte–analyte interactions could have occurred.

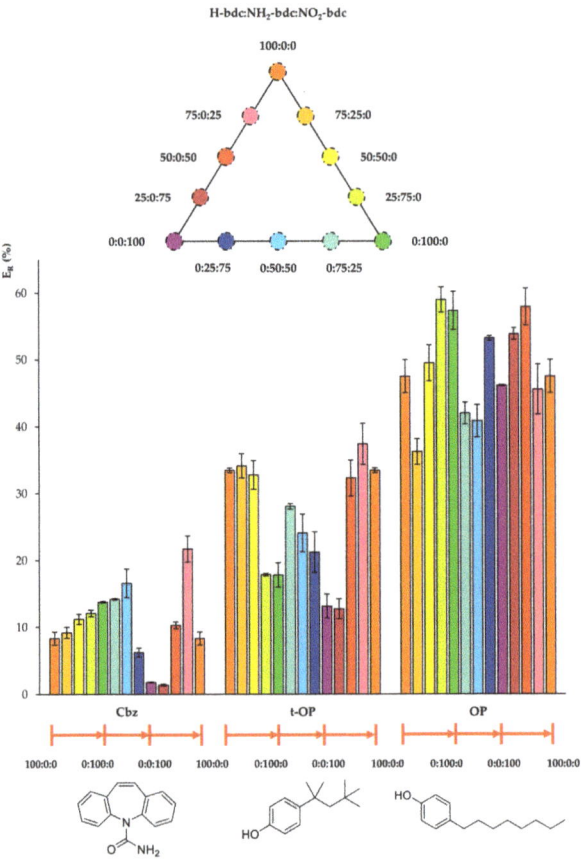

Figure 4. Extraction efficiencies (E_R in %) using the twelve UiO-66-based MOFs as sorbents in D-µSPE-HPLC-DAD for the analytes carbamazepine (Cbz), 4-tert-octylphenol (t-OP), and 4-octylphenol (OP) (selected for presenting similar structures).

Regarding Cbz, Figure 4 shows that the best performance occurs when there is a slight inclusion of NO_2 functionalization in the bare H-bdc UiO-66. Although the nitro group seems to favor the extraction of Cbz, the further increase in nitro groups produces a decrease in E_R, indicating that another effect, most likely the decrease of the pore window, reduces the diffusion through the pores. In the case of the t-OP, the inclusion of functional groups in the UiO-66 decreases the extraction performance, with the best values being obtained for MOFs with little or no functionalization. For OP, its extraction efficiencies fluctuate, but without a clear trend, and the best performances are obtained using 0:100:0 (neat NH_2-UiO-66) and 50:50:0 (mixed nitro and bare UiO-66).

Figure 5 includes the behavior of CuP, which is similar to that of t-OP, but even more drastic. Its E_R value using the MOF 100% NH_2-bdc UiO-66 in D-µSPE-HPLC-DAD is lower than 4%, but when using the neat UiO-66 without any functional group, it is higher than 40%. Although CuP exhibits a hydroxyl H-bond donor, the functionalization of the MOF with polar groups is not key for its extraction, and it is possible that the decrease of pore aperture plays a more significant role. In the case of BP-3, the incorporation of both NH_2 and NO_2 functional groups in the UiO-66 MOF produces better E_R values. The best performance is achieved using 0:25:75 NH_2/NO_2 mixed UiO-66. The presence of H-bond acceptor and donor groups in the same MOF favors the extraction of BP-3, an analyte with H-bond donor (hydroxyl) and acceptor (ketone) groups. For Tr, the best performance occurs when utilizing 0:50:50 and 50:0:50 UiO-MOFs, indicating that the presence of nitro groups contributes to more efficient extractions, but the excess of this group produces a final decrease in E_R. As it has been abovementioned, the combination of amino and nitro groups has good effects on the overall extraction performance of the method for the analyte.

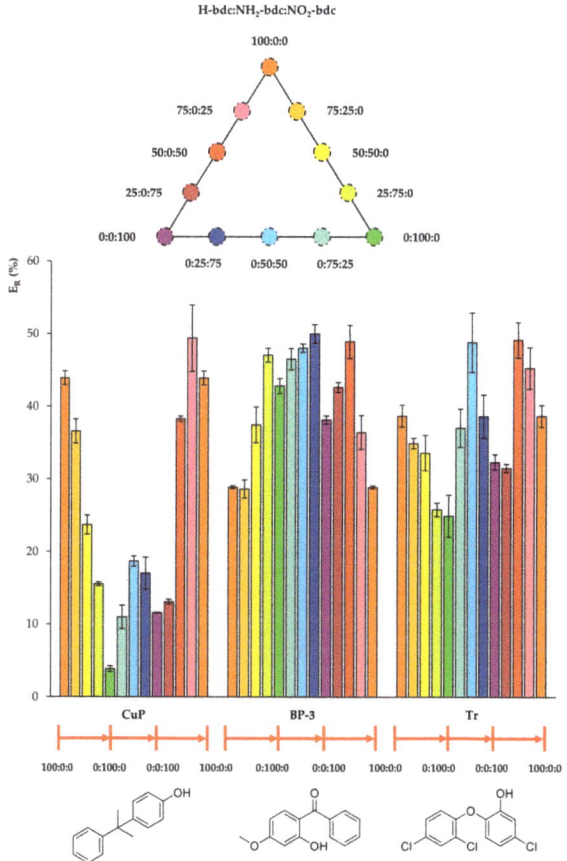

Figure 5. Efficiencies (E_R in %) using the twelve UiO-66-based MOFs as sorbents in D-μSPE-HPLC-DAD for the analytes 4-cumylphenol (CuP), benzophenone-3 (BP-3), and triclosan (Tr) (selected for presenting similar structures).

In Figure 6, it can be observed that any functionalization of the bare UiO-66 increases the extraction efficiency for Chy, and the best performances occur for mixed NH_2/NO_2 UiO-66. Regarding Ind, similarly to Chy, the functionalization of UiO-66 produces better E_R values, and the 50:50:0, 0:50:50 mixed UiO-66 led to the best values.

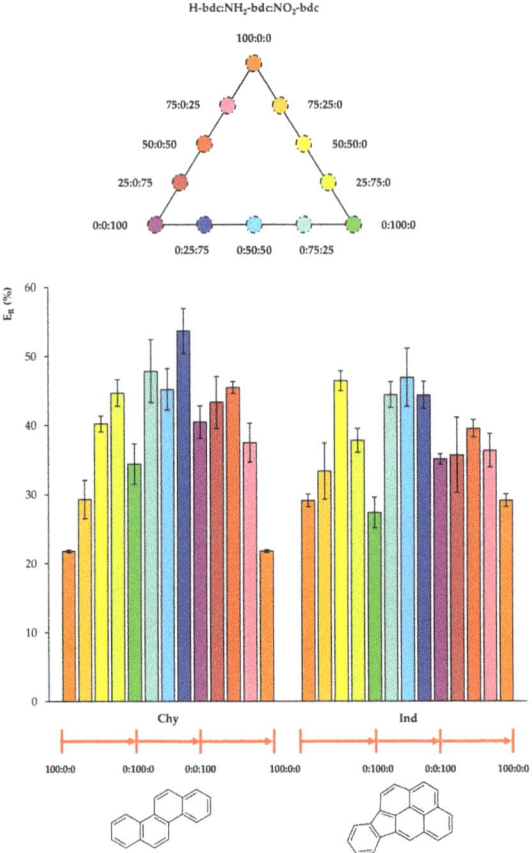

Figure 6. Extraction efficiencies (E_R in %) using the twelve UiO-66-based MOFs as sorbents in D-μSPE-HPLC-DAD for the analytes chrysene (Chy) and indeno(1,2,3-cd)pyrene (Ind) (selected for presenting similar structures).

4. Conclusions

The mixed-ligand strategy has been demonstrated to be a useful and simple tool for incorporating different functionalization groups and thus different pore environments into the MOF UiO-66. The strategy permitted the tailoring of UiO-66, and the resulting MOFs containing a variety of polar functional groups showed a high efficiency for the microextraction of target pollutants from waters when used as sorbents in D-μSPE.

Different trends were observed in the obtained extraction efficiency, depending on the structure of the target analyte and on the type and degree of functionalization of UiO-66. Therefore, the presence of H-bond donor groups in UiO-66 improves the analytical performance of the D-μSPE-HPLC-DAD method for those target compounds containing groups in their structures able to participate in H-bonds. However, the presence of polar groups in UiO-66 can significantly reduce the extraction efficiency for analytes with bulky hydrophobic groups in their structures. Clearly, proper control of the structure of the MOF is possible by carefully considering the type of target analytes intended.

It is not possible to quantitatively estimate the amount of amino and nitro groups to be included in UiO-66 to ensure an improved efficiency for the entire group of target analytes selected in the current study. However, the use of mixed amounts of both groups in UiO-66 seems to represent an adequate selection.

Ongoing studies intend to investigate the application of this mixed functionalization strategy to other MOFs, in order to develop tailored analytical microextraction methods.

Supplementary Materials: The following are available online, Figure S1: Scheme of the D-µSPE-HPLC-DAD method using optimum conditions; Figure S2–S4: XRD patterns; Figure S5, S6: Infrared spectra; Figure S7–S9: Analytical performances; Table S1: Adsorption data for all the synthesized UiO-66-based MOFs; Table S2: Elemental analysis data for all the synthesized UiO-66-based MOFs.

Author Contributions: Conceptualization: J.P., A.B.L., and V.P.; formal analysis: A.B.L. and J.P.; funding acquisition: V.P.; investigation: G.G.-R., A.B.L., J.H.A., I.T.-M., J.P., and V.P.; methodology: G.G.-R.; resources: J.P., V.P., and J.H.A.; software: J.H.A.; supervision: J.P., A.B.L., J.H.A., and V.P.; validation: G.G.-R., I.T.-M., J.H.A., and V.P.; writing—original draft: G.G.-R., I.T.-M., and A.B.L.; Writing—review and editing: J.P., J.H.A., and V.P.

Funding: This research was funded by the Spanish Ministry of Economy (MINECO) project ref. MAT2017-89207-R. A.B.L. and DIAD Group ES for financial support.

Acknowledgments: I.T.-M. is thankful for his collaboration fellowship with the Spanish Ministry of Education (MEC) during the MS studies at ULL. V.P. acknowledges funding from the Spanish Ministry of Economy (MINECO) project ref. MAT2017-89207-R. A.B.L. thanks the DIAD Group ES for financial support.

Conflicts of Interest: The authors declare no conflict of interest.

References

1. Rocío-Bautista, P.; González-Hernández, P.; Pino, V.; Pasán, J.; Afonso, A.M. Metal-organic frameworks as novel sorbents in dispersive-based microextraction approaches. *Trac Trends Anal. Chem.* **2017**, *90*, 114–134. [CrossRef]
2. Maya, F.; Cabello, C.P.; Frizzarin, R.M.; Estela, J.M.; Palomino, G.T.; Cerdà, V. Magnetic solid-phase extraction using metal-organic frameworks (MOFs) and their derived carbons. *Trac Trends Anal. Chem.* **2017**, *90*, 142–152. [CrossRef]
3. Hashemi, B.; Zohrabi, P.; Raza, N.; Kim, K.-H. Metal-organic frameworks as advanced sorbents for the extraction and determination of pollutants from environmental, biological, and food media. *Trac Trends Anal. Chem.* **2017**, *97*, 65–82. [CrossRef]
4. González-Hernández, P.; Gutiérrez-Serpa, A.; Rocío-Bautista, P.; Pasán, J.; Ayala, J.H.; Pino, V. Micro-Solid-Phase Extraction Using Metal-Organic Frameworks. In *Metal Organic Frameworks*; Mittal, V., Ed.; Hardcover Central West Publishing: Orange, Australia, 2019; pp. 99–136.
5. Gutiérrez-Serpa, A.; Pacheco-Fernández, I.; Pasán, J.; Pino, V. Metal-organic frameworks as key materials for solid-phase microextraction devices—A review. *Separations* **2019**, *6*, 47, in press. [CrossRef]
6. Lv, S.-W.; Liu, J.-M.; Wang, Z.-H.; Ma, H.; Li, C.-Y.; Zhao, N.; Wang, S. Recent advances on porous organic frameworks for the adsorptive removal of hazardous materials. *J. Environ. Sci.* **2019**, *80*, 169–185. [CrossRef] [PubMed]
7. Burtch, N.C.; Jasuja, H.; Walton, K.S. Water Stability and Adsorption in Metal-Organic Frameworks. *Chem. Rev.* **2014**, *114*, 10575–10612. [CrossRef] [PubMed]
8. Chisvert, A.; Cárdenas, S.; Lucena, R. Dispersive micro-solid phase extraction. *Trac Trends Anal. Chem.* **2019**, *112*, 226–233. [CrossRef]
9. Khezeli, T.; Daneshfar, A. Development of dispersive micro-solid phase extraction based on micro and nano sorbents. *Trac Trends Anal. Chem.* **2017**, *89*, 99–118. [CrossRef]
10. Wen, Y.; Chen, L.; Li, J.; Liu, D.; Chen, L. Recent advances in solid-phase sorbents for sample preparation prior to chromatographic analysis. *Trac Trends Anal. Chem.* **2014**, *59*, 26–41. [CrossRef]
11. Rocío-Bautista, P.; Pino, V.; Pasán, J.; López-Hernández, I.; Ayala, J.H.; Ruiz-Pérez, C.; Afonso, A.M. Insights in the analytical performance of neat metal-organic frameworks in the determination of pollutants of different nature from waters using dispersive miniaturized solid-phase extraction and liquid chromatography. *Talanta* **2018**, *179*, 775–783. [CrossRef] [PubMed]
12. Taima-Mancera, I.; Rocío-Bautista, P.; Pasán, J.; Ayala, J.H.; Ruiz-Pérez, C.; Afonso, A.M.; Lago, A.B.; Pino, V. Influence of Ligand Functionalization of UiO-66-Based Metal-Organic Frameworks When Used as Sorbents in

12. Dispersive Solid-Phase Analytical Microextraction for Different Aqueous Organic Pollutants. *Molecules* **2018**, *23*, 2869. [CrossRef]
13. Lirio, S.; Shih, Y.-H.; Hsiao, S.-Y.; Chen, J.-H.; Chen, H.-T.; Liu, W.-L.; Lin, C.-H.; Huang, H.-Y. Monitoring the Effect of Different Metal Centers in Metal-Organic Frameworks and Their Adsorption of Aromatic Molecules using Experimental and Simulation Studies. *Chem. Eur. J.* **2018**, *24*, 14044–14047. [CrossRef] [PubMed]
14. Boontongto, T.; Siriwong, K.; Burakham, R. Amine-Functionalized Metal-Organic Framework as a New Sorbent for Vortex-Assisted Dispersive Micro-Solid Phase Extraction of Phenol Residues in Water Samples Prior to HPLC Analysis: Experimental and Computational Studies. *Chromatographia* **2018**, *81*, 735–747. [CrossRef]
15. Jiang, J. Computational screening of metal-organic frameworks for CO_2 separation. *Curr. Opin. Green Sust. Chem.* **2019**, *16*, 57–64. [CrossRef]
16. Wen, H.-M.; Li, B.; Li, L.; Lin, R.-B.; Zhou, W.; Qian, G.; Chen, B. A Metal-Organic Framework with Optimized Porosity and Functional Sites for High Gravimetric and Volumetric Methane Storage Working Capacities. *Adv. Mater.* **2018**, *30*, 1704792–1704797. [CrossRef] [PubMed]
17. Ye, J.; Gagliardi, L.; Cramer, C.J.; Truhlar, D.G. Computational screening of MOF-supported transition metal catalysts for activity and selectivity in ethylene dimerization. *J. Catal.* **2018**, *360*, 160–167. [CrossRef]
18. Gao, G.; Li, S.; Li, S.; Wang, Y.; Zhao, P.; Zhang, X.; Hou, X. A combination of computational-experimental study on metal-organic frameworks MIL-53(Al) as sorbent for simultaneous determination of estrogens and glucocorticoids in water and urine samples by dispersive micro-solid-phase extraction coupled to UPLC-MS/MS. *Talanta* **2018**, *180*, 358–367. [CrossRef] [PubMed]
19. Rocío-Bautista, P.; Martínez-Benito, C.; Pino, V.; Pasán, J.; Ayala, J.H.; Ruiz-Pérez, C.; Afonso, A.M. The metal-organic framework HKUST−1 as efficient sorbent in a vortex-assisted dispersive micro solid-phase extraction of parabens from environmental waters, cosmetic creams, and human urine. *Talanta* **2015**, *139*, 13–20. [CrossRef] [PubMed]
20. González-Sálamo, J.; González-Curbelo, M.A.; Hernández-Borges, J.; Rodríguez-Delgado, M.A. Use of Basolite® F300 metal-organic framework for the dispersive solid-phase extraction of phthalic acid esters from water samples prior to LC-MS determination. *Talanta* **2019**, *195*, 236–244. [CrossRef]
21. Zou, D.; Liu, D. Understanding the modifications and applications of highly stable porous frameworks via UiO-66. *Mater. Today Chem.* **2019**, *12*, 139–165. [CrossRef]
22. Bai, Y.; Dou, Y.; Xie, L.-H.; Rutledge, W.; Li, J.-R.; Zhou, H.-C. Zr-based metal-organic frameworks: Design, synthesis, structure, and applications. *Chem. Soc. Rev.* **2016**, *45*, 2327–2367. [CrossRef] [PubMed]
23. Hu, Z.; Zhao, D. *De facto* methodologies toward the synthesis and scale-up production of UiO-66-type metal-organic frameworks and membrane materials. *Dalton Trans.* **2015**, *44*, 19018–19040. [CrossRef] [PubMed]
24. Rocío-Bautista, P.; Taima-Mancera, I.; Pasán, J.; Pino, V. Metal-Organic Frameworks in Green Analytical Chemistry. *Separations* **2019**, *6*, 33. [CrossRef]
25. Mohyuddin, A.; Hussain, D.; Fatima, B.; Athar, M.; Ashiq, M.N.; Najam-ul-Haq, M. Gallic acid functionalized UiO-66 for the recovery of ribosylated metabolites from human urine samples. *Talanta* **2019**, *201*, 23–32. [CrossRef]
26. Lv, S.-W.; Liu, J.-M.; Ma, H.; Wang, Z.-H.; Li, C.-Y.; Zhao, N.; Wang, S. Simultaneous adsorption of methyl orange and methylene blue from aqueous solution using amino functionalized Zr-based MOFs. *Microporous Mesoporous Mat.* **2019**, *282*, 179–187. [CrossRef]
27. Zhang, W.; Yang, J.-M.; Yang, R.-N.; Yang, B.-C.; Quan, S.; Jiang, X. Effect of free carboxylic acid groups in UiO-66 analogues on the adsorption of dyes from water: Plausible mechanisms for adsorption and gate-opening behavior. *J. Mol. Liq.* **2019**, *283*, 160–166. [CrossRef]
28. Burrows, A.D. Mixed-component metal-organic frameworks (MC-MOFs): Enhancing functionality through solid solution formation and surface modifications. *CrystEngComm.* **2011**, *13*, 3623–3642. [CrossRef]
29. Taddei, M.; Tiana, D.; Casati, N.; van Bokhoven, J.A.; Smit, B.; Ranocchiari, M. Mixed-linker UiO-66: Structure-property relationships revealed by a combination of high-resolution powder X-ray diffraction and density theory calculations. *Phys. Chem. Chem. Phys.* **2017**, *19*, 1551–1559. [CrossRef]
30. Garibay, S.J.; Cohen, S.M. Isoreticular synthesis and modification of frameworks with the UiO-66 topology. *Chem. Commun.* **2010**, *46*, 7700–7702. [CrossRef]

31. Chavan, S.M.; Shearer, G.C.; Svelle, S.; Olsbye, U.; Bonino, F.; Ethiraj, J.; Lillerud, K.P.; Bordiga, S. Synthesis and Characterization of Amine-Functionalized Mixed-Ligand Metal-Organic Frameworks of UiO-66 Topology. *Inorg. Chem.* **2014**, *53*, 9509–9515. [CrossRef]
32. Katz, M.J.; Brown, Z.J.; Colón, Y.J.; Siu, P.W.; Scheidt, K.A.; Snurr, R.Q.; Hupp, J.T.; Farha, O.K. A facile synthesis of UiO-66, UiO-67 and their derivatives. *Chem. Commun.* **2013**, *49*, 9449–9451. [CrossRef] [PubMed]
33. Cavka, J.H.; Jakobsen, S.; Olsbye, U.; Guillou, N.; Lamberti, C.; Bordiga, S.; Lillerud, K.P. A New Zirconium Inorganic Building Brick Forming Metal Organic Frameworks with Exceptional Stability. *J. Am. Chem. Soc.* **2008**, *130*, 13850–13851. [CrossRef]
34. Nakamoto, K. Infrared and raman spectra of inorganic and coordination compounds, theory and applications in inorganic chemistry. In *Infrared and Raman Spectra of Inorganic and Coordination Compounds*, 5th ed.; Wiley: Weinheim, Germany, 2006.

Sample Availability: Samples of the compounds are not available from the authors.

© 2019 by the authors. Licensee MDPI, Basel, Switzerland. This article is an open access article distributed under the terms and conditions of the Creative Commons Attribution (CC BY) license (http://creativecommons.org/licenses/by/4.0/).

Article

Dual Emission in a Ligand and Metal Co-Doped Lanthanide-Organic Framework: Color Tuning and Temperature Dependent Luminescence

Despoina Andriotou [1], Stavros A. Diamantis [1], Anna Zacharia [2], Grigorios Itskos [2], Nikos Panagiotou [3], Anastasios J. Tasiopoulos [3] and Theodore Lazarides [1,*]

1. Department of Chemistry, Aristotle University of Thessaloniki, 54124 Thessaloniki, Greece; despoina.andriotou95@gmail.com (D.A.); sdiamant@chem.auth.gr (S.A.D.)
2. Department of Physics, University of Cyprus, 1687 Nicosia, Cyprus; zacharia.anna@ucy.ac.cy (A.Z.); itskos@ucy.ac.cy (G.I.)
3. Department of Chemistry, University of Cyprus, 1687 Nicosia, Cyprus; panagiotou.nikos@ucy.ac.cy (N.P.); atasio@ucy.ac.cy (A.J.T.)
* Correspondence: tlazarides@chem.auth.gr; Tel.: +30-2310-997853

Received: 1 December 2019; Accepted: 15 January 2020; Published: 25 January 2020

Abstract: In this study, we report the luminescence color tuning in the lanthanide metal-organic framework (LnMOF) ([La(bpdc)Cl(DMF)] (**1**); bpdc^{2-} = [1,1′-biphenyl]-4,4′-dicarboxylate, DMF = N,N-dimethylformamide) by introducing dual emission properties in a La^{3+} MOF scaffold through doping with the blue fluorescent 2,2′-diamino-[1,1′-biphenyl]-4,4′-dicarboxylate (dabpdc^{2-}) and the red emissive Eu^{3+}. With a careful adjustment of the relative doping levels of the lanthanide ions and bridging ligands, the color of the luminescence was modulated, while at the same time the photophysical characteristics of the two chromophores were retained. In addition, the photophysical properties of the parent MOF (**1**) and its doped counterparts with various dabpdc^{2-}/bpdc^{2-} and Eu^{3+}/La^{3+} ratios and the photoinduced energy transfer pathways that are possible within these materials are discussed. Finally, the temperature dependence study on the emission profile of a doped analogue containing 10% dabpdc^{2-} and 2.5% Eu^{3+} (**7**) is presented, highlighting the potential of this family of materials to behave as temperature sensors.

Keywords: metal-organic frameworks; luminescence; lanthanides; color tuning; doping; temperature sensors

---

1. Introduction

The unique luminescence properties of trivalent lanthanide ions (Ln^{3+}), including sharp atomic-like emission spectra, which are largely independent of the metal's coordination environment and long lifetimes, reaching the order of a few milliseconds in the cases of Eu^{3+} and Tb^{3+}, make them well suited as luminophores for a diverse range of applications spanning the fields of biotechnology, telecommunications, sensors and lighting [1–4]. Despite their favorable properties, the Laporte forbidden nature of f-f transitions makes luminescence through direct excitation of Ln^{3+} ions extremely inefficient. However, this shortcoming can be tackled through the coordination of Ln^{3+} ions to strongly absorbing chromophores which act as antennae by sensitizing metal-based emission through photoinduced energy transfer [5]. Recently, a considerable amount of research effort has been directed towards the development of lanthanide ratiometric thermometers [6] which are based on the temperature-induced changes in the photophysical behavior of at least two emission centers, thereby providing a more reliable and accurate self-referenced signal with reduced dependence on the experimental conditions. The majority lanthanide luminescent thermometers reported in the literature are based on measuring the ratio between the emission intensities of Tb^{3+} and Eu^{3+} centers at different temperatures [7–10] while those that involve the emission of a bridging ligand [11] or an encapsulated organic dye [12] are relatively rare.

In this contribution, we report the preparation and study of a homologous series of ligand and metal co-doped lanthanide-organic frameworks where a parent framework [La(bpdc)Cl(DMF)] (**1**) [13] is doped with the strongly fluorescent diamino derivative of the $bpdc^{2-}$ bridging ligand $dabpdc^{2-}$ and with the luminescent lanthanide ion Eu^{3+}. The lanthanide MOF (LnMOF) ([La(bpdc)Cl(DMF)] (**1**) was chosen as a doping platform because: (i) its highly reproducible synthesis and chemical robustness, as dry crystals of **1** can be left in air for several months without showing any sign of deterioration; (ii) the $bpdc^{2-}$ bridging ligand has been demonstrated to be a good sensitizer for the luminescence of the Eu^{3+} ion [14–16]; (iii) the possibility to obtain strong Ln^{3+}-based emission due to the absence of water from the coordination sphere of the Ln^{3+} ion which would provide an efficient non-radiative deactivation pathway for f-f excited states through vibrational coupling with O–H oscillators [17]. Thus, following the above mentioned doping procedure, we prepared an isostructural series of materials with the formula $[La_{1-x}Eu_x(bpdc)_{1-y}(dabpdc)_yCl(DMF)]$ (x = 0–0.025; y = 0 or 0.1) which show emission from both chromophores. With careful adjustment of the Eu^{3+} doping percentage while keeping the $dabpdc^{2-}$ doping level at 10%, luminescence color tuning from blue to red through purple was achieved. In addition, a temperature dependent luminescence study of material **7** (x = 0.025; y = 0.1) shows that good temperature sensing action can be obtained in the region from 80 to 180 K with the sensitivity parameter reaching the maximum value 2.51 %K^{-1} at 80 K.

2. Results and Discussion

2.1. Synthesis and Structural Studies

The reaction of the bridging ligand H_2bpdc with $LaCl_3 \cdot xH_2O$ in a 1:1 molar ratio in DMF at 110 °C, afforded a crystalline product **1** with the formula [La(bpdc)Cl(DMF)] which is isostructural to the compound of the same formula reported by Hou et al. in 2013 [13]. Compound **1** crystallizes in the orthorombic Pnma space group and features one crystallographically unique lanthanum cation with a coordination number of nine while its coordination polyhedron can be best described as a tricapped trigonal prism. As seen in Figure 1, the structure of **1** features an infinite rod secondary building unit (SBU) consisting of a zig-zag chain of La^{3+} ions bridged by μ_2 Cl^- anions and by μ_2-η^2:η^1 carboxylate units. Each bridging ligand is connected to two different chains, thus forming a three-dimensional framework with rhombic channels along the crystallographic axis which are occupied by terminally coordinated DMF molecules displaying two-fold positional disorder around the crystallographic mirror plane. Selected bond lengths and bond angles for **1** are listed in Supplementary Table S1.

In agreement with the findings of Hou et al. [13], frameworks isostructural to **1** could only be obtained with early lanthanide ions such as La^{3+}, Pr^{3+} and Nd^{3+} while our attempts to prepare a luminescent Eu^{3+} analog of **1** were met with failure. We therefore decided to follow the strategy of metal doping in order to introduce the luminescent Eu^{3+} ion within the framework of **1**. Thus, a series of reactions in the presence 1–2.5 mol% of Eu^{3+} afforded crystalline products **2**–**9** which are isostructural to **1**, as confirmed by powder X-ray diffraction (pxrd) studies (Figure 2). The Eu^{3+} doped materials displayed the characteristic red luminescence of the Eu^{3+} ion upon being illuminated with a standard laboratory UV lamp (vide infra) thus showing that Eu^{3+} is successfully incorporated within the parent structure. This observation encouraged us to attempt further doping of the parent framework with the intensely blue fluorescent diamino derivative the H_2bpdc bridging ligand 2,2′-diamino-[1,1′-biphenyl]-4,4′-dicarboxylic acid (H_2dabpdc). Indeed, we found that the presence of up to 33 mol% of H_2dabpdc in the initial reaction mixture leads to crystalline products which are isostructural to **1** (Figure 2). In order to gain better insight on the degree of incorporation of $dabpdc^{2-}$ within the parent framework of **1**, we subjected a sample of **2** (0 mol% Eu^{3+} and 10 mol% of H_2dabpdc in the reaction feed) to ^{1}H-NMR analysis after it was digested in a mixture of D_2O/NaOH.

From the ratio of the peak integrals corresponding to bpdc^{2-} and dabpdc^{2-} (Supplementary Figure S1), we calculated a molar fraction of 12% of dabpdc^{2-} within **2** which is close to the molar percentage of dabpdc^{2-} present in the reaction feed. This finding suggests that, in the employed experimental conditions, dabpdc^{2-} is similar to bpdc^{2-} in terms of reactivity towards La^{3+} and at least relatively low percentages of dabpdc^{2-} in the reaction feed result in a statistical distribution of the amino substituted bridging ligand within the product.

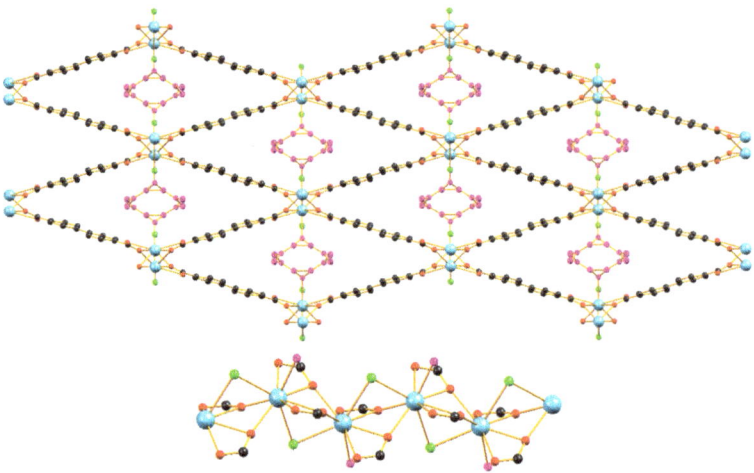

Figure 1. The crystal structure of the parent framework **1** [13] viewed along the a axis. The infinite rod metal SBU of **1** is shown below the main structure. Color code: Lanthanum: turqoise, Carbon: black, Oxygen: red, Chlorine: green. The atoms of the coordinated DMF molecules are shown in magenta.

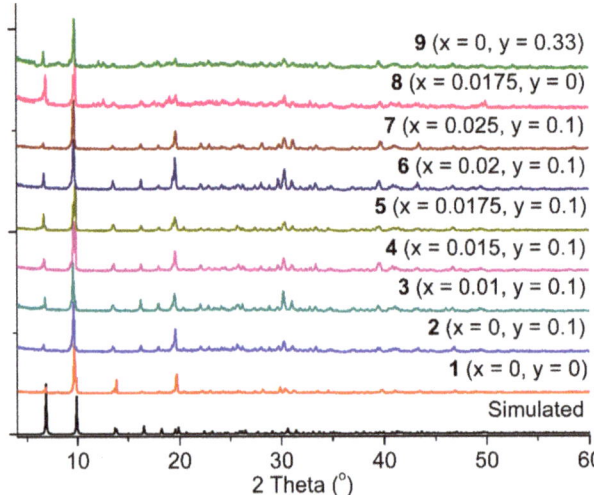

Figure 2. Powder X-ray diffraction paterns of the doped analogues **1–9** with the general formula [La$_{1-x}$Eu$_x$(bpdc)$_{1-y}$(dabpdc)$_y$Cl(DMF)]; the values of x and y corresponding to each doped analogue are shown on the graph.

In addition to the ^{1}H-NMR study, we carried out single crystal X-ray structural analysis on a sample of **9** (0 mol% Eu^{3+} and 33 mol% of H$_2$dabpdc in the reaction feed). Crystal and refinement data can be found in Supplementary Table S2. The overall structure of **9** is virtually identical to that of **1** with the difference that the bridging ligand shows significantly greater disorder and had to be refined in two positions (Figure 3). We were also able to locate one of the two nitrogen atoms of dabpdc^{2-} which refined well with a given site occupancy of ca. 17%. In addition, several constraints were applied in order to keep the C–N distance and the angles around the C–N bond within chemically acceptable values. The disorder of the bridging ligand in **9** is possibly a result of the different conformations adopted by bpdc^{2-} and dabpdc^{2-}. In the parent compound **1**, the bpdc^{2-} ligand adopts a conformation where the two phenylene groups of the biphenyl spacer are virtually co-planar [13] a feature that is commonly found in structures containing the bpdc^{2-} ligand [18–26] while, as observed by us [27] and others [28], the dabpdc^{2-} bridging ligand tends to adopt a staggered syn conformation where the dihedral angle between the two phenylene groups is in the order of 60–70°. It is therefore reasonable to expect that the presence of about one third dabpdc^{2-} at the sites normally occupied by bpdc^{2-} in the parent framework would induce some additional disorder to the diphenylene spacer. Consequently the ^{1}H-NMR and crystallographic data indicate that even though the dabpdc^{2-} moiety seems to have slightly different stereochemical demands than the bpdc^{2-} ligand of the parent framework, its incorporation does not induce a big distortion to the overall structure.

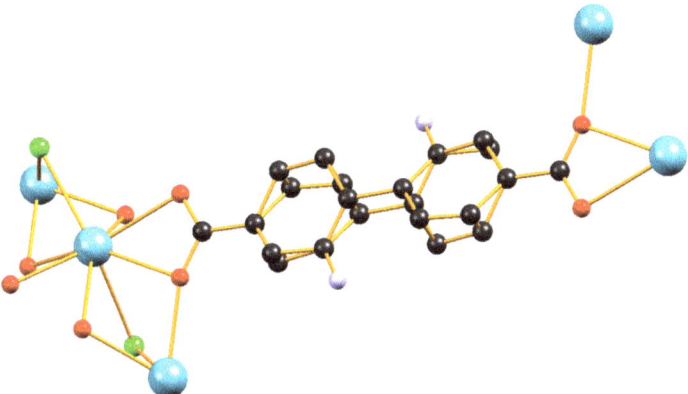

Figure 3. Partial view of the crystal structure of **9** highlighting the disorder of the biphenyl bridging unit. The nitrogen atoms were refined with a site occupancy of ca. 17%. Hydrogen atoms and the coordinated DMF molecules are omitted for clarity. Color code: Lanthanum: turqoise, Carbon: black, Oxygen: red, Nitrogen: blue, Chlorine: green.

2.2. Thermogravimetric Analysis

The thermal stability of **1**, **2** and **6** was studied by thermogravimetric analysis (TGA) under air (see Supplementary Figures S2 and S3). All the analogues show essentially identical behavior and for this reason only the thermograph of **6** (Figure 4) shall be discussed. In particular, the weight loss in **6** was observed in two steps. The first step is observed in the temperature range 240–330 °C and the corresponding weight loss is attributed to the coordinated DMF molecules (experimental loss: 16.83%, theoretically estimated loss of 14.9%). The framework remains thermally stable up to ~500 °C and then the second mass loss step appears, corresponding to the decomposition of the framework, which is completed up to ~525 °C.

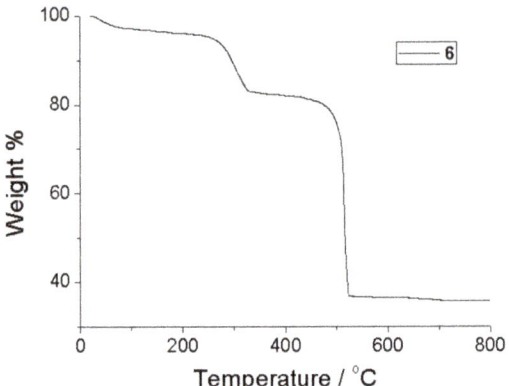

Figure 4. The TGA curve of compound **6**.

2.3. Luminescence Properties

Photophysical studies on microcrystalline powders of the parent framework **1** and its metal and ligand doped counterparts were carried out by emission spectroscopy. Excitation of compound 1 at λ_{exc} = 365 nm gives rise to a fluorescence band with maximum at ca. 440 nm (Figure 5), which is attributed to the radiative deactivation of the lowest energy $^1\pi$-π^* excited state of the bpdc^{2-} bridging ligand [13,14]. The excitation spectrum of **1** (monitored at 450 nm) shows that the lowest energy absorption feature is at 375 nm and tails off rapidly after 400 nm, thereby showing virtually no absorption in the visible region (Figure 5). When the parent framework of **1** is doped 10 mol% with the diamino derivative dabpdc^{2-} (**2**), a rather small red shift in the fluorescence peak which maximizes at ca. 463 nm was observed. Based on a comparison with our previous work on the fluorescence properties of Ca^{2+} and Sr^{2+} MOFs featuring (NH$_2$)$_2$bpdc^{2-} as bridging ligand, we attribute the red shifted emission signal of **2** predominantly to the fluorescence from the dabpdc^{2-} chromophore [27]. From the onsets of the emission peaks of the two chromophores in **1** and **2** the energies of the lowest lying $^1\pi$-π^* of excited states of bpdc^{2-} and dabpdc^{2-} were estimated at ca. 25,000 and 23,500 cm^{-1} respectively [29]. It therefore follows that initial excitation of predominantly the bpdc^{2-} moiety of **2** (mainly due to its much higher abundance within the material's framework) is followed by energy transfer to the dabpdc^{2-} chromophore most possibly through a mechanism involving exciton diffusion to a position adjacent to a dabpdc^{2-} group and subsequent coulombic (Förster) energy transfer to the latter [30–34]. It is important to mention that the emission profiles of samples doped with significantly larger percentages of dabpdc^{2-} (such as sample **9**) are virtually identical to that of **2**, thus confirming that in the latter material interchromophore energy transfer reaches its maximum efficiency.

The emission spectrum of compound **8** (λ_{exc} = 365 nm), where the parent framework of **1** is doped only with 1.75 mol% Eu^{3+}, is dominated by the Eu^{3+}-based $^5D_0 \rightarrow {}^7F_J$ (J = 0–4) emission peaks which are located at 580, 596, 620, 654 and 704 nm respectively. On the high energy region of the spectrum, we observe the relatively weak residual emission of the bpdc^{2-} bridging ligand which maximizes at ca. 440 nm indicating that, even at this relatively low doping level of 1.75 mol%, ligand-to-Eu^{3+} energy transfer is quite efficient [1,35]. The fact that the bpdc^{2-} bridging ligand is a good sensitizer for both dabpdc^{2-} and Eu^{3+} prompted us to synthesize and study frameworks where both energy acceptors are present within the parent framework of **1**.

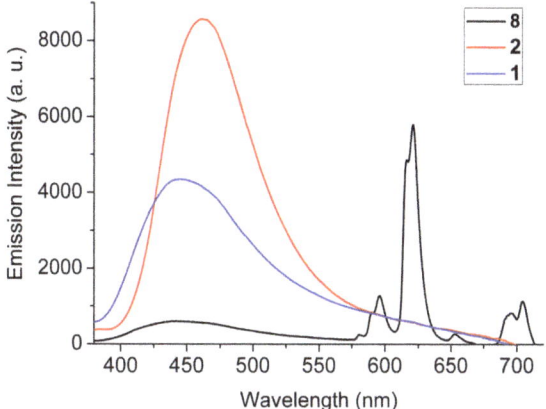

Figure 5. The solid-state emission spectra of compounds **1**, **2** and **8** upon excitation at 365 nm. See main text for details.

In the case of **6**, where both dabpdc^{2-} and Eu^{3+} are doped into the parent framework of **1**, excitation at 365 nm results in emission from both the amino substituted organic chromophore and the luminescent lanthanide ion (Figure 6). The maxima of the $^5D_0 \rightarrow {}^7F_J$ (J = 0–4) peaks of Eu^{3+} in **6** are in the expected positions while the fluorescence from the dabpdc^{2-} chromophore occupies the blue region of the spectrum showing a maximum at ca. 470 nm (vide supra). The excitation spectra of **6** were measured monitoring at both the ligand (470 nm) and Eu^{3+} (620 nm) emissions and are shown in Figure 6. We observe that upon monitoring the dabpdc^{2-} ligand emission, the excitation spectrum of **6** is dominated by the absorption features of the bpdc^{2-} chromophore, while monitoring at the Eu^{3+} emission results in an excitation spectrum showing a relatively weak albeit clear shoulder at ca. 425 nm which tails off above 445 nm. This spectral feature is attributed to the absorption of the dabpdc^{2-} chromophore [27] and indicates that the latter may also sensitize Eu^{3+} emission.

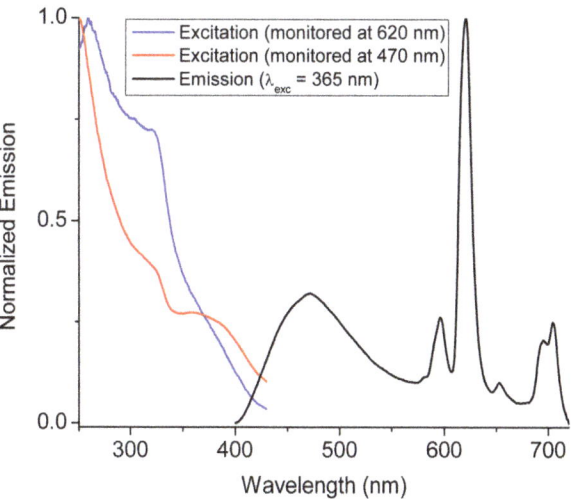

Figure 6. The solid-state emission and excitation spectra of compound **6**. The excitation spectra are monitored both at the dabpdc^{2-} (470 nm) fluorescence and the Eu^{3+} emission (620 nm).

Starting from **2** and progressively doping the framework with increased levels of Eu^{3+} (*vide supra*) results in increased lanthanide-based emission with a concomitant decrease of the contribution from the organic chromophore. The change in color of the materials doped with 10 mol% of $dabpdc^{2-}$ and increasing levels of Eu^{3+} (0–2.5 mol%) is demonstrated by the CIE (Commission Internationale de l'Éclairage) coordinates of the chromaticity diagram of Figure 7, where we see that the emission color gradually shifts from the blue (x = 0.172, y = 0.173 for **2**) to the purple-red region of the spectrum (x = 0.425, y = 0.289 for **8**). However, the absence of a strong yellow-green component in the emission spectra of this series of materials does not allow sufficient color tuning in order to achieve entry in the white region (x and y values of 0.3 and above) [36]. Instead, a further increase of the Eu^{3+} doping levels results in the emission color traversing the purple region and eventually entering the red region.

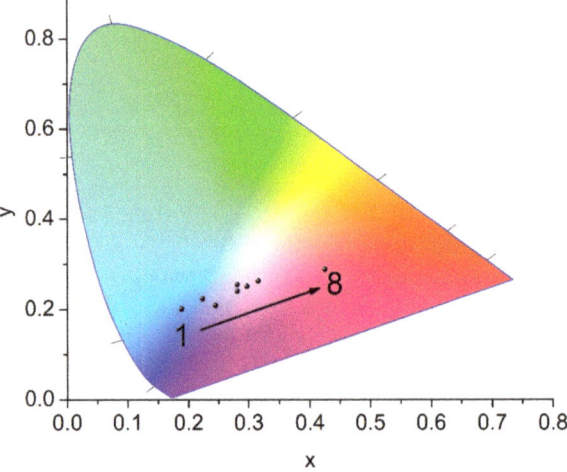

Figure 7. Chromaticity coordinates (CIE 1931) calculated from the corrected emission profiles of materials **1–8** showing the gradual shift from blue to red upon increasing the Eu^{3+} content.

Nonetheless, the presence of two clear emission components in **7**, arising from an organic chromophore and a lanthanide, prompted us to study the effect of temperature on its emission profile in order to explore the potential of the material to perform as a ratiometric luminescence thermometer [8,37–40]. Figure 8 shows the emission spectra of **7** at various temperatures from 80 to 300 K. By examining the spectra of Figure 8, we observe that the ligand component shows a steady decrease in intensity with rising temperature while the Eu^{3+} luminophore shows distinctly different behavior in two different temperature regions. In the region between 80 and ca. 140 K, we observe an increase of Eu^{3+} emission intensity, while in the region between 150 and 300 K, the emission intensity of Eu^{3+} follows the steady decrease of that of the organic component. The decrease in the fluorescence intensity of the organic component as the temperature rises can be mainly attributed to the increasing participation of non-radiative pathways in the decay process of the $dabpdc^{2-}$ chromophore [29]. The initial rise of the Eu^{3+} component between 80 and 150 K possibly indicates that, at that temperature range, ligand-to-metal energy transfer is a major non-radiative deactivation pathway for the $dabpdc^{2-}$ excited state. At higher temperatures, increased molecular and lattice vibrations render thermal deactivation pathways dominant, and thereby lead to the observed reduction in the intensities of both the organic and Eu^{3+} emission components.

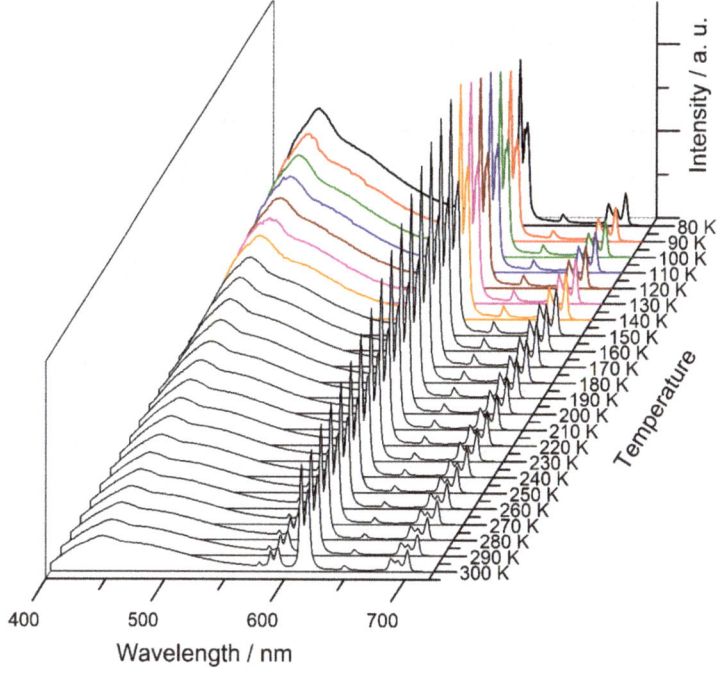

Figure 8. Temperature dependent emission spectra of **7** upon excitation at 375 nm. The spectra in the temperature region between 80 and 140 K are highlighted to emphasize the increase in Eu^{3+}-based emission with rising temperature (see main text).

If we define the ratio of the integrated intensities of the Eu^{3+}-based $^5D_0 \rightarrow {}^7F_2$ emission peak (I_{Eu}) to the ligand-based fluorescence signal (I_L) as the thermometric parameter Δ and plot the result against the temperature T, we obtain the diagram of Figure 9. The data can be fitted satisfactorily (correlation coefficient $R^2 = 0.984$) to a second-degree polynomial (Equation (1)).

$$\Delta = -1.05 \times 10^{-5}\, T^2 + 4.90 \times 10^{-3}\, T - 0.21 \tag{1}$$

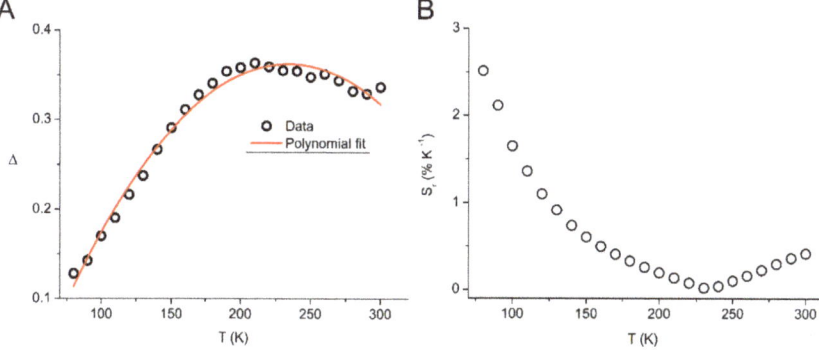

Figure 9. (**A**) The thermometric parameter versus temperature for material **7**. The red line represents a polynomial fit to the experimental data. (**B**) The S_r parameter versus temperature for material **7**. See main text for details.

The performance of a temperature sensor is often reported in terms of its relative sensitivity [41], S_r, which serves as a figure of merit to allow comparison between different temperature sensors reported in the literature and is defined in Equation (2) [8,42].

$$S_r = \frac{1}{\Delta}\left|\frac{\partial \Delta}{\partial T}\right| \quad (2)$$

The S_r parameter of **7** as %K^{-1} is plotted against the temperature in Figure 9B. The maximum value of the relative sensitivity S_m = 2.51 %K^{-1} is obtained at 80 K. These results indicate that **7** can function as a ratiometric luminescence thermometer in the tested region from 80 to 300 K showing its best performance in the 80 to 150 K region. The maximum relative sensitivity of 2.51 %K^{-1} shown by **7** compares well with the typical values obtained with many luminescent thermometers based on lanthanide organic frameworks [8–10,43]. Therefore, combined ligand and metal doping can be an effective route for the preparation of improved luminescent temperature sensors.

3. Conclusions

We demonstrated that the parent framework of [La(bpdc)Cl(DMF)] (**1**) [13] can be easily doped with the bridging ligand, dabpdc^{2-} (the diamino derivative of bpdc^{2-} ligand present in **1**), and Eu^{3+} to produce a range of mixed ligand and mixed metal analogues. Doping of ca. 10 mol% with dabpdc^{2-} (**2**) results in a moderate red shift of the ligand-based fluorescence due to interligand photoinduced energy transfer. Further doping with various amounts of Eu^{3+} yields materials which display sensitized Eu^{3+} emission along with the blue dabpdc^{2-} fluorescence. Emission color tuning was achieved by varying the Eu^{3+} doping percentage from 0 to 2.5 mol%, from blue to red through purple. However, the absence of a yellow-green component in the emission spectra of this series did not allow the achievement of white light luminescence. Finally, we studied the temperature dependence of the emission profile of **7** (10 mol% dabdc^{2-}, 2.5 mol% Eu^{3+}) in the region from 80 to 300 K. While the ligand component shows a steady decrease in fluorescence intensity with increasing temperature, the Eu^{3+} luminescence shows an initial enhancement from 80 to ca. 150 K before following the trend of the organic portion of the emission spectrum of **7**. We attribute the initial enhancement of Eu^{3+} luminescence to the dominance of ligand-to-metal energy transfer over thermal decay pathways at relatively low temperatures. Compound **7** shows good potential as a ratiometric luminescence thermometer displaying its best performance in the 80 to 150 K region with maximum sensitivity of 2.51 %K^{-1} at 80 K. This result shows that combined ligand and metal doping can be a viable route to produce new luminescence-based temperature sensors. Our group is currently working towards the construction of mixed lanthanide–organic frameworks exhibiting white luminescence and luminescence-based temperature sensing properties by following the above described strategy. The results obtained from the current study can be a stepping-stone towards the construction of superior optical materials.

4. Materials and Methods

4.1. Synthesis

Starting materials and solvents were purchased from the usual commercial sources (Sigma-Aldrich and Alfa Aesar) and were used as received.

4.1.1. Synthesis of H$_2$dabpdc

Dimethyl-2,2'-dinitro-[1,1'-biphenyl]-4,4'-dicarboxylate. Dimethyl-biphenyl-4,4'-dicarboxylate (1.00 g, 3.7 mmol) was added into concentrated H$_2$SO$_4$ (10 mL). The mixture was stirred at room temperature for 10 min. Nitric acid (760 µL, 3 eq.) was added into concentrated H$_2$SO$_4$ (2 mL). This solution was added dropwise into the first mixture at room temperature over a period of 20 min. The mixture was stirred at room temperature for 4 h and then poured into ice (300 mL) to form a beige solid. The resulting solid was dissolved in dichloromethane and the aqueous phase was extracted with dichloromethane (3 × 70 mL). The combined organic layers were dried over Na_2SO_4 and evaporated under reduced pressure to afford a beige solid. The crude mixture was recrystallized from 2-propanol and washed with diethyl ether to give the pure product. Yield: 1.2 g (3.33 mmol, 90%). ^{1}H NMR (500 MHz, $CDCl_3$): δ (ppm): 8.90 (s, 2H), 8.37 (d, J = 8.4 Hz, 2H), 7.40 (d, J = 8 Hz, 2H), 4.02 (s, 6H).

Dimethyl-2,2'-diamino-[1,1'-biphenyl]-4,4'-dicarboxylate. Dimethyl-2,2'-dinitro-[1,1'-biphenyl]-4,4'-dicarboxylate (1.2 g, 3.33 mmol) was dissolved in 20 mL acetic acid and the solution was stirred under Ar atmosphere for 10 min. To this solution was added iron powder (3.7 g, 10 eq.) and the resulting mixture was stirred at room temperature for 24 h. The suspension was filtered through celite, washed with 40 mL acetic acid and the filtrate was concentrated under reduced pressure. The solid was dissolved in ethyl acetate (80 mL) and was extracted with saturated aqueous sodium carbonate solution (2 × 80 mL) and H_2O (3 × 80 mL). The combined organic layers were dried over Na_2SO_4 and the solution was evaporated under vacuum to yield the product as a yellow solid. Yield 900 mg (3 mmol, 90%). ^{1}H NMR (500 MHz, DMSO-d_6): δ (ppm): 7.42 (s, 2H), 7.21 (d, J = 7.8 Hz, 2H), 7.06 (d, J = 7.8 Hz, 2H), 4.97 (s, 4H), 3.81 (s, 6H).

2,2'-Diamino-[1,1'-biphenyl]-4,4'-dicarboxylic acid (H_2dabpdc). Dimethyl 2,2'-diamino-[1,1'-biphenyl]-4,4'-dicarboxylate (900 mg, 3.0 mmol) was dissolved in THF (20 mL) and 20 mL of aqueous NaOH (0.6 M) were added dropwise under vigorous stirring. The mixture was stirred overnight at room temperature. The organic solvent was removed under vacuum and the aqueous solution was acidified with acetic acid to yield a light brown solid (735 mg, 2.7 mmol, 90%). ^{1}H-NMR (500 MHz, DMSO-d_6): δ (ppm) = 12.64 (br, 2H), 7.40 (s, 2H), 7.20 (d, J = 7.6 Hz, 2H), 7.40 (d, J = 7.8 Hz, 2H), 4.90 (br, 4H).

4.1.2. Synthesis of MOFs

Synthesis of [La(bpdc)Cl(DMF)] (**1**). $LaCl_3·7H_2O$ (74.8 mg, 0.2 mmol) and biphenyl-4,4'-dicarboxylic acid (48.4 mg, 0.2 mmol) were added in DMF (3 mL) and stirred until the solids were fully dissolved. The resulting solution was sealed in a screw cap 23 mL scintillation vial and placed in a preheated oven at 110 °C where it remained undisturbed for 24 h before being cooled to room temperature. Colorless needle-like crystals were isolated by filtration, washed with DMF (5 × 3 mL), and dried under vacuum overnight. Yield 32 mg (45%).

Synthesis of the [$La_{1-x}Eu_x$(bpdc)$_{1-y}$(dabpdc)$_y$Cl(DMF)] series. $LaCl_3·7H_2O$ (74.8 mg, 0.2 mmol) and biphenyl-4,4'-dicarboxylic acid (48.4 mg, 0.2 mmol) were added in DMF (3 mL) and stirred until the solids were fully dissolved. Standard solutions of $EuCl_3·6H_2O$ (10^{-2} M) and H_2dabdc (10^{-2} M) in DMF were prepared and calculated volumes were added in the reaction feed using a volumetric pipette, while the mixture was magnetically stirred, to achieve the desired La^{3+}/Eu^{3+} and H_2bpdc/H_2dabpdc molar ratio. Otherwise, the procedure was identical to that for the synthesis of **1**. Colorless or pale-yellow needle-like crystals were isolated by filtration, washed with DMF (5 × 3 mL) and dried under vacuum overnight. Exact percentages of Eu^{3+} and dabpdc^{2-} in each doped analogue and reaction yields are shown in Table 1.

Table 1. Percentages of H_2dabpdc and Eu^{3+} doping [1] and yields.

Compound	mol% H_2dabpdc	mol% Eu^{3+}	Yield
2	10	0	42
3	10	1.00	41
4	10	1.50	40
5	10	1.75	45
6	10	2.00	41
7	10	2.50	44
8	0	1.75	39
9	33	0	32

[1] Molar percentage of dopant in the reaction mixture.

4.2. Physical Measurements and Crystallogtraphy

Photoluminescence spectra. The emission spectra were measured on a Horiba fluorescence spectrometer equipped with a powder sample holder. The light source was a 450 W Xenon Arc Lamp (220–1000 nm) and the detector a red sensitive Hamamatsu R928 photomultiplier tube. All spectra were corrected for instrument response using the correction function generated after calibration of the instrument with a standard light source. Appropriate long pass filters were used to remove scattering from the sample and the monochromators. Temperature-dependent photoluminescence measurements were carried out in the 80–300 K range by placing the samples in the cold finger of a Janis VPF liquid nitrogen optical cryostat.

^{1}H-NMR. ^{1}H-NMR spectra were recorded at room temperature on NMR Agilent 500 MHz, with the use of the solvent proton as an internal standard and on an Avance Brucker NMR spectrometer (500 MHz).

PXRD measurements. PXRD diffraction patterns were recorded on a Shimadzu 6000 Series X-ray diffractometer with a Cu K$_\alpha$ source (λ = 1.5418 Å).

X-ray Crystal Structure Determination. Single crystal X-ray diffraction data were collected on a Rigaku Oxford-Diffraction Supernova diffractometer, equipped with a CCD area detector utilizing Cu Kα (λ = 1.5418 Å) radiation. A suitable crystal was mounted on a Hampton cryoloop with Paratone-N oil and transferred to a goniostat where it was cooled for data collection. Empirical absorption corrections (multiscan based on symmetry-related measurements) were applied using CrysAlis RED software [44]. The structure was solved by direct methods using SIR2004 [45] and refined on F^2 using full-matrix least-squares with SHELXL-2014/7 [46] within the WinGX [47] platform. Software packages used were as follows: CrysAlis CCD for data collection [44], CrysAlis RED for cell refinement and data reduction [44], and MERCURY [48] for molecular graphics. The non-H atoms were treated anisotropically, except for those that belong to disordered parts. The aromatic H atoms were placed in calculated, ideal positions and refined depending on their respective carbon atoms. Selected crystal data for **9** are summarized in Table S2.

Thermogravimetric Analysis (TGA). Thermal stability studies were performed with a Shimadzu TGA 50 thermogravimetric analyzer. Thermal analysis was conducted from 25 to 800 °C under air with a heating rate of 10 °C min^{-1}.

Supplementary Materials: The following are available online, Table S1: Selected bond lengths and angles for 1., Table S2: Crystal and refinement data for 9, Figure S1: 1H-NMR spectrum of a digested sample of 2 in D2O/NaOH., Figure S2: The TGA curve for 1, Figure S3: The TGA curve for **2**.

Author Contributions: T.L. designed research and wrote the paper; D.A. and S.A.D. preformed the syntheses; D.A., A.Z., N.P., A.J.T., G.I. and T.L. contributed in photophysical studies, structural characterization and data interpretation. All authors have read and agreed to the published version of the manuscript.

Funding: This research is co-financed by Greece and the European Union (European Social Fund—ESF) through the Operational Programme «Human Resources Development, Education and Lifelong Learning» in the context of the project "Strengthening Human Resources Research Potential via Doctorate Research-2nd Cycle" (MIS-5000432), implemented by the State Scholarships Foundation (IKΥ).

Acknowledgments: S.A.D. wishes to thank the IKY foundation for a PhD scholarship.

Conflicts of Interest: The authors declare no conflict of interest.

References

1. Bünzli, J.-C.G. Chapter 287—Lanthanide Luminescence: From a Mystery to Rationalization, Understanding, and Applications. In *Handbook on the Physics and Chemistry of Rare Earths*; Bünzli, J.-C.G., Pecharsky, V.K., Eds.; Elsevier: Amsterdam, The Netherlands, 2016; Volume 50, pp. 141–176.
2. Bünzli, J.-C.G. On the design of highly luminescent lanthanide complexes. *Coord. Chem. Rev.* **2015**, *293*, 19–47. [CrossRef]

3. Aletti, A.B.; Gillen, D.M.; Gunnlaugsson, T. Luminescent/colorimetric probes and (chemo-) sensors for detecting anions based on transition and lanthanide ion receptor/binding complexes. *Coord. Chem. Rev.* **2018**, *354*, 98–120. [CrossRef]
4. Armelao, L.; Quici, S.; Barigelletti, F.; Accorsi, G.; Bottaro, G.; Cavazzini, M.; Tondello, E. Design of luminescent lanthanide complexes: From molecules to highly efficient photo-emitting materials. *Coord. Chem. Rev.* **2010**, *254*, 487–505. [CrossRef]
5. Bünzli, J.-C.G.; Piguet, C. Taking advantage of luminescent lanthanide ions. *Chem. Soc. Rev.* **2005**, *34*, 1048–1077. [CrossRef] [PubMed]
6. Dramicanin, M.D. Sensing temperature via downshifting emissions of lanthanide-doped metal oxides and salts. A review. *Methods Appl. Fluoresc.* **2016**, *4*, 042001. [CrossRef]
7. Cui, Y.J.; Zhu, F.L.; Chen, B.L.; Qian, G.D. Metal-organic frameworks for luminescence thermometry. *Chem. Commun.* **2015**, *51*, 7420–7431. [CrossRef]
8. Rocha, J.; Brites, C.D.S.; Carlos, L.D. Lanthanide Organic Framework Luminescent Thermometers. *Chem. Eur. J.* **2016**, *22*, 14782–14795. [CrossRef]
9. Abdelhameed, R.M.; Ananias, D.; Silva, A.M.S.; Rocha, J. Luminescent Nanothermometers Obtained by Post-Synthetic Modification of Metal-Organic Framework MIL-68. *Eur. J. Inorg. Chem.* **2019**, *2019*, 1354–1359. [CrossRef]
10. Zhao, D.; Yue, D.; Jiang, K.; Zhang, L.; Li, C.; Qian, G. Isostructural Tb^{3+}/Eu^{3+} Co-Doped Metal–Organic Framework Based on Pyridine-Containing Dicarboxylate Ligands for Ratiometric Luminescence Temperature Sensing. *Inorg. Chem.* **2019**, *58*, 2637–2644. [CrossRef]
11. D'Vries, R.F.; Alvarez-Garcia, S.; Snejko, N.; Bausa, L.E.; Gutierrez-Puebla, E.; de Andres, A.; Monge, M.A. Multimetal rare earth MOFs for lighting and thermometry: Tailoring color and optimal temperature range through enhanced disulfobenzoic triplet phosphorescence. *J. Mater. Chem. C* **2013**, *1*, 6316–6324. [CrossRef]
12. Cui, Y.J.; Song, R.J.; Yu, J.C.; Liu, M.; Wang, Z.Q.; Wu, C.D.; Yang, Y.; Wang, Z.Y.; Chen, B.L.; Qian, G.D. Dual-Emitting MOF superset of Dye Composite for Ratiometric Temperature Sensing. *Adv. Mater.* **2015**, *27*, 1420–1425. [CrossRef] [PubMed]
13. Jia, L.-N.; Hou, L.; Wei, L.; Jing, X.-J.; Liu, B.; Wang, Y.-Y.; Shi, Q.-Z. Five sra Topological Ln(III)-MOFs Based on Novel Metal-Carboxylate/Cl Chain: Structure, Near-Infrared Luminescence and Magnetic Properties. *Cryst. Growth Des.* **2013**, *13*, 1570–1576. [CrossRef]
14. Han, Y.-F.; Zhou, X.-H.; Zheng, Y.-X.; Shen, Z.; Song, Y.; You, X.-Z. Syntheses, structures, photoluminescence, and magnetic properties of nanoporous 3D lanthanide coordination polymers with 4,4′-biphenyldicarboxylate ligand. *CrystEngComm* **2008**, *10*, 1237–1242. [CrossRef]
15. Amghouz, Z.; García-Granda, S.; García, J.R.; Ferreira, R.A.S.; Mafra, L.; Carlos, L.D.; Rocha, J. Series of Metal Organic Frameworks Assembled from Ln(III), Na(I), and Chiral Flexible-Achiral Rigid Dicarboxylates Exhibiting Tunable UV–vis–IR Light Emission. *Inorg. Chem.* **2012**, *51*, 1703–1716. [CrossRef] [PubMed]
16. Chatenever, A.R.K.; Warne, L.R.; Matsuoka, J.E.; Wang, S.J.; Reinheimer, E.W.; LeMagueres, P.; Fei, H.; Song, X.; Oliver, S.R.J. Isomorphous Lanthanide Metal–Organic Frameworks Based on Biphenyldicarboxylate: Synthesis, Structure, and Photoluminescent Properties. *Cryst. Growth Des.* **2019**, *19*, 4854–4859. [CrossRef]
17. Horrocks, W.D.; Sudnick, D.R. Lanthanide ion probes of structure in biology. Laser-induced luminescence decay constants provide a direct measure of the number of metal-coordinated water molecules. *J. Am. Chem. Soc.* **1979**, *101*, 334–340. [CrossRef]
18. Gao, M.-L.; Wang, W.-J.; Liu, L.; Han, Z.-B.; Wei, N.; Cao, X.-M.; Yuan, D.-Q. Microporous Hexanuclear Ln(III) Cluster-Based Metal–Organic Frameworks: Color Tunability for Barcode Application and Selective Removal of Methylene Blue. *Inorg. Chem.* **2017**, *56*, 511–517. [CrossRef]
19. Wang, J.; Lu, G.; Liu, Y.; Wu, S.-G.; Huang, G.-Z.; Liu, J.-L.; Tong, M.-L. Building Block and Directional Bonding Approaches for the Synthesis of {DyMn₄} n (n = 2, 3) Metallacrown Assemblies. *Cryst. Growth Des.* **2019**, *19*, 1896–1902. [CrossRef]
20. Volkringer, C.; Marrot, J.; Férey, G.; Loiseau, T. Hydrothermal Crystallization of Three Calcium-Based Hybrid Solids with 2,6-Naphthalene- or 4,4′-Biphenyl-Dicarboxylates. *Cryst. Growth Des.* **2008**, *8*, 685–689. [CrossRef]
21. Qiao, Y.; Li, Z.-M.; Wang, X.-B.; Guan, W.-S.; Liu, L.-H.; Liu, B.; Wang, J.-K.; Che, G.-B.; Liu, C.-B.; Lin, X. Thermal behaviors and adsorption properties of two Europium(III) complexes based on 2-(4-carboxyphenyl) imidazo [4, 5-f]-1,10-phenanthroline. *Inorg. Chim. Acta* **2018**, *471*, 397–403. [CrossRef]

22. Yan, W.; Zhang, C.; Chen, S.; Han, L.; Zheng, H. Two Lanthanide Metal–Organic Frameworks as Remarkably Selective and Sensitive Bifunctional Luminescence Sensor for Metal Ions and Small Organic Molecules. *ACS Appl. Mater. Interfaces* **2017**, *9*, 1629–1634. [CrossRef] [PubMed]
23. Zhou, J.-M.; Shi, W.; Li, H.-M.; Li, H.; Cheng, P. Experimental Studies and Mechanism Analysis of High-Sensitivity Luminescent Sensing of Pollutional Small Molecules and Ions in Ln4O4 Cluster Based Microporous Metal–Organic Frameworks. *J. Phys. Chem. C* **2014**, *118*, 416–426. [CrossRef]
24. Zhang, S.; Yang, Y.; Xia, Z.-Q.; Liu, X.-Y.; Yang, Q.; Wei, Q.; Xie, G.; Chen, S.-P.; Gao, S.-L. Eu-MOFs with 2-(4-Carboxyphenyl) imidazo [4,5-f]-1,10-phenanthroline and Ditopic Carboxylates as Coligands: Synthesis, Structure, High Thermostability, and Luminescence Properties. *Inorg. Chem.* **2014**, *53*, 10952–10963. [CrossRef] [PubMed]
25. Min, Z.; Singh-Wilmot, M.A.; Cahill, C.L.; Andrews, M.; Taylor, R. Isoreticular Lanthanide Metal-Organic Frameworks: Syntheses, Structures and Photoluminescence of a Family of 3D Phenylcarboxylates. *Eur. J. Inorg. Chem.* **2012**, *2012*, 4419–4426. [CrossRef]
26. Liu, C.; Eliseeva, S.V.; Luo, T.-Y.; Muldoon, P.F.; Petoud, S.; Rosi, N.L. Near infrared excitation and emission in rare earth MOFs via encapsulation of organic dyes. *Chem. Sci.* **2018**, *9*, 8099–8102. [CrossRef]
27. Diamantis, S.A.; Pournara, A.D.; Hatzidimitriou, A.G.; Manos, M.J.; Papaefstathiou, G.S.; Lazarides, T. Two new alkaline earth metal organic frameworks with the diamino derivative of biphenyl-4,4′-dicarboxylate as bridging ligand: Structures, fluorescence and quenching by gas phase aldehydes. *Polyhedron* **2018**, *153*, 173–180. [CrossRef]
28. Dikhtiarenko, A.; Olivos Suarez Alma, I.; Pustovarenko, A.; García-Granda, S.; Gascon, J. Crystal structure of 2,2′-diamino-[1,1′-biphenyl]-4,4′-dicarboxylic acid dihydrate, $C_{14}H_{16}N_2O_6$. *Z. Kristallogr. NCS* **2016**, *231*, 65–67. [CrossRef]
29. Lakowicz, J.R. *Principles of Fluorescence Spectroscopy*; Springer Science: New York, NY, USA, 2006.
30. Kent, C.A.; Mehl, B.P.; Ma, L.Q.; Papanikolas, J.M.; Meyer, T.J.; Lin, W.B. Energy Transfer Dynamics in Metal-Organic Frameworks. *J. Am. Chem. Soc.* **2010**, *132*, 12767–12769. [CrossRef]
31. Lin, J.X.; Hu, X.Q.; Zhang, P.; Van Rynbach, A.; Beratan, D.N.; Kent, C.A.; Mehl, B.P.; Papanikolas, J.M.; Meyer, T.J.; Lin, W.B.; et al. Triplet Excitation Energy Dynamics in Metal-Organic Frameworks. *J. Phys. Chem. C* **2013**, *117*, 22250–22259. [CrossRef]
32. So, M.C.; Wiederrecht, G.P.; Mondloch, J.E.; Hupp, J.T.; Farha, O.K. Metal-organic framework materials for light-harvesting and energy transfer. *Chem. Commun.* **2015**, *51*, 3501–3510. [CrossRef]
33. Son, H.J.; Jin, S.; Patwardhan, S.; Wezenberg, S.J.; Jeong, N.C.; So, M.; Wilmer, C.E.; Sarjeant, A.A.; Schatz, G.C.; Snurr, R.Q.; et al. Light-harvesting and ultrafast energy migration in porphyrin-based metal-organic frameworks. *J. Am. Chem. Soc.* **2013**, *135*, 862–869. [CrossRef] [PubMed]
34. Yu, J.; Park, J.; Van Wyk, A.; Rumbles, G.; Deria, P. Excited-State Electronic Properties in Zr-Based Metal-Organic Frameworks as a Function of a Topological Network. *J. Am. Chem. Soc.* **2018**, *140*, 10488–10496. [CrossRef] [PubMed]
35. Shavaleev, N.M.; Eliseeva, S.V.; Scopelliti, R.; Bünzli, J.-C.G. Influence of Symmetry on the Luminescence and Radiative Lifetime of Nine-Coordinate Europium Complexes. *Inorg. Chem.* **2015**, *54*, 9166–9173. [CrossRef]
36. Kotova, O.; Comby, S.; Lincheneau, C.; Gunnlaugsson, T. White-light emission from discrete heterometallic lanthanide-directed self-assembled complexes in solution. *Chem. Sci.* **2017**, *8*, 3419–3426. [CrossRef] [PubMed]
37. Cui, Y.J.; Xu, H.; Yue, Y.F.; Guo, Z.Y.; Yu, J.C.; Chen, Z.X.; Gao, J.K.; Yang, Y.; Qian, G.D.; Chen, B.L. A Luminescent Mixed-Lanthanide Metal-Organic Framework Thermometer. *J. Am. Chem. Soc.* **2012**, *134*, 3979–3982. [CrossRef] [PubMed]
38. Rao, X.T.; Song, T.; Gao, J.K.; Cui, Y.J.; Yang, Y.; Wu, C.D.; Chen, B.L.; Qian, G.D. A Highly Sensitive Mixed Lanthanide Metal-Organic Framework Self-Calibrated Luminescent Thermometer. *J. Am. Chem. Soc.* **2013**, *135*, 15559–15564. [CrossRef]
39. Li, L.; Zhu, Y.L.; Zhou, X.H.; Brites, C.D.S.; Ananias, D.; Lin, Z.; Paz, F.A.A.; Rocha, J.; Huang, W.; Carlos, L.D. Visible-Light Excited Luminescent Thermometer Based on Single Lanthanide Organic Frameworks. *Adv. Funct. Mater.* **2016**, *26*, 8677–8684. [CrossRef]
40. Liu, X.; Akerboom, S.; Jong, M.D.; Mutikainen, I.; Tanase, S.; Meijerink, A.; Bouwman, E. Mixed-Lanthanoid Metal–Organic Framework for Ratiometric Cryogenic Temperature Sensing. *Inorg. Chem.* **2015**, *54*, 11323–11329. [CrossRef]

41. Wade, S.A.; Collins, S.F.; Baxter, G.W. Fluorescence intensity ratio technique for optical fiber point temperature sensing. *J. Appl. Phys.* **2003**, *94*, 4743–4756. [CrossRef]
42. Brites, C.D.S.; Lima, P.P.; Silva, N.J.O.; Millán, A.; Amaral, V.S.; Palacio, F.; Carlos, L.D. Thermometry at the nanoscale. *Nanoscale* **2012**, *4*, 4799–4829. [CrossRef]
43. Qiu, L.; Yu, C.; Wang, X.; Xie, Y.; Kirillov, A.M.; Huang, W.; Li, J.; Gao, P.; Wu, T.; Gu, X.; et al. Tuning the Solid-State White Light Emission of Postsynthetic Lanthanide-Encapsulated Double-Layer MOFs for Three-Color Luminescent Thermometry Applications. *Inorg. Chem.* **2019**, *58*, 4524–4533. [CrossRef] [PubMed]
44. *CrysAlis CCD and CrysAlis RED*; Oxford Diffraction Ltd: Abingdon, UK, 2008.
45. Burla, M.C.; Caliandro, R.; Camalli, M.; Carrozzini, B.; Cascarano, G.L.; Caro, L.D.; Giacovazzo, C.; Polidori, G.; Spagna, R.J. SIR2004: An improved tool for crystal structure determination and refinement. *Appl. Crystallogr.* **2005**, *38*, 381–388. [CrossRef]
46. Sheldrick, G.M. SHELXT–Integrated space-group and crystal-structure determination. *Acta Cryst.* **2015**, *C71*, 3–8. [CrossRef] [PubMed]
47. Farrugia, L.J.J. *J.* WinGX and ORTEP for Windows: an update. *Appl. Cryst.* **2012**, *45*, 849–854. [CrossRef]
48. Macrae, C.F.; Edgington, P.R.; McCabe, P.; Pidcock, E.; Shields, G.P.; Taylor, R.; Towler, M.; van de Streek, J.J. Mercury: visualization and analysis of crystal structures. *Appl. Cryst.* **2006**, *39*, 453–457. [CrossRef]

Sample Availability: Samples of the compounds are available from the authors.

© 2020 by the authors. Licensee MDPI, Basel, Switzerland. This article is an open access article distributed under the terms and conditions of the Creative Commons Attribution (CC BY) license (http://creativecommons.org/licenses/by/4.0/).

Article

Facile Preparation of Metal-Organic Framework (MIL-125)/Chitosan Beads for Adsorption of Pb(II) from Aqueous Solutions

Xue-Xue Liang, Nan Wang, You-Le Qu *, Li-Ye Yang, Yang-Guang Wang and Xiao-Kun Ouyang *

School of Food and Pharmacy, Zhejiang Ocean University, Zhoushan 316022, China; 13665804509@163.com (X.-X.L.); ynwangnan@163.com (N.W.); liyey@zjou.edu.cn (L.-Y.Y.); ygw0510@sohu.com (Y.-G.W.)
* Correspondence: youle1960@163.com (Y.-L.Q.); xkouyang@163.com (X.-K.O.);
  Tel.: +86-580-255-4781 (X.-K.O.); Fax: +86-580-255-4781 (X.-K.O.)

Academic Editors: Victoria F. Samanidou, Eleni Deliyanni and Derek J. McPhee
Received: 20 May 2018; Accepted: 23 June 2018; Published: 25 June 2018

Abstract: In this study, novel composite titanium-based metal-organic framework (MOF) beads were synthesized from titanium based metal organic framework MIL-125 and chitosan (CS) and used to remove Pb(II) from wastewater. The MIL-125-CS beads were prepared by combining the titanium-based MIL-125 MOF and chitosan using a template-free solvothermal approach under ambient conditions. The surface and elemental properties of these beads were analyzed using scanning electron microscopy, Fourier transform infrared and X-ray photoelectron spectroscopies, as well as thermal gravimetric analysis. Moreover, a series of experiments designed to determine the influences of factors such as initial Pb(II) concentration, pH, reaction time and adsorption temperature was conducted. Notably, it was found that the adsorption of Pb(II) onto the MIL-125-CS beads reached equilibrium in 180 min to a level of 407.50 mg/g at ambient temperature. In addition, kinetic and equilibrium experiments provided data that were fit to the Langmuir isotherm model and pseudo-second-order kinetics. Furthermore, reusability tests showed that MIL-125-CS retained 85% of its Pb(II)-removal capacity after five reuse cycles. All in all, we believe that the developed MIL-125-CS beads are a promising adsorbent material for the remediation of environmental water polluted by heavy metal ions.

Keywords: metal-organic framework; chitosan beads; adsorption; Pb(II)

1. Introduction

With the development of modern industry, the standards of living have been continuously improving. However, this has also led to many environmental problems, including the heavy metal pollution of water, which creates risks for human health, the environment and ecological systems [1]. Most specifically, lead can enter the body through contaminated food and the respiratory tract in the forms of vapor, dust and chemicals [2]. Moreover, the amount of lead absorbed by the body from food and water increases with age [3]. Lead poisoning mainly damages the nervous and hematopoietic systems, as well as the kidneys, but can also affect the functions of the circulatory and reproductive systems, and can even cause cancer. In addition, lead has been reported as teratogenic and mutagenic [4]. Since the lead pollution problem is growing, it is important to develop a highly efficient adsorbent for the remediation of Pb(II) [5]. Many methods, mostly based on physical and chemical processes, for the removal of Pb(II) from polluted environmental water have been described in the literature [6]. Among these, the adsorption of Pb(II) has proven to be the best treatment approach owing to several significant advantages that include design simplicity, cost efficiency, ease of operation and absence of secondary pollution [7].

With the rapid development of new materials, metal-organic frameworks (MOFs) have received an increasing amount of attention in recent years [8]. MOFs are crystalline porous materials that have periodic network structures formed by the self-assembly of transition metal ions and organic ligands [9]. They not only have many excellent features such as high porosity, low density, large surface area and adjustable aperture, but are also topologically diverse and scalable [10]. In addition, MOFs provide significant advantages to the fields of gas storage, small-molecule separation and catalysis due to their special structural properties and ability to change their internal structures [11]. Indeed, MOFs have been shown to favorably adsorb species that include heavy metals and drugs.

The development of titanium-based MIL-125 represents an important breakthrough in the use of metal nodes as functional moieties [12]. Various modifications of MIL-125 have been reported in recent years. One example is the highly crystalline NH_2-MIL-125, which exhibits good sorption-isotherm-model behavior and high water capacity for an adsorbent-heat-transformation system [13]. Although titanium based metal organic framework MIL-125 can remove Pb(II) well and separate it from waste-water simply by centrifugation, this method has proven to be troublesome, which has restricted its application [14].

Chitosan (CS) is a natural and completely biodegradable polymeric material that is very attractive to researchers because of its versatile chemical and physical properties. Chitin is the raw material for chitosan and is produced from chitinous solid waste from the food industry [15]. Chitosan is obtained by chitin deacylation and has broader application prospects than chitin since it contains a larger number of chelating amino groups that can be modified [16]. Even though this valuable structure contains reactive hydroxyl and amino groups that can potentially bind heavy metals, as well as various polymers, chitosan derivatives have proven to be more effective than the pure form [17,18]. Reportedly, chitosan can easily be processed into membranes, nanofibers, beads, microparticles and nanoparticles. Moreover, the outstanding biological properties of this molecule have led to its enormous importance in a variety of pharmaceutical and biomedical applications [19]. However, despite all of its attractive features, in its pure form, chitosan adsorbs Pb(II) poorly [20].

In this study, we synthesized MIL-125 using the hydrothermal-solvent method and mixed it with chitosan to form solidified beads in a sodium tripolyphosphate solution. The synthesized MIL-125-CS beads contained carboxyl and hydroxyl groups derived from chitosan, and its bead structure facilitated the separation from water in subsequent experiments. Furthermore, chitosan wraps on the surface of MIL-125 were found to enhance the stability of the beads in water. Finally, the abilities of the MIL-125-CS beads to adsorb Pb(II) from polluted water were tested. Analysis of the Pb(II) levels revealed that the beads exhibited good adsorption capacities and solid-liquid separation characteristics. In addition, filtration or centrifugation was not required during the recovery of the MIL-125-CS beads.

2. Results and Discussion

2.1. Synthesis and Characterization of MIL-125-CS

SEM pictures of MIL-125, the MIL-125-CS beads and Pb(II) loaded on MIL-125-CS (MIL-125-CS-Pb) are displayed in Figure 1. The SEM image in Figure 1a revealed that MIL-125 had an octahedral structure with a size range of 5–20 um and smooth surfaces [21], thereby verifying that MIL-125 had been successfully synthesized. A comparison of the SEM pictures of pure MIL-125 (Figure 1a) and the MIL-125-CS beads (Figure 1b,c) revealed that the surface of MIL-125 had undergone some changes. Notably, the surface of the MIL-125-CS beads was rougher and denser than that of MIL-125, which was attributed to the assembly of chitosan on the MIL-125 layers. Figure 2c displays a half-cut view of a dry MIL-125-CS bead that shows Pb(II) successfully loaded on the MIL-125-CS surface. This result clearly indicates that the surface of MIL-125-CS becomes rougher after the Pb(II) loading.

Figure 1. SEM pictures of (**a**) MIL-125(Ti), (**b**,**c**) MIL-125-CS and (**d**) MIL-125-CS-Pb.

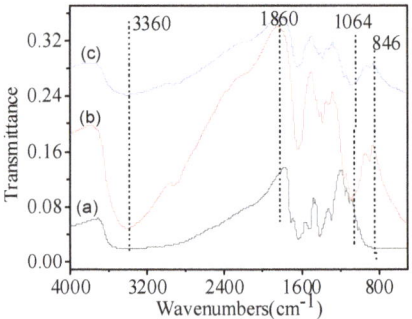

Figure 2. FTIR spectra of (**a**) MIL-125, (**b**) MIL-125-CS and (**c**) MIL-125-CS-Pb.

FTIR spectra of MIL-125, MIL-125-CS and MIL-125-CS-Pb are shown in Figure 2. The band observed in Figure 2a at 3360 cm^{-1} corresponded to the O–H stretching vibrations, while the band at 1860 cm^{-1} was ascribed to the carboxyl C=O moiety. Similarly, the bands at 3360 cm^{-1} in both spectra shown in Figure 2b were attributed to the O–H vibrations. In addition, the adsorption peak at 2900 cm^{-1} corresponded to the C–H stretching vibrations based on MIL-125-CS, while those at 1162 and 1064 cm^{-1} were assigned to the C–C stretching vibrations, proving that MIL-125-CS contained MIL-125. In addition, the band 1210 cm^{-1} was ascribed to the O-Ti-O vibration [22] . Furthermore, as can be seen from Figure 2a,b, the key peaks in the spectrum of MIL-125 can also be detected in that of MIL-125-CS, and the band between 1000 and 846 was ascribed to -NH of chitosan, which indicates that MIL-125-CS was successfully synthesized. Furthermore, the peak shift at 1665–1600 cm^{-1} (-NH$_2$ bending mode) reflected the interaction between the Fe^{3+} ion and -NH$_2$ group [23]. Moreover, the hydroxyl and carboxyl bands in the spectrum of MIL-125-CS-Pb were better defined than those in the spectrum of pure MIL-125-CS, which implied that Pb(II) was embedded through interactions with both the O–H and

COO− units [24]. Reportedly, the stretching and torsional vibrations of these functional groups weaken with the increasing ionic volume, thus resulting in an altered adsorption maximal [25]. Through the aforementioned analysis, we concluded that Pb(II) reacted chemically with MIL-125-CS and caused changes in the infrared adsorption peaks, thereby demonstrating that Pb(II) was successfully adsorbed onto the surfaces of the MIL-125-CS beads.

XPS was used to investigate the chemical compositions of MIL-125-CS and MIL-125-CS-Pb. In particular, Ti, C, O, N and Pb were found to be present in the adsorbent exposed to Pb(II). The XPS survey, Pb 4f and C 1s spectra of MIL-125-CS and Pb(II)-loaded MIL-125-CS are shown Figure 3. The Pb(II) adsorption peak that appears in the survey spectrum of MIL-125-CS-Pb (Figure 3a) confirms that Pb(II) was indeed present on MIL-125-CS. Moreover, the Pb 4f spectrum exhibits a binding energy of 138.8 eV (Figure 3b) assigned to Pb $4f_{7/2}$, further verifying that Pb(II) was loaded on MIL-125-CS and suggesting that lead carbonate and lead oxide were possibly formed during the adsorption process [26]. In contrast, binding energies of 284.8, 286.581, 285.521 and 288.185 eV were observed in the C 1s spectrum of MIL-125-CS shown in Figure 3c, which were ascribed to the C–C, C–O, C–N and C=O groups, respectively, of MIL-125-CS. Meanwhile, Figure 3d revealed important differences in the signals corresponding to the main functional groups such as C–O and C=O, with the peaks corresponding to these groups shifted to 286.468 and 288.254 eV, respectively. All in all, the XPS analysis clearly indicates that carboxylate groups played a significant role in the Pb(II) adsorption process.

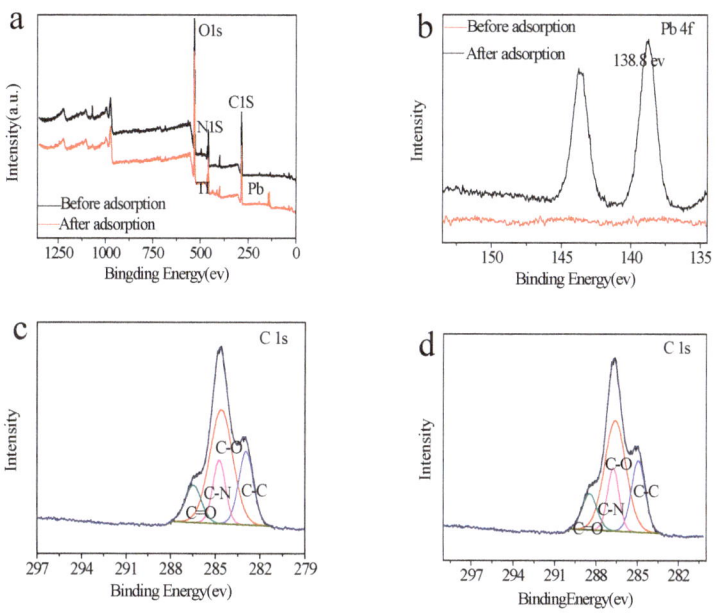

Figure 3. XPS survey spectra of (**a**) MIL-125/CS and MIL-125/CS-Pb, (**b**) Pb 4f, (**c**) C1s of MIL-125/CS and (**d**) C1s of MIL-125/CS-Pb.

TGA examines the stabilities and compositions of the materials being studied, in this case MIL-125 or MIL-125-CS, through programmed temperature changes. Figure 4 displays TGA traces of MIL-125-CS and MIL-125, which revealed that temperature increases led to weight losses. The first weight-loss step was associated with the vaporization of water from the sample at ambient temperature (0–180 °C). The weight losses of MIL-125-CS and MIL-125 were about 11%, which was due to the decomposition of guest elements adsorbed on the adsorbent. The further degradation step was associated with a series of processes. Overall, the weight of the MIL-125-CS sample declined faster than that of MIL-125, which was ascribable to the large number of hydroxyl functional groups in chitosan that are easily dehydrated with the loss of water at higher temperatures. However, by the time both samples had reached 500 °C, they had lost the same amount of weight. Moreover, no further weight losses were observed with the increasing temperature, since chitosan had formed carbide [27], and MIL-125 had transformed into an ultrafine TiO_2 powder that was not easy to crack at higher temperatures (>500 °C) [28]. The main weight losses observed between 300 and 500 °C were attributed to the degradation of each MOF through the decomposition of the aminoterephthalic acid units in MIL-125, ultimately producing an amorphous TiO_2 residue [29].

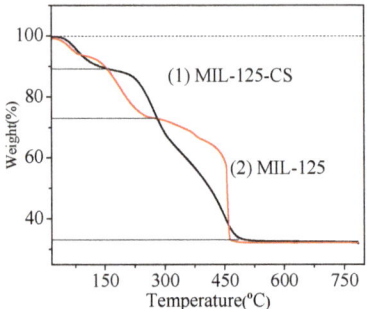

Figure 4. TGA curves of (1) MIL-125-CS and (2) MIL-125.

The XRD patterns of pure MIL-125 and MIL-125-CS are shown in Figure 5. The diffraction peaks of MIL-125 are consistent with those reported in previous studies [30], which indicated that MIL-125 was prepared successfully. In addition, the XRD pattern of MIL-125-CS was almost the same as that of MIL-125, except for a few changes in the (20.2) diffraction, which may correspond to $-NH_2$, and this similarity was due to the fact that CS and MIL-125 have polymerized and that the structure of MIL-125 had not changed.

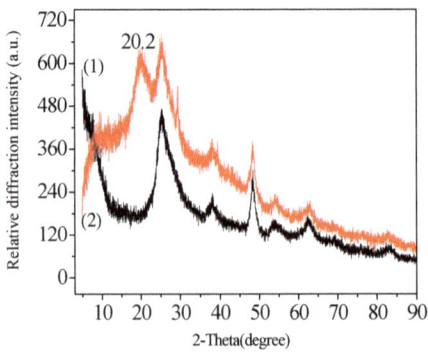

Figure 5. XRD curves of (1) MIL-125 and (2) MIL-125-CS.

2.2. Adsorption Studies

2.2.1. Effect of the Contact Time

Contact time experiments were conducted in the time range of 5–300 min during the Pb(II) removal using the MIL-125-CS beads, and the respective adsorption capacities were determined for different adsorption times. As can be observed in Figure 6a, q_t increases rapidly with the increasing adsorption time during the initial stages of adsorption, which was ascribable to unoccupied MIL-125-CS. However, the adsorption of Pb(II) into the MIL-125-CS beads gradually reached equilibrium within 180 min, as evidenced by the fact that no obvious adsorption-capacity changes were noticed after 180 min. Consequently, a contact time of 180 min was applied to the following experiments.

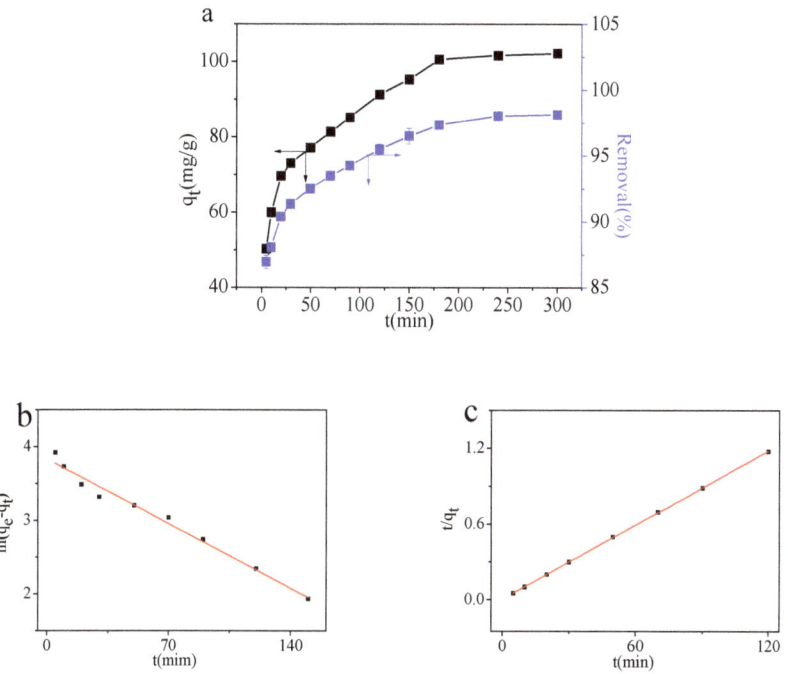

Figure 6. (a) Effect of contact time on Pb(II) adsorption capacity of MIL-125-CS. Data fitted to (b) pseudo-first-order and (c) pseudo-second-order kinetic models.

Additionally, we studied the kinetics of the adsorption of Pb(II) by the MIL-125-CS beads. Experiments were conducted in which MIL-125-CS beads (0.5 g) were added to 20-mL aliquots of Pb(II) solutions with different initial concentrations (10–1000 mg/g). During the analysis of the adsorption data, we assumed either a pseudo-first-order or pseudo-second-order kinetics model, the specific rate equations of which are:

$$\ln(q_e - q_t) = \ln q_e - k_1 t \qquad (1)$$

$$\frac{t}{q_t} = \frac{1}{k_2 q_e^2} + \frac{t}{q_e} \qquad (2)$$

In these equations, k_1 (min^{-1}) and k_2 (g mg^{-1} min^{-1}) are the adsorption rate constants from the two kinetics models, respectively. Figure 6b,c shows the experimental data fitted to these models, with the resulting parameters provided in Table 1.

The rate constant data for the adsorption process fit the pseudo-second-order kinetic model (R^2 = 0.9999) better than the pseudo-first-order model (R^2 = 0.97978). In addition, the pseudo-second-order-calculated q_e value of 102.04 mg/g is in good agreement with the experimentally-determined value of 99.94 mg/g. Hence, we concluded that the Pb(II) adsorption onto the MIL-125-CS beads follows the pseudo-second-order kinetics, which is consistent with the chemical-adsorption process [31].

Table 1. Kinetic parameters for Pb(II) adsorption on MIL-125-CS at different contact times.

$q_{e,exp}$	Pseudo-First-Order Model			Pseudo-Second-Order Model		
	$q_{e,cal}$	k_1	R^2	$q_{e,cal}$	k_2	R^2
99.94	46.45	0.01256	0.97978	102.04	0.0098	0.9999

2.2.2. Effect of pH

Additionally, we investigated the effect of the initial Pb(II)-solution pH on the adsorption by the MIL-125-CS beads and summarize the results in Figure 7a. As can be seen, the pH greatly influenced the adsorption properties of the beads toward Pb(II). The q_e for the Pb(II) adsorption was observed to increase from 40.1 ± 0.57–101.75 ± 0.67 mg/g, as the pH increased from 2–6, respectively. Therefore, it is clearly more beneficial to remove Pb(II) with MIL-125-CS at a higher pH. This phenomenon was attributed to the prolific H^+ ions in the solution that compete with Pb(II) for loading onto the MIL-125-CS beads [32]. At the same time, the functional groups on the surfaces of the MIL-125-CS beads were protonated, thus leading to a positively-charged adsorbent surface that resulted in a decreased tendency to adsorb activated Pb(II). However, the adsorption capacity clearly dropped sharply as the pH was increased from 6–7. This observation was attributed to the hydroxide (OH^-) and ferric (Fe^{2+}) ions reacting to form a precipitate. Finally, we conclude that at low values, the pH had a significant impact on the Pb(II)-adsorption process, with a maximum q_e observed at pH 6. As a result, subsequent tests were carried out at pH 6, at which hydrolysis and Pb(II) sediment formation can be avoided.

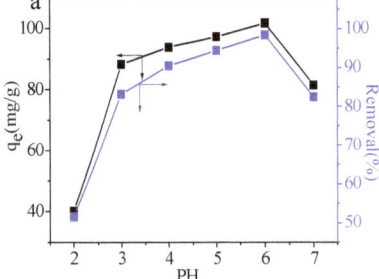 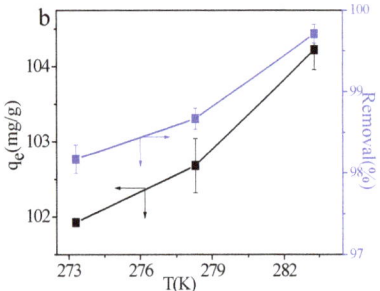

Figure 7. The effect of the Pb(II)-solution pH (a) and reaction temperature (b) on Pb(II) adsorption on MIL-125-CS beads.

2.2.3. Effect of the Temperature

The effect of the temperature on the Pb(II)-adsorption capacity of the MIL-125-CS beads is illustrated in Figure 7b. In this experiment, MIL-125-CS (0.5 g) was mixed into 20 mL of a 200-mg/g Pb(II) solution, after which it was shaken at the required temperature until reaching an adsorption equilibrium. The relationship between the adsorption capacity and temperature was clearly evident, with the adsorption capacity increasing from 101.9–104.1 mg/g with the increasing temperature. Therefore, we concluded that the process of loading Pb(II) onto the MIL-125-CS beads was endothermic [33].

In order to provide insight into the mechanism of the adsorption process, the changes in C_e/q_e were used to determine the changes in Gibbs free energy (ΔG, kJ/mol), enthalpy (ΔH, kJ/mol) and entropy (ΔS, J/mol.K). These quantities are related by the following formula:

$$\Delta G = -RT \ln \frac{q_e}{C_e} = -RT(-\frac{\Delta H}{RT} + \frac{\Delta S}{R}) \qquad (3)$$

In this equation, T (K) indicates the adsorption temperature and represents the common gas constant. Moreover, ΔG was used to establish if the adsorption process is spontaneous and if the thermodynamic temperature is conducive to the Pb(II) loading on the MIL-125-CS beads. Notably, ΔH was affected by temperature, with the positive values confirming the spontaneous nature of the experimental process [34]. These values are listed in Table 2.

Table 2. The parameters of thermodynamics for Pb(II) removal by the MIL-125-CS beads.

ΔG (kJ/mol) T (K)			ΔH (kJ/mol)	ΔS (J/mol·K)
293.2	298.2	303.2	8.343	28.00
−5.78	−5.92	−6.13		

2.2.4. Effect of Pb(II) Concentration

Batch experiments were carried out in order to determine the impact of the Pb(II) concentration on the adsorption capacity of the MIL-125-CS beads. As illustrated in Figure 8, the adsorption capacity of the MIL-125-CS beads increased from 49.18 ± 0.30% (100 mg/g) to 360.05 ± 3.36% (1000 mg/g) as the initial Pb(II) concentration was increased. While the heavy metal ions were still easily captured by the adsorption sites at higher Pb(II) concentrations, the increase in adsorption capacity gradually decreased, which was attributable to the relatively fewer adsorption sites with the larger C_0 values.

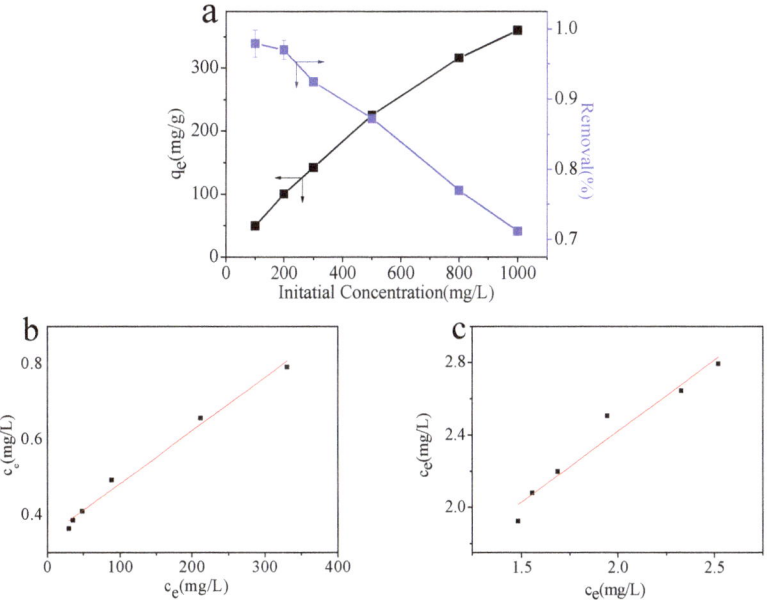

Figure 8. (a) The impact of initial Pb(II)-solution concentration (C_0) on the adsorption capacity of the MIL-125-CS beads. Linear fits to the (b) Langmuir and (c) Freundlich isotherm models.

Equilibrium adsorption isotherms are an efficient means of confirming the mechanism associated with the adsorption process of an adsorbent. In this study, we employed the isotherm model to analyze the adsorption-equilibrium curve for Pb(II) interacting with the MIL-125-CS beads. The Langmuir and Freundlich isotherm models are described by the following formulas [35]:

$$\frac{C_e}{q_e} = \frac{1}{bq_m} + \frac{C_e}{q_m} \quad (4)$$

$$\lg q_e = \lg K_F + \frac{1}{n}\lg C_e \quad (5)$$

where q_{max} (mg/g) indicates the maximum removal capacity for Pb(II) loaded onto the MIL-125-CS beads, b is the Langmuir constant, K_F (mg/g) is the Freundlich adsorption constant and n is the Freundlich adsorption constant that is related to the adsorption intensity. These values are listed in Table 3.

Figure 8b,c shows the degree of coincidence between the experimental curves and the two isotherm models. It can be clearly observed that the Langmuir model better describes how Pb(II) interacts with the MIL-125-CS beads. The Langmuir model supposes that Pb(II) is foremost adsorbed as a monolayer on the surface of the MIL-125-CS bead and that the adsorption energies were uniformly distributed over the adsorbent surface [36].

R_L is a dimensionless separation factor that can be determined from the Langmuir model according to the following formula:

$$R_L = \frac{1}{1 + bC_0} \quad (6)$$

The resulting R_L values (0–1, Table 4) indicate that Pb(II) was effectively adsorbed on the surfaces of the MIL-125-CS beads.

Table 3. Langmuir and Freundlich model parameters for Pb(II) loaded onto MIL-125-CS beads.

T (K)	Langmuir Isotherm			Freundlich Isotherm		
	q_m (mg/g)	b (L/mg)	R^2	K_F	n	R^2
298.2	406.50	0.02	0.99077	27.53	1.12	0.94637

Table 4. R_L data for Pb(II) adsorption on the MIL-125-CS beads based on the Langmuir model.

C_0 (mg/L)	100	200	300	500	800	1000
R_L	0.33	0.20	0.14	0.090	0.058	0.047

2.3. Comparison of the Adsorption Capacities of the Adsorbents

Comparison experiments involving the absorption of Pb(II) on MIL-125, the chitosan beads and the MIL-125-CS beads revealed that MIL-125-CS exhibited a q_e of 100.03 mg·g^{-1}, which is higher than that of the chitosan beads (60.97 mg·g^{-1}) and MIL-125 (94.72 mg·g^{-1}). The results also indicated that MIL-125 and the chitosan beads played the same significant role during the adsorption of Pb(II), especially when combined in the composite MIL-125-CS beads. Consequently, the MIL-125-CS beads were used as the adsorbent in subsequent experiments.

2.4. Reusability of MIL-125-CS Beads

In order to investigate the recycling characteristics of the MIL-125-CS beads, 0.1 mol/L NaOH (desorbent) were added to a mixture of the MIL-125-CS beads in the Pb(II) solution [37]. Following the Pb(II) desorption, Pb(II) was re-adsorbed onto the MIL-125-CS beads. This process was repeated five times (Figure 9b), with the adsorption capacity measured after each cycle. While the adsorption capacity of the MIL-125-CS beads was slightly lower, having dropped from 100.02 ± 0.15%– 87.70 ± 0.14% mg/g, after five continuous usage cycles, the Pb(II)-elimination rate was retained at

83.85 ± 0.28%. As a result of the series of experiments and analyses presented in this study, we suggest that the MIL-125-CS beads are prospective materials for efficient Pb(II) removal.

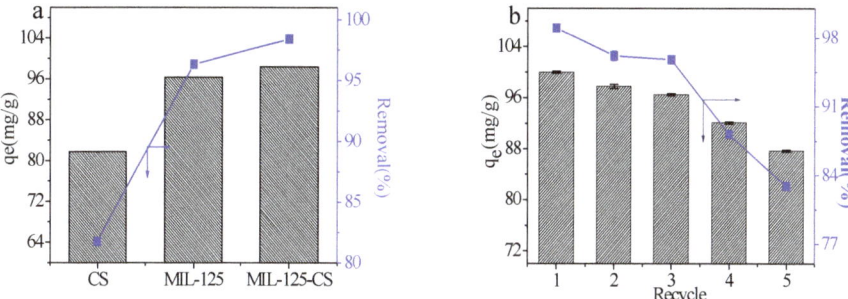

Figure 9. Adsorption capacities and removal ratios (%) of CS, MIL-125 and the MIL-125-CS beads (**a**) and the reusability of the MIL-125-CS beads for Pb(II) adsorption (**b**).

2.5. Stability of MIL-125

The structure stability of MIL-125 in a water environment with a pH range of 2–7 was investigated by FTIR and XRD analyses. As can be seen from the FTIR spectra (shown in Figure 10a,b), no obvious difference was observed in MIL-125 after acidity treatment. A few changes were detected in the crystal form of MIL-125 after the acid treatment (shown in Figure 10c,d). Based on these results, we concluded that the crystal structure did not change after the acid treatment, and MIL-125 maintained its stability in a water environment with a pH ranging from 2–7.

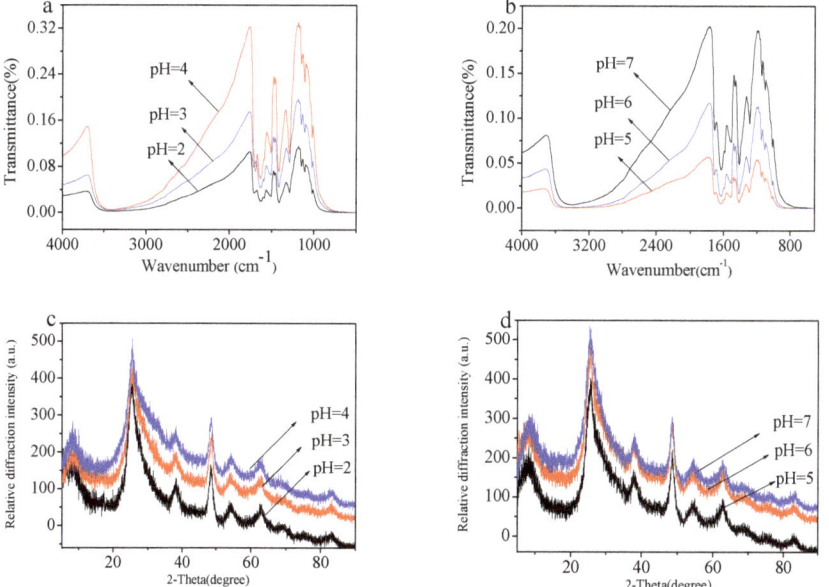

Figure 10. FTIR spectra (**a**,**b**) and XRD patterns (**c**,**d**) of MIL-125 after treatment with solution of different pH.

3. Materials and Methods

3.1. Materials

Titanium isopropoxide ($C_{12}H_{28}O_4Ti$, 95%), ferric chloride hexahydrate ($FeCl_3 \cdot 6H_2O$), benzene-1,4-dicarboxylicacid (BDC), N,N-dimethylformamide (DMF), sodium tripolyphosphate ($Na_5P_3O_{10}$) solution and methanol (CH_3OH) were provided by Aladdin Reagent Co., Ltd. (Shanghai, China), while chitosan with a viscosity of 100–200 mpa.s and a degree of deacetylation ≥95% was furnished by Aladdin Reagent Co., Ltd. (Shanghai, China). Deionized water was prepared by our laboratory. All reagents were used without further purification.

3.2. Preparation of MIL-125 and the MIL-125-CS Beads

MIL-125(Ti) was formed according to a previously-reported method [38]. $C_{12}H_{28}O_4Ti$ (14.64 g) and BDC (13.38 mL) were added to a mixture of methanol (24 mL) and DMF (216 mL) and stirred at room temperature until a homogeneous solution was obtained. Then, the mixed solution was placed in a Teflon-lined stainless steel autoclave and heated at 80 °C for 48 h. The reaction was cooled to room temperature and collected by filtration, after which the white precipitate was washed three times with DMF to remove the remaining unreacted titanium isopropoxide from the porous framework. The resulting solid was washed several times with methanol and dried at 80 °C under vacuum for 5 h, and the prepared white solid powder was triturated for subsequent use.

The synthesis of the MIL-125-CS beads is depicted schematically in Scheme 1. First, $FeCl_3$ (1.35 g) was mixed with DI water (50 mL) and stirred until evenly dispersed. Then, sodium chitosan (1.5 g) was evenly added to the solution. Finally, MIL-125 (1.5 g) was added into the above solution until well mixed. This mixed solution was added dropwise into a pre-prepared 200 mL solution of 3% Na5P3O10 with constant stirring at room temperature for 2 h. The synthesized MIL-125-CS beads were collected and washed three times with DI water. These beads were then comprehensively characterized using multiple physicochemical techniques to confirm the formation of MIL-125-CS.

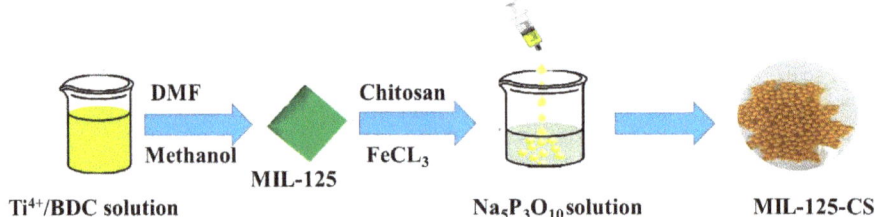

Scheme 1. Synthesis of the MIL-125-CS beads. BDC, benzene-1,4-dicarboxylicacid.

3.3. Characterization

The surface morphologies of the beads were analyzed using scanning electron microscopy (SEM-S4800, Hitachi, Japan). The functional groups and elemental composition of the adsorbent were monitored by Fourier transform infrared (FTIR) spectroscopy (Tensor II, Bruker, Germany) and X-ray photoelectron spectroscopy (XPS, ESCALAB 250, Shimadzu, Japan). Additionally, thermal gravimetric analysis (TGA) was performed using a thermal gravimetric analyzer (PTC-10A, Rigaku, Lorentz, Japan).

3.4. Adsorption Studies

A 0.5-g aliquot of MIL-125-CS was added to 20 mL of a 200-mg/L Pb(II) solution in a 100-mL conical flask. A set of initial Pb(II) concentrations (100–1000 mg/L) was used in the following experiments. In order to study the effect of pH on the adsorption reaction, the pH of the Pb(II) solution was adjusted to 2–7 with HCl (0.1 M) or NaOH (0.1 M). The flask with the adsorbent and Pb(II) solution was shaken at 150 rpm, while maintaining a temperature of 298.2 K during the adsorption process. After the adsorption, the solution was poured into a small brown bottle, and the post-adsorption concentration of Pb(II) was determined.

The removal capacity (q_e, mg/g) of MIL-125-CS was determined by Equation (7):

$$q_e = \frac{(C_0 - C_e)V}{m} \tag{7}$$

where m (g) is the weight of MIL-125-CS, C_0 (mg/L) is the initial concentration of Pb(II), V (L) is the volume of the Pb(II) solution and C_e (mg/L) is the Pb(II) concentration at equilibrium.

The value of q_t (mg/g) for MIL-125-CS at time t was calculated according to Equation (8):

$$q_t = \frac{(C_0 - C_t)V}{m} \tag{8}$$

where C_t (mg/L) indicates the resulting concentration of Pb(II).

The rate (R) for the elimination of Pb(II) from the polluted water by MIL-125-CS was calculated using Equation (9):

$$R = \frac{C_0 - C_e}{C_0} \times 100\% \tag{9}$$

All experimental results show the average values of three parallel experiments. In order to compare the adsorption performances of MIL-125, chitosan and MIL-125-CS toward Pb(II), aqueous Pb(II) solutions (200 mg/g, 20 mL) were treated with each adsorbent (0.5 g).

4. Conclusions

In this study, we hydrothermally prepared MIL-125 following a literature procedure. Then, we mixed it with chitosan and dropped it into a $Na_5P_3O_{10}$ solution to form beads under ambient conditions. The synthetic polymer beads were analyzed using SEM, FTIR, XPS and TGA, and a thorough study of the surface characteristics of MIL-125-CS confirmed its successful preparation. Furthermore, the formed beads, as novel adsorbents, were used to adsorb Pb(II) from simulated wastewater. A series of adsorption kinetic studies, adsorption-isotherm modeling and thermodynamic studies led us to conclude that the adsorption process was spontaneous. Notably, 180 min were required to reach equilibrium, eventually leading to a q_{max} of the attached Pb(II) of 406.5 mg/g at a pH of 6. After batch experiments and regeneration testing, we finally concluded that the MIL-125-CS beads were an effective and reusable material for the adsorption of Pb(II) from polluted water.

Author Contributions: Y.-L.Q. and X.K.O. conceived and designed the experiments; X.-X.L. and N.W. performed the experiments; X.-X.L., L.-Y.Y. and Y.-G.W. analyzed the data; X.-X.L. wrote the paper.

Funding: This work was financially supported by the National Natural Science Foundation of China (21476212) and the Key research and development plan of Zhejiang Province (2018C02038).

Conflicts of Interest: The authors declare that there are no conflicts of interest.

References

1. Hu, Z.H.; Omer, A.M.; Ouyang, X.K.; Yu, D. Fabrication of carboxylated cellulose nanocrystal/sodium alginate hydrogel beads for adsorption of Pb(II) from aqueous solution. *Int. J. Biol. Macromol.* **2018**, *108*, 149–157. [CrossRef] [PubMed]
2. OuYang, X.K.; Jin, R.N.; Yang, L.P.; Wang, Y.G.; Yang, L.Y. Bamboo-derived porous bioadsorbents and their adsorption of Cd(II) from mixed aqueous solutions. *RSC Adv.* **2014**, *4*, 28699–28706. [CrossRef]
3. Bindler, R.; Renberg, I.; Klaminder, J.; Emteryd, O. Tree rings as Pb pollution archives? A comparison of $^{206}Pb/^{207}Pb$ isotope ratios in pine and other environmental media. *Sci. Total Environ.* **2004**, *319*, 173–183. [CrossRef]

4. Yang, H.; Linge, K.; Rose, N. The Pb pollution fingerprint at Lochnagar: The historical record and current status of Pb isotopes. *Environ. Pollut.* **2007**, *145*, 723–729. [CrossRef] [PubMed]
5. Wu, Z.; Cheng, Z.; Ma, W. Adsorption of Pb(II) from glucose solution on thiol-functionalized cellulosic biomass. *Bioresour. Technol.* **2012**, *104*, 807–809. [CrossRef] [PubMed]
6. Younas, M.; Shahzad, F.; Afzal, S.; Khan, M.I.; Ali, K. Assessment of Cd, Ni, Cu, and Pb pollution in Lahore, Pakistan. *Environ. Int.* **1998**, *24*, 761–766. [CrossRef]
7. Odukoya, O.O.; Arowolo, T.A.; Bamgbose, O. Pb, Zn, and Cu levels in tree barks as indicator of atmospheric pollution. *Environ. Int.* **2000**, *26*, 11–16. [CrossRef]
8. Burtch, N.C.; Jasuja, H.; Walton, K.S. Water Stability and Adsorption in Metal-Organic Frameworks. *Chem. Rev.* **2014**, *114*, 10575–10612. [CrossRef] [PubMed]
9. Stock, N.; Biswas, S. ChemInform Abstract: Synthesis of Metal—Organic Frameworks (MOFs): Routes to Various MOF Topologies, Morphologies, and Composites. *Chem. Rev.* **2012**, *43*, 933–969. [CrossRef] [PubMed]
10. Hermes, S.; Schröter, M.K.; Schmid, R.; Khodeir, L.; Muhler, M.; Tissler, A.; Fischer, R.W.; Fischer, R.A. Metal@MOF: Loading of Highly Porous Coordination Polymers Host Lattices by Metal Organic Chemical Vapor Deposition. *Angew. Chem. Int. Ed.* **2005**, *44*, 6237–6241. [CrossRef] [PubMed]
11. Wang, N.; Jin, R.N.; Omer, A.M.; Ouyang, X.K. Adsorption of Pb(II) from fish sauce using carboxylated cellulose nanocrystal: Isotherm, kinetics, and thermodynamic studies. *Int. J. Biol. Macromol.* **2017**, *102*, 232–240. [CrossRef] [PubMed]
12. Hendon, C.H.; Tiana, D.; Fontecave, M.; Sanchez, C.; D'Arras, L.; Sassoye, C.; Rozes, L.; Mellot-Draznieks, C.; Walsh, A. Engineering the optical response of the titanium-MIL-125 metal-organic framework through ligand functionalization. *J. Am. Chem. Soc.* **2013**, *135*, 10942–10945. [CrossRef] [PubMed]
13. Kim, S.N.; Kim, J.; Kim, H.Y.; Cho, H.Y.; Ahn, W.S. Adsorption/catalytic properties of MIL-125 and NH 2-MIL-125. *Catal. Today* **2013**, *204*, 85–93. [CrossRef]
14. Guo, H.; Lin, F.; Chen, J.; Li, F.; Weng, W. Metal–organic framework MIL-125(Ti) for efficient adsorptive removal of Rhodamine B from aqueous solution. *Appl. Organomet. Chem.* **2015**, *29*, 12–19. [CrossRef]
15. Krajewska, B. Application of chitin- and chitosan-based materials for enzyme immobilizations: A review. *Enzyme Microb. Technol.* **2004**, *35*, 126–139. [CrossRef]
16. Pillai, C.K.S.; Paul, W.; Sharma, C.P. Chitin and chitosan polymers: Chemistry, solubility and fiber formation. *Prog. Polym. Sci.* **2009**, *34*, 641–678. [CrossRef]
17. Takeshita, S.; Konishi, A.; Takebayashi, Y.; Yoda, S.; Otake, K. Aldehyde Approach to Hydrophobic Modification of Chitosan Aerogels. *Biomacromolecules* **2017**, *18*, 2172–2178. [CrossRef] [PubMed]
18. Rezvani, H.; Riazi, M.; Tabaei, M.; Kazemzadeh, Y.; Sharifi, M. Experimental investigation of interfacial properties in the EOR mechanisms by the novel synthesized Fe_3O_4@Chitosan nanocomposites. *Colloids Surf. A* **2018**, *544*, 15–27. [CrossRef]
19. Guibal, E. Interactions of metal ions with chitosan-based sorbents: A review. *Sep. Purif. Technol.* **2004**, *38*, 43–74. [CrossRef]
20. Dash, M.; Chiellini, F.; Ottenbrite, R.M.; Chiellini, E. Chitosan—A versatile semi-synthetic polymer in biomedical applications. *Prog. Polym. Sci.* **2011**, *36*, 981–1014. [CrossRef]
21. Yang, T.B.; Sun, L.X.; Xu, F.; Wang, Z.Q.; Zou, Y.J.; Chu, H.L. Synthesis of MIL-125/Graphene Oxide Composites and Hydrogen Storage Properties. *Key Eng. Mater.* **2017**, *727*, 683–687. [CrossRef]
22. Martis, M.; Mori, K.; Fujiwara, K.; Ahn, W.S.; Yamashita, H. Amine-Functionalized MIL-125 with Imbedded Palladium Nanoparticles as an Efficient Catalyst for Dehydrogenation of Formic Acid at Ambient Temperature. *J. Phys. Chem. C* **2013**, *117*, 22805–22810. [CrossRef]
23. McNamara, N.D.; Neumann, G.T.; Masko, E.T.; Urban, J.A.; Hicks, J.C. Catalytic performance and stability of (V) MIL-47 and (Ti) MIL-125 in the oxidative desulfurization of heterocyclic aromatic sulfur compounds. *J. Catal.* **2013**, *305*, 217–226. [CrossRef]
24. Naceur, B.; Abdelkader, E.; Nadjia, L.; Sellami, M.; Noureddine, B. Synthesis and characterization of Bi1.56Sb1.48Co0.96O7 pyrochlore sun-light-responsive photocatalyst. *Mater. Res. Bull.* **2016**, *74*, 491–501. [CrossRef]
25. Kordić, B.; Kovačević, M.; Sloboda, T.; Vidović, A.; Jović, B. FT-IR and NIR Spectroscopic Investigation of Hydrogen Bonding in Indole-Ether Systems. *J. Mol. Struct.* **2017**, *1144*, 159–165. [CrossRef]

26. Guo, X.; Du, B.; Qin, W.; Jian, Y.; Hu, L.; Yan, L.; Xu, W. Synthesis of amino functionalized magnetic graphenes composite material and its application to remove Cr(VI), Pb(II), Hg(II), Cd(II) and Ni(II) from contaminated water. *J. Hazard. Mater.* **2014**, *278*, 211–220. [CrossRef] [PubMed]
27. Archana, D.; Dutta, J.; Dutta, P.K. Evaluation of chitosan nano dressing for wound healing: Characterization, in vitro and in vivo studies. *Int. J.Biol. Macromol.* **2013**, *57*, 193–203. [CrossRef] [PubMed]
28. Liu, Z.; Wu, Y.; Chen, J.; Li, Y.; Zhao, J.; Gao, K.; Na, P. Effective elimination of As(III) via simultaneous photocatalytic oxidation and adsorption by a bifunctional cake-like TiO_2 derived from MIL-125(Ti). *Catal. Sci. Technol.* **2018**, *8*, 1936–1944. [CrossRef]
29. Zhang, S.Y.; Xiao, F.P.; Wu, L.Y.; Li, C.Y.; Chen, Z.H. Effect of Heat Treatment on the Microstructure of TiO_2 Nanofibers Prepared by Refluxing in Basic Solution. *Key Eng. Mater.* **2008**, *368–372*, 800–802. [CrossRef]
30. Yang, Z.; Xu, X.; Liang, X.; Lei, C.; Cui, Y.; Wu, W.; Yang, Y.; Zhang, Z.; Lei, Z. Construction of heterostructured MIL-125/Ag/g-C_3N_4 nanocomposite as an efficient bifunctional visible light photocatalyst for the organic oxidation and reduction reactions. *Appl. Catal. B Environ.* **2017**, *205*, 42–54. [CrossRef]
31. Badiea, A.M.; Mohana, K.N. Corrosion Mechanism of Low-Carbon Steel in Industrial Water and Adsorption Thermodynamics in the Presence of Some Plant Extracts. *J. Mater. Eng. Perform.* **2009**, *18*, 1264–1271. [CrossRef]
32. Ivanchikova, I.D.; Lee, J.S.; Maksimchuk, N.V.; Shmakov, A.N.; Chesalov, Y.A.; Ayupov, A.B.; Hwang, Y.K.; Chang, J.S.; Kholdeeva, O.A. Highly Selective H_2O_2-Based Oxidation of Alkylphenols to *p*-Benzoquinones Over MIL-125 Metal–Organic Frameworks. *Eur. J. Inorg. Chem.* **2014**, *2014*, 132–139. [CrossRef]
33. Auta, M.; Hameed, B.H. Modified mesoporous clay adsorbent for adsorption isotherm and kinetics of methylene blue. *Chem. Eng. J.* **2012**, *198–199*, 219–227. [CrossRef]
34. Chaudhry, S.A.; Zaidi, Z.; Siddiqui, S.I. Isotherm, kinetic and thermodynamics of arsenic adsorption onto Iron-Zirconium Binary Oxide-Coated Sand (IZBOCS): Modelling and process optimization. *J. Mol. Liq.* **2017**, *229*, 230–240. [CrossRef]
35. Mojzsis, S.J. Confirmation of mass-independent isotope effects in archent(2.5 3.8 GA) sedimentary sulfides as determined by ion microprobe analysis. *J. Chromatogr. A* **2008**, *1189*, 19–31.
36. Zhang, Y.; Hidajat, K.; Ray, A.K. Determination of competitive adsorption isotherm parameters of pindolol enantiomers on α_1-acid glycoprotein chiral stationary phase. *J.Chromatogr. A* **2006**, *1131*, 176–184. [CrossRef] [PubMed]
37. Wu, Y.; Zhang, Y.; Qian, J.; Xin, X.; Hu, S.; Zhang, S.; Wei, J. An exploratory study on low-concentration hexavalent chromium adsorption by Fe(III)-cross-linked chitosan beads. *R. Soc. Open Sci.* **2017**, *4*, 170905. [CrossRef] [PubMed]
38. Danhardi, M.; Serre, C.; Frot, T.; Rozes, L.; Maurin, G.; Sanchez, C.; Férey, G. A new photoactive crystalline highly porous titanium(IV) dicarboxylate. *J. Am. Chem. Soc.* **2009**, *131*, 10857–10859. [CrossRef] [PubMed]

Sample Availability: Samples of the compounds are not available from the authors.

© 2018 by the authors. Licensee MDPI, Basel, Switzerland. This article is an open access article distributed under the terms and conditions of the Creative Commons Attribution (CC BY) license (http://creativecommons.org/licenses/by/4.0/).

Communication

Hydrogen-Bonding Linkers Yield a Large-Pore, Non-Catenated, Metal-Organic Framework with pcu Topology

Mohammad S. Yazdanparast [1], Victor W. Day [2] and Tendai Gadzikwa [1,*]

[1] Department of Chemistry, Kansas State University, Manhattan, KS 66506, USA; yazdanparast@ksu.edu
[2] Department of Chemistry, University of Kansas, Lawrence, KS 66045, USA; vwday@ku.edu
* Correspondence: gadzikwa@ksu.edu

Academic Editors: Victoria Samanidou, Eleni Deliyanni and Liudmil Antonov
Received: 22 December 2019; Accepted: 3 February 2020; Published: 6 February 2020

Abstract: Pillared paddle-wheel-based metal-organic framework (MOF) materials are an attractive target as they offer a reliable method for constructing well-defined, multifunctional materials. A drawback of these materials, which has limited their application, is their tendency to form catenated frameworks with little accessible volume. To eliminate this disadvantage, it is necessary to investigate strategies for constructing non-catenated pillared paddle-wheel MOFs. Hydrogen-bonding substituents on linkers have been postulated to prevent catenation in certain frameworks and, in this work, we present a new MOF to further bolster this theory. Using 2,2'-diamino-[1,1'-biphenyl]-4,4'-dicarboxylic acid, BPDC-$(NH_2)_2$, linkers and dipyridyl glycol, DPG, pillars, we assembled a MOF with **pcu** topology. The new material is non-catenated, exhibiting large accessible pores and low density. To the best of our knowledge, this material constitutes the **pcu** framework with the largest pore volume and lowest density. We attribute the lack of catenation to the presence of H-bonding substituents on both linkers.

Keywords: metal-organic framework; mixed-ligand; pillared; paddle-wheel; non-catenated; large-pore; hydrogen-bonding

1. Introduction

While there are a variety of ways to assemble well-defined, multifunctional metal-organic framework (MOF) materials [1], the construction of mixed-linker, pillared paddle-wheel MOFs is the most efficacious (Figure 1) [2]. Their assembly provides a reliable strategy for introducing two different organic linkers into an MOF, allowing for the chemical pore environment to be tuned with high fidelity [3]. Despite the advantages that they offer, following their first introduction, pillared paddle-wheel frameworks have received much less attention than their potential would warrant. This is owing to two limitations: the M^{2+}-paddle-wheel secondary building unit (SBU, Figure 1) is not as chemically stable as many other clusters [4] and, due to the small size of the SBU, the frameworks are prone to catenation. Though the challenges of relatively poor stability and low porosity due to catenation can be addressed post MOF assembly, via transmetallation [5–7] and solvent-assisted linker exchange (SALE) [8,9], the de novo synthesis of such materials would be preferable. Thus, there is a need to investigate strategies to incorporate preferred cations and to prevent catenation in the solvothermal synthesis of pillared paddle-wheel MOFs. In this report, we present an unusual, non-catenated, large pore, pillared paddle-wheel MOF, providing an additional datapoint to support current postulation on the factors that may influence catenation in these frameworks.

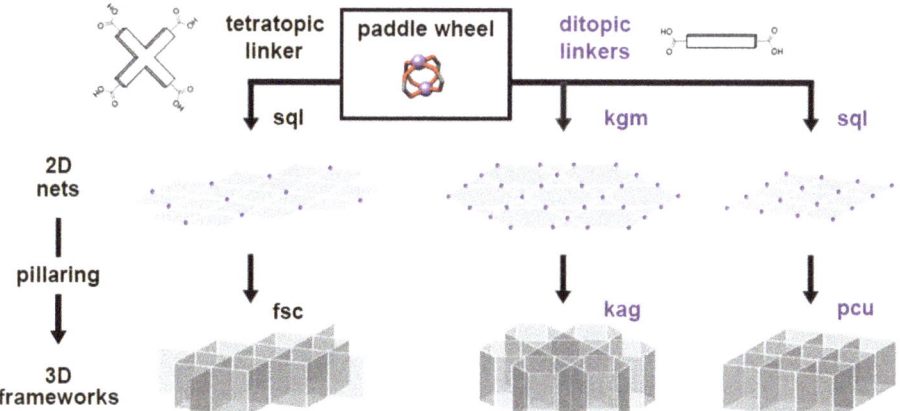

Figure 1. Schematic representation of the possible topologies for pillared, paddle-wheel metal organic frameworks (MOFs). The 2D nets are pillared to form 3D MOFs. MOFs of **pcu** and **fsc** are derived from **sql** nets, and **kag** MOFs are derived from **kgm** nets.

A major focus of our group is the uniform multifunctionalization of MOFs. To this end, we have been synthesizing pillared paddle-wheel MOFs where the two different linkers bear reactive groups that can be addressed independently post MOF assembly [10,11]. In this work, we specifically target non-catenated frameworks that can accommodate additional functionality. Specifically, we sought pillared frameworks with the **kag** topology as, unlike the more common **pcu**-based structures, they are non-catenated with large pores (Figure 1). For our ligands, we employed dipyridyl glycol, DPG, together with either 2-amino-1,4-benzenediacarboxylic acid (BDC-NH_2 or 2-azido-1,4-benzenediacarboxylic acid (BDC-N_3). With the intent of constructing a symmetric version of such MOFs, we then attempted the construction of a **kag** MOF composed of Zn^{2+}, 2,2'-diamino-[1,1'-biphenyl]-4,4'-dicarboxylic acid, BPDC-$(NH_2)_2$, and DPG. Gratifyingly, we obtained a non-catenated structure. Unexpectedly, however, we found the structure to have the **pcu** topology.

2. Results

Combining BPDC-$(NH_2)_2$, and DPG under the low-temperature nucleation conditions generally employed to obtain **kag** MOFs [12,13], we obtained pale-yellow, block-like crystals that were suitable for single-crystal X-ray analysis. Following refinement of the diffraction data, we found that we had obtained a **pcu** framework, **KSU-100**, that is non-catenated (Figure 2b).

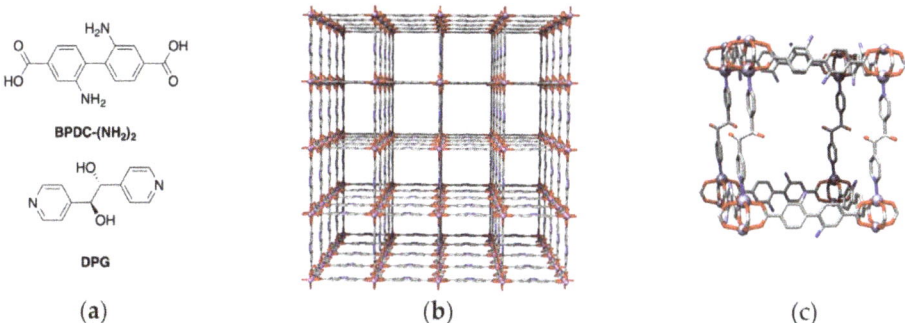

Figure 2. (a) MOF linkers; (b) **KSU-100** viewed down the c-axis; (c) Network unit of **KSU-100**.

Crystal data for C10H7N1.5O2.5Zn0.50 (M = 220.86 g/mol): monoclinic, space group P4, a = 15.1970(5) Å, b = 15.1970(5) Å, c = 16.2095(5) Å, V = 3743.6(3) Å^3, Z = 4, T = 200(2) K, µ(CuKα) = 0.542 mm^{-1}, D_{calc} = 0.392 g/cm^3, 33,836 reflections measured (2.908° ≤ 2Θ ≤ 68.403°), 6586 unique (R_{int} = 0.0490) which were used in all calculations. The final R1 was 0.1026 (I > 2σ(I)) and wR2 was 0.2860 (all data).

The new MOF, **KSU-100**, has the BPDC-(NH$_2$)$_2$ linkers connected by Zn paddle-wheel clusters, defining the *xy*-plane in a **sql** net. This 2D net is then pillared together by the DPG ligand to form a **pcu** framework with large pore dimensions of 11Å × 11Å × 9Å, and a low calculated density of 0.392 g/cm^3. Powder X-ray diffraction (PXRD) of bulk samples of single-crystals of the material confirmed the purity of the structure (Figure 3a, and Figure S1 in Supporting Information for the indexed pattern). Note that large crystals were used instead of powders, as the powders lost solvent rapidly and did not produce adequate diffraction patterns. Thermogravimetric analysis (TGA) indicates that **KSU-100** loses 60% of its weight as solvent. Such a significant loss confirms that the material has a large solvent-accessible volume and supports that the bulk material is indeed non-catenated (Figure 3b).

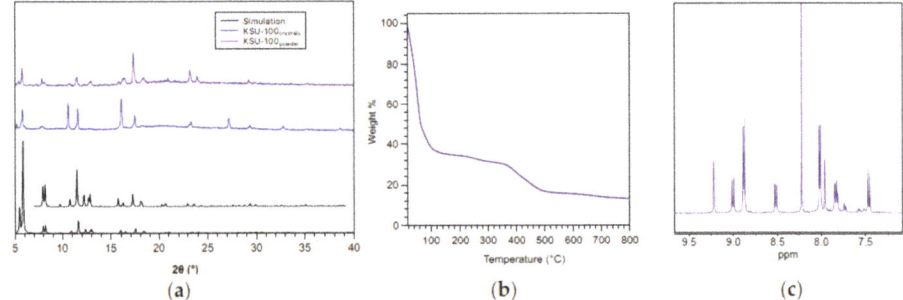

Figure 3. Characterization of **KSU-100**: (**a**) Powder diffraction patterns of the simulated pattern based on the single-crystal data, single crystals, and powder of **KSU-100**. The insert is a magnification of the smaller peaks of the simulation; (**b**) Thermogravimetric analysis (TGA) trace of the MOF solvated with DMF; (**c**) ^{1}H-NMR of **KSU-100** digested in a TFA-d_1 and d_6-DMSO mixture.

To confirm the composition of the material, we performed proton nuclear magnetic resonance (^{1}H-NMR) spectroscopy of **KSU-100** digested in a mixture of deuterated trifluoroacetic acid, TFA-d_1, and deuterated dimethylsulfoxide, DMSO-d_6 (Figure 3c). Integration of the ligand peaks (Figure S2) indicates a BPDC-(NH$_2$)$_2$:DPG ratio of 2:1, as is to be expected for a pillared paddle-wheel MOF. Note that there are two sets of protons corresponding to BPDC-(NH$_2$)$_2$. Literature precedence indicates that the ligand undergoes multiple transformations in the presence of metal cations and strong acid [14]. We found that using TFA as the digestion acid reduced the number of complexes formed, allowing us to identify and integrate the peaks.

3. Discussion

Pillared paddle-wheel MOFs comprise M^{2+}-acetate centers connected by multicarboxylate linkers to form two possible two-dimensional (2D) nets: **sql** and **kgm** (Figure 1) [15]. When the paddle-wheel nets are pillared together by ditopic linkers they form 3D frameworks. Of the three pillared paddle-wheel MOF topologies that can be formed, only the **kag** topology, formed from the **kgm** net, has consistently been non-catenated. While low temperature nucleation has been suggested as a method for generating these wide-channel MOFs [13], this topology is relatively rare, with a handful of reports in the last several years [10–13,16,17]. The more common **sql** net, formed by di- and tetratopic dicarboxylate ligands, is pillared to form the **pcu** and **fsc** nets, respectively [15].

Of the **sql**-based structures, the **fsc** frameworks constructed with tetracarboxy linkers have thus far provided the most reliable route to large-pore, non-catenated, pillared paddle-wheel MOFs.

The topology has been primarily reported with the tetrakis(4-carboxyphenyl)porphine (TCPP) [18–21] and tetrakis(4-carboxyphenyl)benzene, TCPB [22]. The TCPP-based MOFs are all non-catenated, but the TCPB linker produces a net with openings that are large enough to accommodate an additional framework, resulting in frequent catenation [22,23]. Catenation has been prevented by the presence of blocking functional groups on the tetracarboxylate [8], or by using dipyridyl pillars that are bulky or have bulky substituents [24–26]. The exception to this trend is **DO-MOF**, a structure where the dipyridyl linker is the seemingly inobtrusive dipyridyl glycol, DPG [27].

The **pcu** topology is generally catenated, unless the MOF is composed of short dicarboxylate linkers or short pillars. The use of short linkers results in frameworks that lack the space to accommodate additional frameworks, thus the lack of catenation comes at the expense of larger pore volumes [28]. The single exception is a non-catenated **pcu** framework, **BMOF-1-bpdc-NO$_2$**, composed of bipyridine (BIPY) pillars, and a long 4,4'-biphenyl dicarboxylate (BPDC) linker bearing a nitro substituent [29]. For the TCPB-based MOF, it has been speculated that the hydrogen-bonding capability of the DPG linker is responsible for preventing catenation in **DO-MOF** [8]. That result, combined with the existence of a large-pore, non-catenated **pcn** framework that is decorated with H-bond accepting nitro groups, prompts the question of whether catenation can be influenced by substituents that participate in hydrogen bonding. The new MOF, **KSU-100**, lends credence to this theory by presenting another large-pore, non-catenated **pcu** MOF, constructed using linkers bearing H-bonding substituents.

While we cannot identify the limits of pore dimensions for the construction of **pcu** type MOFs that are non-catenated, it is extraordinary that a **pcu** MOF with a BPDC-based linker is non-catenated. For the small dicarboxylate linker benzene dicarboxylate (BDC), catenated **pcu** MOFs are formed when sufficiently long pillars are used, including the relatively short BIPY [30–35]. The same is true for the slightly longer naphthalene dicarboxylate (NDC) linker [36], which still forms catenated structures when the dipyridyl liker has bulky (trimethylsilyl)ethynyl substituents [37]. Even the small, 4-carbon fumarate (FMA) linker forms a catenated **pcu** MOF with the one-ring pyrazine (PYZ) pillar [38]. Given the prevalence of catenation even with short dicarboxylates, it is expected that **pcu** structures of the longer BPDC linker should be catenated.

There is only one report of a non-catenated **pcu** structure of BPDC and a non-hydrogen-bonding pillar, and it is one where the co-linker is the short and bulky DABCO [39]. All other **pcu** structures of BPDC are, at minimum, 2-fold catenated, even when the dipyridyl linker has bulky substituents. Sterically demanding substituents on pillars include anthracene [40] and a 24-member interlocking ring [41], and these pillars have formed 2-fold catenated **pcu** MOFs with BPDC. A pillar containing the bulky triptycene moiety results in a 4-fold catenated MOF with BPDC [35]. Given these examples, and the lack of catenation in **KSU-100** and in **BMOF-1-bpdc-NO$_2$**, it is reasonable to assume that it is the electronic nature of the substituents, not their size, that prevents catenation.

Hupp and co-workers [23], and others [42], have suggested that H-bonding between linker substituents and solvent molecules can increase the steric requirements of linkers, preventing catenation. It should be noted that that there is a **pcu** structure that is 3-fold catenated despite having DPG as a pillar [43]. The dicarboxylate ligand in this case is azobenzene-4,4'-dicarboxylic acid, a linker that is only ~2 Å longer than **BPDC**. This result suggests that this may be where the threshold void volume exists, i.e., where the H-bonding capability of DPG is no longer sufficient to prevent catenation. In **KSU-100**, the BPDC-(NH$_2$)$_2$ linkers and the DPG pillar each have two H-bonding substituents, resulting in a dense concentration of H-bond donors and acceptors in the framework. Such an environment is conducive to the creation of a dense network of H-bonded solvent molecules in the pores. We presume that this is why catenation does not take place despite the significant void volume.

4. Materials and Methods

All chemicals were used as received from commercial sources unless otherwise noted. Meso-α,β-di(4-pyridyl) glycol (DPG) was purchased from TCI America (Portland, OR, USA). N,N-dimethylformamide (DMF) was purchased from Fisher Scientific (Pittsburgh, PA, USA), and zinc nitrate hexahydrate from Strem Chemicals (Newburyport, MA, USA). Dimethyl sulfoxide-d_6 (d_6-DMSO, 99.9 atom % D) was purchased from Cambridge Isotope Laboratories (Tewksbury, MA, USA), while trifluoroacetic acid-d (TFA-*d1* 99.5 atom % D) was purchased from Sigma-Aldrich (St Louis, MO, USA). 2,2′-Diaminobiphenyl-4,4′-dicarboxylic acid (BPDC-(NH$_2$)$_2$) was synthesized following a literature procedure [14].

Synthesis of **KSU-100**: in a 500 mL round-bottom flask, Zn(NO$_3$)$_2$·6H$_2$O (400.0 mg, 0143 mmol) and DPG (160.0 mg, 0.74 mmol) were added to 250 mL DMF and stirred at RT for 30 min. BPDC-(NH$_2$)$_2$ (400.0 mg, 0.147 mmol) was added to the mixture and left to stir at room temperature for 10 min. The flask was then incubated at 60 °C. After 14 h, the flask was removed from the heating block and left at room temperature for 30 h. Pale yellow crystals (300 mg, 30% yield) of the product were collected by filtration and stored in fresh DMF.

Details of single-crystal X-ray analysis are available in the Supporting Information. CCDC 1972127 contains the supplementary crystallographic data for this paper. These data are provided free of charge by the Cambridge Crystallographic Data Centre.

Powder X-ray diffraction (PXRD) patterns were recorded on a Bruker AXS D8 Advance Phaser diffractometer (Bruker AXS GmbH, Karlsruhe, Germany) with Cu Kα radiation (λ = 1.5418 Å) over a range of 5° < 2θ < 40° in 0.02° steps, with a 0.5 s counting time per step. Samples were collected from the bottom of the reaction vial as a thick suspension in DMF and spread on a Si-Einkristalle plate immediately before PXRD measurements.

Thermogravimetric analysis (TGA) was performed on a TGA-Q50 (TA Instruments, New Castle, DE, USA) interfaced with a PC using TA Universal Analysis software. Samples were heated at a rate of 10 °C/min under a nitrogen atmosphere. All samples were extensively solvent-exchanged with fresh DMF prior to analysis.

The proton NMR spectrum of **KSU-100** was recorded on a Bruker Avance NEO spectrometer (400 MHz for ^{1}H, Bruker BioSpin, Billerica, MA, USA). NMR chemical shifts are reported in ppm against a residual solvent resonance as the internal standard ($\delta(d_6$-DMSO) = 2.5 ppm). In a typical analysis, MOF materials were washed thoroughly with DMF. The sample was isolated and dried under vacuum at 60 °C for minimum of 2 h. The dry MOF sample (~5 mg) was digested in a mixture of 0.400 mL d_6-DMSO (0.1 mL) and TFA-d_1 (0.100 mL) and then transferred into an NMR tube.

Fourier-transform infrared spectroscopy of **KSU-100** was performed on an Agilent Cary 630 spectrometer (Agilent Technologies, Santa Clara, CA, USA). The MOF sample (~1 mg) was combined with five mass equivalents (~5 mg) of KBr and ground together to a fine powder.

5. Conclusions

In our own work of covalently functionalizing MOFs post-synthesis, it has been crucial to synthesize non-catenated frameworks that have enough space to accommodate additional functionality. Doubtless, accessible pore volume is necessary for a variety of other MOF applications. In this work, we have provided an additional datapoint to support the assertion that hydrogen-bonding substituents on linkers can prevent catenation in pillared, paddle-wheel MOFs. With this information, MOF chemists who are interested in the well-defined multifunctionality of these materials now have a potential avenue for constructing non-catenated variants of these pillared frameworks.

Supplementary Materials: The following are available online, Crystallographic Information File (CIF) for **KSU-100** and Supporting Information.

Author Contributions: Conceptualization, T.G.; investigation, M.S.Y.; crystallography, V.W.D.; writing—review and editing, M.S.Y. and T.G.; funding acquisition, T.G. All authors have read and agreed to the published version of the manuscript.

Funding: This study was supported by a National Science Foundation grant, CHE-1800517, and NSF-MRI grants, CHE-0923449 to the University of Kansas to purchase the X-ray diffractometer and software used in this study, and CHE-1826982 to Kansas State University for the NMR spectrometer used in this study. The authors also acknowledge the Aakeröy Lab at KState for use of their TGA.

Acknowledgments: We acknowledge the Aakeröy lab at KState for use of their TGA instrument, and Kanchana P. Samarakoon for assistance with NMR studies.

Conflicts of Interest: The authors declare no conflict of interest.

References

1. Qin, J.-S.; Yuan, S.; Wang, Q.; Alsalme, A.; Zhou, H.-C. Mixed-linker strategy for the construction of multifunctional metal–organic frameworks. *J. Mater. Chem. A* **2017**, *5*, 4280–4291. [CrossRef]
2. Hashemi, L.; Morsali, A. *Pillared Metal-Organic Frameworks: Properties and Applications*; John Wiley & Sons: Hoboken, NJ, USA, 2019; ISBN 978-1-119-46024-4.
3. ZareKarizi, F.; Joharian, M.; Morsali, A. Pillar-layered MOFs: Functionality, interpenetration, flexibility and applications. *J. Mater. Chem. A* **2018**, *6*, 19288–19329. [CrossRef]
4. Burtch, N.C.; Walton, K.S. Modulating adsorption and stability properties in pillared metal–organic frameworks: A model system for understanding ligand effects. *Acc. Chem. Res.* **2015**, *48*, 2850–2857. [CrossRef] [PubMed]
5. Lalonde, M.; Bury, W.; Karagiaridi, O.; Brown, Z.; Hupp, J.T.; Farha, O.K. Transmetalation: Routes to metal exchange within metal–organic frameworks. *J. Mater. Chem. A* **2013**, *1*, 5453–5468. [CrossRef]
6. Karagiaridi, O.; Bury, W.; Fairen-Jimenez, D.; Wilmer, C.E.; Sarjeant, A.A.; Hupp, J.T.; Farha, O.K. Enhanced gas sorption properties and unique behavior toward liquid water in a pillared-paddlewheel metal–organic framework transmetalated with Ni(II). *Inorg. Chem.* **2014**, *53*, 10432–10436. [CrossRef]
7. Xu, Y.; Howarth, A.J.; Islamoglu, T.; da Silva, C.T.; Hupp, J.T.; Farha, O.K. Combining solvent-assisted linker exchange and transmetallation strategies to obtain a new non-catenated nickel (II) pillared-paddlewheel MOF. *Inorg. Chem. Commun.* **2016**, *67*, 60–63. [CrossRef]
8. Farha, O.K.; Malliakas, C.D.; Kanatzidis, M.G.; Hupp, J.T. Control over catenation in metal–organic frameworks via rational design of the organic building block. *J. Am. Chem. Soc.* **2010**, *132*, 950–952. [CrossRef]
9. Karagiaridi, O.; Bury, W.; Tylianakis, E.; Sarjeant, A.A.; Hupp, J.T.; Farha, O.K. Opening metal–organic frameworks Vol. 2: Inserting longer pillars into pillared-paddlewheel structures through solvent-assisted linker exchange. *Chem. Mater.* **2013**, *25*, 3499–3503. [CrossRef]
10. Samarakoon, K.P.; Satterfield, C.S.; McCoy, M.C.; Pivaral-Urbina, D.A.; Islamoglu, T.; Day, V.W.; Gadzikwa, T. Uniform, binary functionalization of a metal–organic framework material. *Inorg. Chem.* **2019**, *58*, 8906–8909. [CrossRef]
11. Samarakoon, K.; Yazdanparast, M.; Day, V.W.; Gadzikwa, T. Uniform and simultaneous orthogonal functionalization of a metal-organic framework material. *ChemRxiv* **2020**. [CrossRef]
12. Kondo, M.; Takashima, Y.; Seo, J.; Kitagawa, S.; Furukawa, S. Control over the nucleation process determines the framework topology of porous coordination polymers. *CrystEngComm* **2010**, *12*, 2350–2353. [CrossRef]
13. Zhou, K.; Chaemchuen, S.; Wu, Z.; Verpoort, F. Rapid room temperature synthesis forming pillared metal-organic frameworks with Kagomé net topology. *Microporous Mesoporous Mater.* **2017**, *239*, 28–33. [CrossRef]
14. Ko, N.; Hong, J.; Sung, S.; Cordova, K.E.; Park, H.J.; Yang, J.K.; Kim, J. A significant enhancement of water vapour uptake at low pressure by amine-functionalization of UiO-67. *Dalton Trans.* **2015**, *44*, 2047–2051. [CrossRef]
15. Guillerm, V.; Kim, D.; Eubank, J.F.; Luebke, R.; Liu, X.; Adil, K.; Soo Lah, M.; Eddaoudi, M. A supermolecular building approach for the design and construction of metal–organic frameworks. *Chem. Soc. Rev.* **2014**, *43*, 6141–6172. [CrossRef]
16. Chun, H.; Moon, J. Discovery, synthesis, and characterization of an isomeric coordination polymer with pillared kagome net topology. *Inorg. Chem.* **2007**, *46*, 4371–4373. [CrossRef]
17. Hungerford, J.; Walton, K.S. Room-Temperature synthesis of metal–organic framework isomers in the tetragonal and kagome crystal structure. *Inorg. Chem.* **2019**, *58*, 7690–7697. [CrossRef] [PubMed]
18. Barron, P.M.; Son, H.-T.; Hu, C.; Choe, W. Highly tunable heterometallic frameworks constructed from paddle-wheel units and metalloporphyrins. *Cryst. Growth Des.* **2009**, *9*, 1960–1965. [CrossRef]

19. Lee, C.Y.; Farha, O.K.; Hong, B.J.; Sarjeant, A.A.; Nguyen, S.T.; Hupp, J.T. Light-Harvesting metal–organic frameworks (MOFs): Efficient strut-to-strut energy transfer in bodipy and porphyrin-based MOFs. *J. Am. Chem. Soc.* **2011**, *133*, 15858–15861. [CrossRef]
20. Park, J.; Feng, D.; Yuan, S.; Zhou, H.-C. Photochromic metal–organic frameworks: Reversible control of singlet oxygen generation. *Angew. Chem. Int. Ed.* **2015**, *54*, 430–435. [CrossRef]
21. Danowski, W.; van Leeuwen, T.; Abdolahzadeh, S.; Roke, D.; Browne, W.R.; Wezenberg, S.J.; Feringa, B.L. Unidirectional rotary motion in a metal–organic framework. *Nat. Nanotechnol.* **2019**, *14*, 488–494. [CrossRef]
22. Mulfort, K.L.; Farha, O.K.; Malliakas, C.D.; Kanatzidis, M.G.; Hupp, J.T. An Interpenetrated Framework Material with Hysteretic CO2 Uptake. *Chem. Eur. J.* **2010**, *16*, 276–281. [CrossRef] [PubMed]
23. Bury, W.; Fairen-Jimenez, D.; Lalonde, M.B.; Snurr, R.Q.; Farha, O.K.; Hupp, J.T. Control over catenation in pillared paddlewheel metal–organic framework materials via solvent-assisted linker exchange. *Chem. Mater.* **2013**, *25*, 739–744. [CrossRef]
24. Gadzikwa, T.; Farha, O.K.; Malliakas, C.D.; Kanatzidis, M.G.; Hupp, J.T.; Nguyen, S.T. Selective bifunctional modification of a non-catenated metal–organic framework material via "Click" chemistry. *J. Am. Chem. Soc.* **2009**, *131*, 13613–13615. [CrossRef] [PubMed]
25. Shultz, A.M.; Farha, O.K.; Hupp, J.T.; Nguyen, S.T. A catalytically active, permanently microporous MOF with metalloporphyrin struts. *J. Am. Chem. Soc.* **2009**, *131*, 4204–4205. [CrossRef] [PubMed]
26. Shultz, A.M.; Farha, O.K.; Adhikari, D.; Sarjeant, A.A.; Hupp, J.T.; Nguyen, S.T. Selective Surface and Near-Surface Modification of a Noncatenated, Catalytically Active Metal-Organic Framework Material Based on Mn(salen) Struts. *Inorg. Chem.* **2011**, *50*, 3174–3176. [CrossRef]
27. Mulfort, K.L.; Farha, O.K.; Stern, C.L.; Sarjeant, A.A.; Hupp, J.T. Post-Synthesis alkoxide formation within metal–organic framework materials: A strategy for incorporating highly coordinatively unsaturated metal ions. *J. Am. Chem. Soc.* **2009**, *131*, 3866–3868. [CrossRef]
28. Ma, B.-Q.; Mulfort, K.L.; Hupp, J.T. Microporous pillared paddle-wheel frameworks based on mixed-ligand coordination of zinc ions. *Inorg. Chem.* **2005**, *44*, 4912–4914. [CrossRef]
29. Dau, P.V.; Cohen, S.M. The influence of nitro groups on the topology and gas sorption property of extended Zn(II)-paddlewheel MOFs. *CrystEngComm* **2013**, *15*, 9304–9307. [CrossRef]
30. Chen, B.; Liang, C.; Yang, J.; Contreras, D.S.; Clancy, Y.L.; Lobkovsky, E.B.; Yaghi, O.M.; Dai, S. A microporous metal–organic framework for gas-chromatographic separation of alkanes. *Angew. Chem. Int. Ed.* **2006**, *45*, 1390–1393. [CrossRef]
31. Du, M.; Zhang, Z.-H.; Wang, X.-G.; Tang, L.-F.; Zhao, X.-J. Structural modulation of polythreading and interpenetrating coordination networks with an elongated dipyridyl building block and various anionic co-ligands. *CrystEngComm* **2008**, *10*, 1855–1865. [CrossRef]
32. Seo, J.; Bonneau, C.; Matsuda, R.; Takata, M.; Kitagawa, S. Soft secondary building unit: Dynamic bond rearrangement on multinuclear core of porous coordination polymers in gas media. *J. Am. Chem. Soc.* **2011**, *133*, 9005–9013. [CrossRef] [PubMed]
33. Liu, X.-M.; Xie, L.-H.; Lin, J.-B.; Lin, R.-B.; Zhang, J.-P.; Chen, X.-M. Flexible porous coordination polymers constructed from 1,2-bis(4-pyridyl)hydrazine via solvothermal in situ reduction of 4,4′-azopyridine. *Dalton Trans.* **2011**, *40*, 8549–8554. [CrossRef] [PubMed]
34. Safarifard, V.; Morsali, A. Influence of an amine group on the highly efficient reversible adsorption of iodine in two novel isoreticular interpenetrated pillared-layer microporous metal–organic frameworks. *CrystEngComm* **2014**, *16*, 8660–8663. [CrossRef]
35. Jiang, X.; Duan, H.-B.; Khan, S.I.; Garcia-Garibay, M.A. Diffusion-Controlled rotation of triptycene in a metal–organic framework (MOF) sheds light on the viscosity of mof-confined solvent. *ACS Cent. Sci.* **2016**, *2*, 608–613. [CrossRef]
36. Chun, H.; Dybtsev, D.N.; Kim, H.; Kim, K. Synthesis, X-ray Crystal Structures, and Gas Sorption Properties of Pillared Square Grid Nets Based on Paddle-Wheel Motifs: Implications for Hydrogen Storage in Porous Materials. *Chem. Eur. J.* **2005**, *11*, 3521–3529. [CrossRef]
37. Gadzikwa, T.; Lu, G.; Stern, C.L.; Wilson, S.R.; Hupp, J.T.; Nguyen, S.T. Covalent surface modification of a metal–organic framework: Selective surface engineering via CuI-catalyzed Huisgen cycloaddition. *Chem. Commun.* **2008**, 5493–5495. [CrossRef]

38. Chen, B.; Ma, S.; Zapata, F.; Fronczek, F.R.; Lobkovsky, E.B.; Zhou, H.-C. Rationally designed micropores within a metal–organic framework for selective sorption of gas molecules. *Inorg. Chem.* **2007**, *46*, 1233–1236. [CrossRef]
39. Dau, P.V.; Kim, M.; Garibay, S.J.; Münch, F.H.L.; Moore, C.E.; Cohen, S.M. Single-Atom ligand changes affect breathing in an extended metal–organic framework. *Inorg. Chem.* **2012**, *51*, 5671–5676. [CrossRef]
40. Qi, Y.; Xu, H.; Li, X.; Tu, B.; Pang, Q.; Lin, X.; Ning, E.; Li, Q. Structure transformation of a luminescent pillared-layer metal–organic framework caused by point defects accumulation. *Chem. Mater.* **2018**, *30*, 5478–5484. [CrossRef]
41. Zhu, K.; Vukotic, V.N.; O'Keefe, C.A.; Schurko, R.W.; Loeb, S.J. Metal–Organic frameworks with mechanically interlocked pillars: Controlling ring dynamics in the solid-state via a reversible phase change. *J. Am. Chem. Soc.* **2014**, *136*, 7403–7409. [CrossRef]
42. Servati-Gargari, M.; Mahmoudi, G.; Batten, S.R.; Stilinović, V.; Butler, D.; Beauvais, L.; Kassel, W.S.; Dougherty, W.G.; VanDerveer, D. Control of interpenetration in two-dimensional metal–organic frameworks by modification of hydrogen bonding capability of the organic bridging subunits. *Cryst. Growth Des.* **2015**, *15*, 1336–1343. [CrossRef]
43. Goswami, R.; Mandal, S.C.; Pathak, B.; Neogi, S. Guest-Induced Ultrasensitive detection of multiple toxic organics and Fe^{3+} Ions in a strategically designed and regenerative smart fluorescent metal–organic framework. *ACS Appl. Mater. Interfaces* **2019**, *11*, 9042–9053. [CrossRef] [PubMed]

Sample Availability: Samples of the compounds are not available from the authors.

© 2020 by the authors. Licensee MDPI, Basel, Switzerland. This article is an open access article distributed under the terms and conditions of the Creative Commons Attribution (CC BY) license (http://creativecommons.org/licenses/by/4.0/).

Review

Novel Approaches Utilizing Metal-Organic Framework Composites for the Extraction of Organic Compounds and Metal Traces from Fish and Seafood

Sofia C. Vardali [1,*], Natalia Manousi [2,*], Mariusz Barczak [3] and Dimitrios A. Giannakoudakis [4,*]

1. Institute of Biological Marine Resources, Hellenic Center of Marine Research, Agios Kosmas, Hellenikon, 16777 Athens, Greece
2. Laboratory of Analytical Chemistry, Department of Chemistry, Aristotle University of Thessaloniki, 54124 Thessaloniki, Greece
3. Department of Theoretical Chemistry, Institute of Chemical Sciences, Faculty of Chemistry, Maria Curie-Sklodowska University in Lublin, 20-031 Lublin, Poland; mbarczak@umcs.pl
4. Institute of Physical Chemistry, Polish Academy of Sciences, Kasprzaka 44/52, 01-224 Warsaw, Poland
* Correspondence: sofvardali@gmail.com (S.C.V.); nmanousi@chem.auth.gr (N.M.); DaGchem@gmail.com (D.A.G.); Tel.: +30-2310-997693 (S.C.V.)

Academic Editor: Rafael Lucena
Received: 2 January 2020; Accepted: 21 January 2020; Published: 24 January 2020

Abstract: The determination of organic and inorganic pollutants in fish samples is a complex and demanding process, due to their high protein and fat content. Various novel sorbents including graphene, graphene oxide, molecular imprinted polymers, carbon nanotubes and metal-organic frameworks (MOFs) have been reported for the extraction and preconcentration of a wide range of contaminants from fish tissue. MOFs are crystalline porous materials that are composed of metal ions or clusters coordinated with organic linkers. Those materials exhibit extraordinary properties including high surface area, tunable pore size as well as good thermal and chemical stability. Therefore, metal-organic frameworks have been recently used in many fields of analytical chemistry including sample pretreatment, fabrication of stationary phases and chiral separations. Various MOFs, and especially their composites or hybrids, have been successfully utilized for the sample preparation of fish samples for the determination of organic (i.e., antibiotics, antimicrobial compounds, polycyclic aromatic hydrocarbons, etc.) and inorganic pollutants (i.e., mercury, palladium, cadmium, lead, etc.) as such or after functionalization with organic compounds.

Keywords: metal-organic frameworks; MOFs; fish; extraction; antibiotics; antimicrobial agents; metal ions

1. Introduction

The determination of organic and inorganic pollutants in fish tissue samples is a demanding procedure due to the complexity of the sample matrix [1]. Fish samples exhibit high protein and high fat content and therefore, their analysis is a challenging step for analytical chemists. The main problem occurring from the complexity of the sample matrix is the potential low recovery of organic and inorganic compounds that can be attributed to interactions of the analyte with endogenous food components and/or interferences from other chemical substances of the sample [2]. In order to overcome this problem, an efficient experimental protocol must be implemented for the extraction of the target analytes prior to their determination with an instrumental technique [3].

Antibiotics (including fluoroquinolones, penicillins, amphenicols, tetracyclines, sulfonamides etc.) have been widely used in farming industries to prevent bacterial infections. These chemical compounds exhibit activity against both Gram-positive and Gram-negative bacteria. Today, antibiotics are widely used for prevention and treatment of fish diseases [1,3–9]. The extensive use of antibiotics is a significant risk for human health which it is associated with the consumption of antibiotic residues that can directly cause allergic hypersensitivity reactions or toxic effects in humans. Moreover, their extensive use can cause an increase in antibiotic resistance in fish pathogens and potential transfer of these resistance determinants to human pathogens [3,4].

Other emerging organic contaminants that can be detected in edible fish samples include polychlorinated biphenyls and polybrominated diphenyl ethers [10], malachite green [11], polycyclic aromatic hydrocarbons [12] and food colorants [2]. The most common sample preparation techniques that are until today widely used for the extraction of organic pollutants from fish tissue are solid-phase extraction (SPE) [3,4] and liquid-liquid extraction (LLE) [13,14]. However, those conventional techniques tend to have many fundamental drawbacks since they include complicated and time-consuming steps. Moreover, they exhibit many difficulties in automation and require relatively large amounts of sample and organic solvents including ethyl acetate, chloroform, n-hexane, dichloromethane, etc. [15].

As an alternative to classical sample preparation approaches, various microextraction techniques including solid-phase microextraction (SPME) or liquid-phase microextraction (LPME) have been proposed. Among the most important benefits of those techniques are the consumption of less organic solvents and sample as well as the reduction of the sample treatments steps [15–17]. Furthermore, a wide variety of novel materials have been employed to prepare efficient sorbents for the extraction of various analytes from fish samples. These include ionic liquids (ILs) and polymeric ILs [18], graphene [19], graphene oxide [20], carbon nanotubes [21], molecularly imprinted polymers [22] and metal-organic frameworks [23].

Metal-organic frameworks (MOFs) are crystalline porous materials that consist of metal ions or clusters coordinated with organic linkers [23]. These materials are the new development on the interface between materials science and molecular coordination chemistry [24]. MOFs exhibit extraordinary properties including high porosity, tunable pore size, adjustable internal surface, high thermal and chemical stability [23–25]. Until today metal-organic frameworks have gained attention in a plethora of applications such as gas storage and separation [26], desulfurization of fuels [27], sensors [28], detoxification [29–31], catalysis [32], drug delivery and molecular imaging [33].

There are various approaches for the synthesis of metal-organic frameworks. Among them, the solvothermal approach is the most frequently used technique due to its simplicity. With this approach, the metal salt, the organic ligand and a proper solvent system are placed into a Teflon-lined vessel which is subjected to high temperature or/and pressure for a certain time span [34,35]. Other synthetic approaches involve alternative power sources like the microwave, electrochemical, mechanochemical (ball milling or ultrasonication), etc. The last year although, the synthesis of MOFs was reported either under milder conditions, for instance at less than 80 °C and near atmospheric pressure [29].

The experimental parameters during the synthesis of MOFs play a significant role to the structure and properties of the obtained material, since the level of the defect sites can be tuned [29,30]. Therefore, by controlling the quantity and the ratio of the selected metals and/or linkers, the nature and the amount of the selected solvent or the reaction temperature, pressure, and duration, it is possible to obtain MOF materials with different properties [36]. Moreover, since there is a huge variety of metal ions and organic linkers that can be used as precursors for the fabrication of MOFs, there is a nearly infinite number of MOFs that can be prepared.

In the field of analytical chemistry, MOFs have been successfully employed as adsorbents for the extraction and preconcentration of organic compounds from a wide range of samples. In the literature there are applications of MOFs for SPE [37], SPME [38], magnetic solid-phase extraction (MSPE) [39], dispersive solid-phase extraction (d-SPE) [40] etc. MOFs have been also utilized as stationary phase for gas chromatography (GC) [41], high performance liquid chromatography (HPLC) [42] and capillary electrophoresis [43]. Chiral separations with chiral MOFs as stationary phases have been also reported [44]. MOFs have also been used for the fabrication of electrochemical and fluorescent sensors for the determination of organic compounds such as antibiotics and antimicrobial agents [45–50].

MOFs have been also utilized for the extraction and preconcentration of metal ions from fish tissue. Inorganic contaminants including mercury, cadmium, lead, chromium and arsenic are dangerous contaminants in the environment, threatening human health and natural ecosystems. Water pollution led to contaminate fish with toxic metals from various sources such as use of fertilizers, discharge of industrial effluents, chemical waste agricultural drainage and domestic wastewater [51,52]. Since, those metals exhibit toxic activity even in low concentrations, it is essential to develop efficient sample preparation techniques to successfully extract and preconcentrate them from fish samples prior to their determination spectroscopically or with spectroscopic technique, including flame atomic absorption spectroscopy (FAAS) [53], electrothermal atomic absorption spectroscopy (ETAAS) [54], cold vapor atomic absorption spectroscopy (CVAAS) [55], inductively coupled plasma optical emission spectrometry (ICP-OES) [56] and inductively coupled plasma mass spectrometry (ICP-MS) [57].

In order to design a MOF material, a proper choice of its constituents should take place. In general, low-toxicity metal ions such as Fe, Mn, and Zr are preferred. Regarding the selection of metal salts, nitrates and perchlorates can be oxidized and are considered potentially explosives, while metal chlorides are usually corrosive compounds. Therefore, metal oxides and hydroxides are preferred, since they are safer and produce less hazardous by-products. Recently, the application of zero-valence metal precursors has also been introduced. As for the organic ligands, low cost carboxylic acid, such as terephthalic acid are widely used. The synthetic procedure should comply with the green chemistry principles (i.e., less steps, lower consumption of organic solvents etc). Chemometric tools are recommended for the optimization of the synthesis procedure. Finally, introduction of suitable functional groups in order to enhance the extraction efficiency and selectivity is also crucial [58].

For the real-world commercialization of MOFs and especially as a tool for analytic processes/methods, various drawbacks exist, with the most crucial to be assumed the poor chemical stability in aquatic environments, the low stability after exposure to acidic or basic solvents/solutions, the limited thermal stability as well as the difficulties in regeneration/reusability/recyclability. To strengthen MOFs' structure and to tune them for specialized practical applications, different strategies for the design and synthesis of MOFs were showed to have a great potential, like changing/modifying the metal ions (change the oxidation state, metal ion doping) or the ligands, with the later to be the most well-explored field. For instance, ligands can be chemically functionalized either by the attachment or the insertion of specific functional groups or can be exchanged with others of different physicochemical properties and size (dimensionality) [59–61]. Regulating the surface properties or the structural architecture, for instance by interpenetration or formation of multi-walled frameworks and by developing inter-connection between the metal ions/clusters, are also innovative and prosperous strategies, elevating the desirable features and increasing the structural resistance against water [62]. However, these strategies can also lead to an extended presence of defect sites, as well as to the alteration of the size and the chemical environment of the pores, facts that are hard to be controlled and analysed in detail. Additionally, this kind of chemical and structural tuning upraise the complexity and the cost towards large-scale synthesis and as a result the potential of a pragmatic commercial use.

Briefly, the main approaches are the formation of composites by coating the MOFs' particles with nanoparticles (NPs) or the growth of the framework together with the NPs. The later aspect involves the core-shell growth of the MOF around the NPs, the encapsulation of the NPs inside the final MOF nanoparticles, the individually nucleation of the MOF phase/particles in between the NPs, and the formation of macrocrystals with the NPs inside the structure as well as on the surface. In order to promote the feasibility of utilization, the establishment of easy ways to separate/obtain the MOF phase after the use are of a great demand. Towards this direction, the development of magnetic composites is a well explored and functional tactic. To achieve so, the usage of magnetic Fe-NPs has been explored in various cases. The introduction of carboxylic surface functional groups, except helping on NPs stability, is crucial for the growth of the framework. Recently, Giannakoudakis and Bandosz showed that the geometry of the framework's structure plays a crucial role on how the growth of the MOF will occur around the nanoparticles, resulting to different effects, like the creation of mesoporosity [30,63].

Alternative approaches that will be also discussed is the coating of MOF with a polymeric layer or with bio-molecules like aptamer. Moreover, covering fibers with a layer of the active phase showed as an alternative functional approach, leading to high dispersion and availability of the active sites. Finally, the formation of graphitic/carbon-based material with incorporated NPs, derived after carbonization of functionalized MOFs, is a prosperous technique which can additionally serve as a potential way for the use of the spent samples for alternative applications [64–66].

A plethora of articles regarding the application of MOFs for the sample pretreatment of food, agricultural, biological, and environmental matrices can be found in the literature. Alternative approaches for increasing the MOF stability and for specifying their practical use for analytical applications is the formation of composite or hybrid materials. By this mean, simple and already widely studied MOFs as well as their composites/hybrids can be post-synthetic modified, functionalized, or immobilized on substances.

Herein, we aim to discuss the applications of MOFs and essentially their composites/hybrids as potential medias for the extraction, detection, or sensing of organic and inorganic pollutants from fish samples, prior to their determination with an instrumental technique. Emphasis will be given on the extraction of antibiotics as well as metals from fish tissue, since they are considered as significant contaminants of the marine environment [23,58,67–70]. All the studied in the literature cases in which MOFs were tested as extraction, detection, or biosensors media are collected in Figure 1.

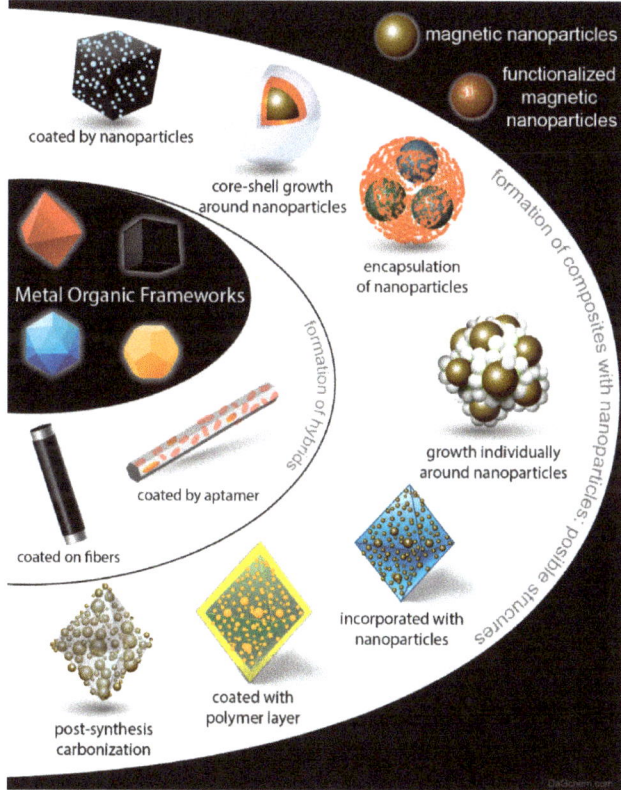

Figure 1. Different approaches for the formation of composite/hybrids as potential medias for the extraction, detection, or sensing of organic and inorganic pollutants from fish samples.

2. Extraction and Detection of Antibiotics and Antimicrobial Agents from Fish and Seafood

2.1. Extraction of Antibiotics with MOFs from Seafood and Fish Samples

Fish and seafood are valued as sustenance of high nutritional value for human consumption because of their valuable fatty acid and amino acid composition. The presence of a wide range of antibiotics and antimicrobial agents has been examined in fish and seafood and a variety of analytical methods and protocols have been developed for their determination in order to satisfy the maximum residue limits (MRLs) established for the safety of consumers. Several research papers have been reported the use of the various MOF based materials (composites or hybrids) not only for the extraction of antibiotics and antimicrobials from fish and seafood and but also for their utility as fluorescent and electrochemical sensors for the sensitive detection of these compounds in fish and seafood. Table 1 summarizes the applications of novel MOF or/and their composites/hybrids for the determination of antibiotics and antimicrobials agents in seafood and fish samples, as well as for different samples for the sake of a comparison.

Table 1. Applications of MOF materials for the extraction and detection of antibiotics and antimicrobial agents in seafood and fish samples.

Matrix	Analyte	Analytical Technique	MOF Material	Sample Preparation Technique	Recovery %	LODs	Ref.
Fish (Tilapia)	Flumequine, nalidixic acid, sulfadimethoxine, sulfaphenazole, tilmicosin & trimethoprim	HPLC-MS/MS	MIL-101(Cr)NH$_2$	SPME	NA	0.2–1.1 ng g^{-1}	[71]
Shrimp, Chicken, pork	Sulfonamides	HPLC-DAD	Fe$_3$O$_4$@JUC-48	MSPE	76.1–102.6	1.73–5.23 ng g^{-1}	[72]
Fish, chicken, water	Fluoroquinolones enrofloxacin, ciprofloxacin, norfloxacin, lomefloxacin	HPLC-UV	Cu based MOF	DSPE	81.3–104.3	0.18–0.58 ng g^{-1}	[73]
Fish, milk, pork	Tetracyclines tetracycline, chlorotetracycline and oxytetracycline	Fluorescence Sensing	In-sbdc	SLE	96.35–102.57	0.28–0.30 nM	[45]
Shrimp	Chloramphenicol	Ratiometric Fluorescence Sensing	PCN-222	SLE	91.25–104.47	0.08 pg mL^{-1}	[46]
Fish, urine samples	Chlortetracycline	Aggregation-Induced Fluorescence (AIF)	Zn-BTEC	SLE	91.5–108.5	28 nM	[47]
Fish, urine samples	Doxycycline	Fluorescence Sensing	Eu-In-BTEC	SLE	105.5–109.5	47 nM	[48]
Fish, milk, urine, serum	Kanamycin and chlortetracycline	Fluoride-Selective Electrodes (FSE)	NMOF-F^{-}@Apt	SLE	91–108	0.35–0.46 nM	[49]
Fish, water samples	Malachite green	UV-Vis Spectroscopy	Tb-MOF	SPE	95.6–104.3	1.66 ng mL^{-1}	[74]
Fish	Malachite green & crystal violet	UHPLC-MS/MS	Fe$_3$O$_4$@PEI-MOF-5	MSPE	83.15–96.53	0.30 ng mL^{-1} & 0.08 ng mL^{-1}	[75]
Fish	Malachite green & crystal violet	UV-Vis Spectroscopy	Fe$_3$O$_4$eNH$_2$@HKUST-1@PDES	MSPE	89.43–100.65	98.19 ng mL^{-1} & 23.19 ngmL^{-1}	[76]
Fish	Malachite green	Differential pulse voltammetry (DPV)	Ag/Cu-MOF-modified electrode	SPE	NA	2.2 nM	[50]

Mondal et al. [71] synthesized a novel polyacrylonitrile fiber coated with amino group modified MOF material, MIL-101(Cr)-NH$_2$, which was used as a solid-phase micro extraction (SPME) fiber for the simultaneous determination of six antibiotics (flumequine, nalidixic acid, sulfadimethoxine, sulfaphenazole, tilmicosin and trimethoprim), representatives of four different antibiotic classes (quinolones, sulfonamides, macrolides, pyrimethamines), in the muscle tissue of living tilapia. Detection was performed using of a high-performance liquid chromatography system coupled with a tandem mass spectrometry detector (LC-MS/MS). MOF was synthesized hydrothermally by mixing chromic nitrate hydrate and 2-aminoterephthalic acid in water followed by thermal treatment (130 °C, 24 h) in an autoclave. For the preparation of the novel SPME fiber, small sized particles of MIL-101(Cr)-NH$_2$ were used to form a slurry for the coating onto the surface by dip and dry of biocompatible polyacrylonitrile quartz fiber. The fiber that gave the optimum results included 50 mg of the new material. The in vivo SPME method was carried out in anesthetized fish. A hypodermic syringe was used to pierce the dorsal epaxial muscle and then the fiber was inserted in the hole for 10 min. The novel fibers were found to be stable for six sampling-desorption cycles. As for the sensitivity of the method the Limit of Detection (LOD) ranged between 0.2 ng g^{-1} to 1.1 ng g^{-1} and limit of quantification (LOQ) ranged between 0.6 ng g^{-1} and 3.7 ng g^{-1} for the six examined antibiotics. The comparison with commercial utilized fibers like C18, PDMS, PDMS/DVB composite or acrylate fiber revealed that the MOF coated fiber shower higher performances. It should be pointed out that the non-amino functionalized MOF showed dramatically lower detection capability. The novel fibers were found to be of low cost, easy to prepare, reproducible in antibiotic determination in fish muscle samples so it was assumed to be an ideal fiber for in vivo experiments [71]. As the authors concluded, the high surface area and mesoporosity, as well as the presence of the amino groups revealed to play the detrimental role.

Xia et al. composed a magnetic and mesoporous MOF-based composite material as a magnetic matrix solid phase sorbent and they used it for the extraction of sulfonamides from shrimps [72]. The determination was performed by high-performance liquid chromatography (HPLC) coupled with a photodiode array detector (DAD). The Fe$_3$O$_4$@JUC-48 nanocomposite material was synthesized by mixing cadmium nitrate tetrahydrate, and 1,4-biphenyldicarboxylic acid with mercaptoacetic acid functionalized Fe$_3$O$_4$ nanoparticles. The iron oxide nanoparticles were coated with the formed rod-shaped JUC-48 crystals, with the carboxylate groups upon the modification of the NPs to act as seeds for the growth of the framework, leading to a micro-porous composite material (Figure 2).

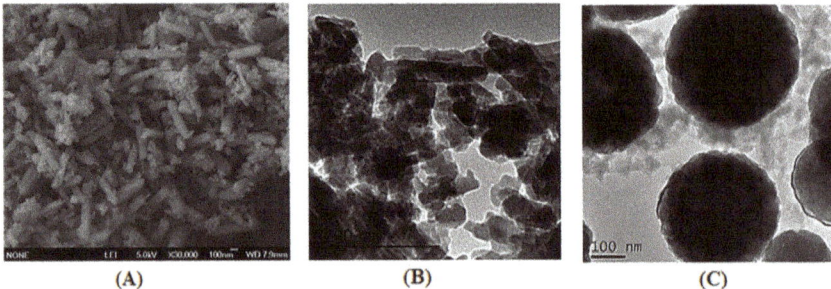

Figure 2. SEM and TEM images of pure JUC-48 (**A**,**B**), and TEM image of Fe$_3$O$_4$@JUC-48 (**C**). Adapted with permission from Reference [72]. Copyright (2017) Elsevier.

The developed method was successfully applied except to shrimps, to a variety of samples such as chicken or pork. The LODs for all matrices were ranged from 1.73 ng g^{-1} to 5.23 ng g^{-1}, while the respective LOQ values ranged between 3.97 and 15.89 ng g^{-1}. Recovery rates for shrimp, pork, and chicken samples were between 76.1% and 102.6%. The novel sorbent was found to be reusable for at least seven times [72].

Wang et al., reported the use of a MOF as a precursor for the synthesis of a three-dimensional (3D) porous Cu@graphitic octahedron carbon cages [73]. After the rapid room-temperature synthesis of a

Cu-based metal–organic framework from copper(II) nitrate trihydrate and 1,3,5-benzene-tricarboxylic acid with the presence of ZnO nanoparticles as nucleation center (Figure 3A), the obtained material was further pyrolyzed at 700 °C under nitrogen. The final obtained material consisted of Cu nanoparticles of size 20-30 nm, encapsulated in a graphitic-carbon phase. The shape of the particles was octahedral (Figure 3B), as the one of the precursor MOF, with an open-pore structure (Figure 3C). The material showed a relatively high surface area of around 224 $m^2\ g^{-1}$.

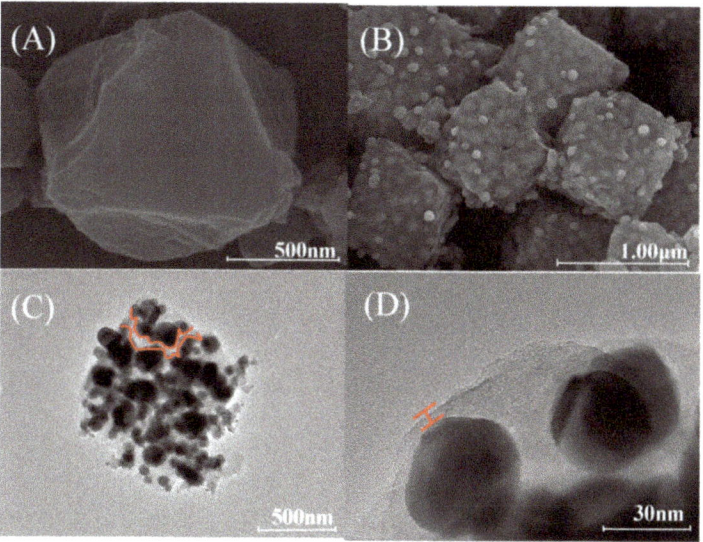

Figure 3. SEM images of the $Cu_3(BTC)_2$ MOF precursor (**A**), and SEM (**B**) and TEM (**C,D**) images of Cu@graphitic carbon composite. Adapted with permission from [73]. Copyright (2018) Elsevier.

This composite material was used for the dispersive solid phase extraction (DSPE) of four fluoroquinolones (FQs) from fish tissue prior to their determination with HPLC coupled with a UV detector. For the extraction of FQs, a portion of 1 g of homogenized fish muscle tissues was used and after the addition of methanol the mixture was sonicated for 10 min. This method was also applied for the detection of the FQs in chicken muscle tissue as well as for water samples. The recoveries of the method in all cases were very satisfactory, ranging between 81.3% and 104.3% while the LODs were found between 0.18 ng g^{-1} and 0.58 ng g^{-1} [73].

2.2. Detection of Antibiotics in Seafood and Fish Samples

Except for the use of MOFs as extraction sorbents enhancing the extraction step of antibiotics, these novel materials have been used in recent years as a novel kind of fluorescent sensing materials.

Liu et al. constructed a MOF-based sensing system which had specific response to tetracyclines TC (tetracycline, chlortetracycline and oxytetracycline) antibiotics [45]. The novel luminescent MOF material (In-sbdc) was synthesized by mixing indium (III) chloride ($InCl_3$) and 4,4'-stilbene-dicarboxylic acid (H_2sbdc) in DMF-H_2O at room temperature. The excitation and emission wavelengths that were chosen were 327 nm and 377 nm, respectively. In-sbdc showed great selectivity/specificity over other classes of antibiotics such as macrolides, chloramphenicols, aminoglycosides, β-lactams, glycopeptides, nitroimidazoles and nitrofurans. The method was applied after an easy pretreatment procedure of fish, milk, pork, or aqueous samples. The extraction of tetracyclines from fish muscle was conducted with acetonitrile by a simple solid liquid extraction (SLE) procedure. In fish, pork and milk samples recoveries ranged between 96.35% and 102.57%, while the LODs were found between 0.28 nM–0.30 nM [45].

The same research group developed a ratiometric fluorescent sensing method for the determination of trace chloramphenicol (CAP) levels in shrimp tissues [46]. They developed a highly stable zirconium-porphyrin MOF (PCN-222) fluorescence quencher with strongly adsorbed dye-labeled Fam-aptamer due to π-π stacking, hydrogen bond and coordination interactions (Figure 4).

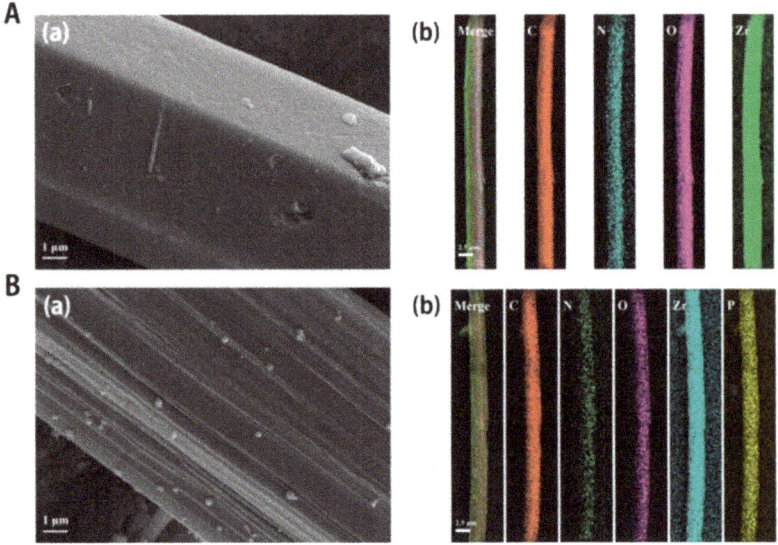

Figure 4. SEM images (**A**(**a**) and **B**(**a**)) and EDS mapping analysis (**A**(**b**) and **B**(**b**))) of PCN-222 (**A**(**a**) and **A**(**b**)) and of PCN-222 adsorbed with FAM-aptamer (**B**(**a**) and **B**(**b**)), with the presence of P characteristic for the homogeneous adsorption of the aptamer. Adapted with permission from [46]. Copyright (2020) Elsevier.

When CAP exists in the sample, dye-labeled aptamers are released from the surface of the novel MOF modified material, resulting in the recovery of fluorescence. The PCN-222 was prepared by dissolving zirconium (IV) chloride $ZrCl_4$, tetrakis(4-carboxyphenyl)porphyrin H_2TCPP and benzoic acid in N,N-dimethylformamide (DMF) by ultrasonication followed by the heating of mixture in an oven at 120 °C for 48 h and then at 130 °C for 24 h. The optimal conditions for the fluorescence detection of CAP was found to be the ratio of intensities $I_{520\,nm}/I_{675\,nm}$. The linear detection range was between 0.1 pg mL^{-1} and 10 ng mL^{-1} while the LOD was found 0.08 pg mL^{-1}. The applicability of the new method for the CAP determination in shrimp samples was evaluated by a comparison with a commercial ELISA kit with the sample pretreatment proposed by the ELISA kit. The recoveries were found (91.25–104.47%) in spiked shrimp tissues and the relative standard deviation values RSD that were found 2.83%–5.02% suggested the fine accuracy and the good precision of the proposed assay [46].

Yu et al. developed an analytical method for the selective and sensitive detection of chlortetracycline (CTC) in fish muscle tissue after developing a zinc-based metal organic framework of pyromellitic acid (Zn-BTEC) [47]. The new material had the ability to enhance the aggregation-induced emission (AIE) of CTC. The new MOF material was synthesized by heating Zn-BTEC and nanometer-sized zinc oxide ZnO in DMF/H_2O (10/1) at 180 °C for 3.5 days. The MOF after the addition CTC showed significant fluorescence enhancement at 446 nm and 540 nm which were different from the peak displacement of other tetracyclines (TCs). For the extraction of CTC by fish muscle tissue, an easy solid-liquid extraction (SLE) was used. The Zn-BTEC MOF material showed great specificity and selectivity over other antibiotics. The method was also applied in urine samples. After spiking of CTC in zebrafish samples and urine samples the recoveries ranged between 91.5% and 108.5% while the LOD of method was found 28 nM. The sensor of MOF was found to be reusable after the removal of CTC but at the cost of tedious washing [47].

The same scientific team reported the synthesis of an europium-based functional MOF with pyromellitic acid as linker, co-doped with indium (Eu-In-BTEC) [48]. The new material was applied in fluorescence sensing of doxycycline (DOX) in fish muscle tissues and urine samples. The material Eu-In-BTEC was synthesized by mixing indium nitrate hydrate, europium chloride hexahydrate and pyrometallitic acid in DMF/H_2O (10/1 v/v) at 180 °C for 84 h. The MOF after the addition of doxycycline showed significant fluorescence enhancement at 526 nm and 617 nm. For the extraction of DOX by fish muscle tissue the same SLE pretreatment as described above was used with a homogenization of fish tissue with methanol for 5 min followed by centrifugation. The supernatants were diluted to the sensing system and mixed well before recording their emission spectra. The new MOF system could discriminate DOX from a plethora of other antibiotics such as TCs showing great specificity and selectivity. After the application of the new method in fish samples and urine samples the recoveries of DOX ranged between 105.5% and 109.5%, while the LOD was estimated 47 nM. The novel sensors indicated the specific and good performance of MOF for this kind of applications [48].

An aptamer-sensing platform was also developed for the detection of small organic molecules kanamycin and chloramphenicol using a portable fluoride-selective electrode (FSE). For this reason, the research group fabricated signal tags of nanometal-organic frameworks (NMOF) encapsulating F^- and labeling aptamers immobilized on one stir-bar. A double stir bar was composed to convert organic small molecules to F^- for signal development. The qualification of the target molecule (kanamycin or chloramfenicol) was achieved after reaction when signal tags from bar-b were washed and F^- was released. The preparation of signal probe (NMOF-F-@S1) was made by mixing UiO-66-COOH nanoparticles and F^- solution in room temperature. The double stir-bars assisted target system was prepared by the use of gold nanoparticles AuNPs. For the extraction of antibiotics from fish tissue, anhydrous sodium sulphate and ethyl acetate were added to fish muscle into a centrifuge tube. The mixture was homogenized and the supernatant was removed and transferred to a round flask. After the second extraction step with ethyl acetate the combined extract was evaporated to dryness. The residue was reconstituted by the addition acetonitrile and n-hexane and the dissolved residue was transferred into a graduated glass stopped reagent bottle and shaken. The n-hexane phase was discarded and this step was repeated with n-hexane. The acetonitrile phase was evaporated to dryness under a stream of dry nitrogen and the dry residue was dissolved in 0.5 mL of PBS (pH 7.4). The new assay was found to be very selective and sensitive in different matrices (water, milk, fish muscle, serum and urine) giving LODs between 0.35 nmol L^{-1} and 0.46 nmol L^{-1} for both antibiotics, while the recoveries ranged between 91 to 108% in all matrices for both compounds [49].

2.3. Extraction and Detection of Antimicrobial Agents in Seafood and Fish Samples

Malachite green (MG) and crystal violet (CV) are triphenylmethane dyes which are widely used in the aquaculture industry as antimicrobial agents due to their antifungal and antiparasitic properties in fish [77,78]. These antimicrobials have a long withdrawal period in fish and can lead to side effects, such as high toxins, high residual, carcinogenic, teratogenic, and mutation. The use of MG and CV in aquaculture remains common despite its prohibition in many countries, because of their highly effective parasiticide and fungicide [79]. A few recent applications of MOFs have been found in literature for the detection of MG and CV in fish muscle tissue.

Mohammadnejad et al. synthesized a terbium metal-organic framework (Tb-MOF), utilized as a solid phase extraction (SPE) sorbent for the extraction of MG from fish muscle tissues and water samples following by detection using a UV-Vis spectrophotometer [74]. Tb-MOF was made using the hydrothermal method by mixing Terbium(III) nitrate hexahydrate, benzene-1,3,5-tricarboxylic acid in DMF and H_2O. Fish tissues were homogenized with the addition of hydroxylamine hydrochloride and ammonium acetate (pH 4.5 adjusted with acetic acid). A portion of the homogenate was used for

MG extraction with the addition of acetonitrile followed by ultrasonication and centrifugation. The supernatant was extracted twice with acetonitrile. The solutions collected were mixed with the MOF sorbent and the mixture was stirred at room temperature for 2 h. The sorbent was separated after centrifugation and then eluted by methanol. Then the MG solution was transferred for UV-Vis analysis. After several extraction cycles Tb-MOF was found to be intact after 10 SPE cycles, showing high regeneration ability. The LOD of the method for water and fish samples was calculated 1.66 ng mL^{-1}, while recoveries for fish samples and water samples ranged between 95.6% and 104.3% [74].

The same year, a magnetic mesoporous metal-organic framework-5 was composed by Zhou et al., for the effective enrichment of MG and CV in fish samples [75]. The developed MOF-5 material was synthesized using the solvothermal method by mixing zinc acetate hydrate, terephthalic acid and polyethyleneimine functionalized Fe$_3$O$_4$ nanoparticles. SEM and TEM analysis (Figure 5) revealed that the cubic particles of MOF were coated with modified Fe$_3$O$_4$ nanoparticles.

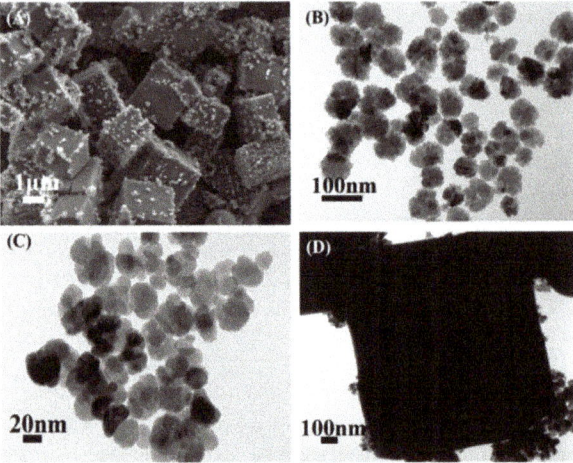

Figure 5. SEM image of Fe$_3$O$_4$@PEI-MOF-5 (**A**); TEM images of Fe$_3$O$_4$ (**B**), Fe$_3$O$_4$@PEI (**C**), Fe$_3$O$_4$@PEI-MOF-5 (**D**). Adapted with permission from [75]. Copyright (2018) Elsevier.

Fish samples were homogenized with acetonitrile and the mixture was sonicated and centrifuged. This extraction step was repeated twice. The extract was dried and dissolved in EtOH and then 10 mg of the MOF composite were added for the MSPE procedure. The mixture was stirred for 40 min and then the material was separated by a magnet and washed with 1 mL of methanol for 3 times. Malachite green and crystal violet were eluted in acidic methanol (1% formic acid) while the desorption time was 20 min. The LODs of the method for MG and CV were estimated to be 0.30 ng mL^{-1} and 0.08 ng mL^{-1}, respectively while recoveries for both compounds were between 83.15% and 96.53%. The novel sorbent exhibited high magnetization, large surface area, good chemical stability and a distinctive morphology [75].

Polymeric deep eutectic solvents (PDES) functionalized amino-magnetic (Fe$_3$O$_4$) MOF (HKUST-1-MOF) composites (Fe$_3$O$_4$-NH$_2$@HKUST-1@PDES) were synthesized by Wei et al. [76] and used for the selective separation of MG and CV coupled with MSPE prior to detection with UV-Vis spectrometry. For the preparation of the composite framework, spheroidal Fe$_3$O$_4$ nanoparticles (FeNPs) of size around 20 nm (Figure 6), were modified with 3-aminopropyltriethoxysilane (APTES). The MOF/FeNPs composite was synthesized under reflux and by mixing the amino-modified FeNPs with benzene-1,3,5- tricarboxylic acid (H$_3$BTC) in a ethanol/DMF solution, followed by the addition of an aqueous copper (II) acetate monohydrate (Cu(OAc$_2$)$_2$·H$_2$O) solution. The polymeric composite, Fe$_3$O$_4$-NH$_2$@HKUST-1@PDES, was prepared following polymerization with deep eutectic solvents

(PDES) based on 3-acrylamidopropyl trimethylammonium chloride and *N,N*-methylene-bisacrylamide through a seeded emulsion polymerization method.

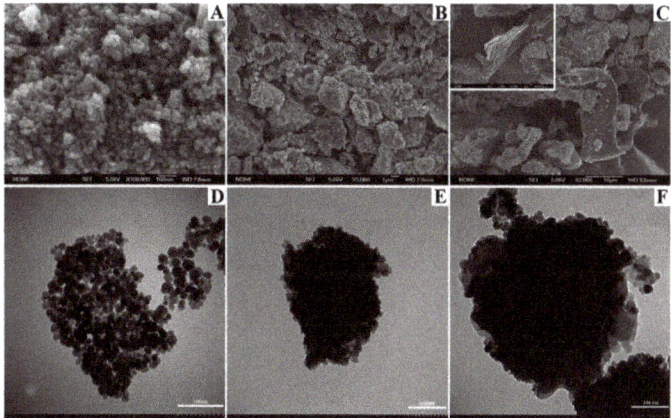

Figure 6. SEM images of Fe_3O_4 (**A**), Fe_3O_4@HKUST-1 (**B**), Fe_3O_4@HKUST-1@PDES (**C**), and TEM images of Fe_3O_4 (**D**), Fe_3O_4@HKUST-1 (**E**), Fe_3O_4@HKUST-1@PDES (**F**). Adapted with permission from [76]. Copyright (2019) Elsevier.

For the sample preparation 5 g of fish samples were homogenized with 1.5 mL of 20% hydroxylamine hydrochloride and 3.5 mL of 50 mmol L^{-1} ammonium acetate for 30 min under vigorous stirring. For the MSPE procedure, the magnetic sorbent was added into the supernatant and the mixture was vortexed at room temperature to extract the dyes. The magnetic material sorbents were removed with a magnet and the supernatant was used for the detection of MG and CV by a UV-Vis spectrophotometer. Limits of detection were found to be 98.19 ng mL^{-1} for MG and 23.97 ng mL^{-1} for CV, respectively, while recoveries from fish samples for both compounds ranged between 89.43% and 100.65% [76].

A Cu-based MOF modified by silver (Ag/Cu-MOF) was fabricated by Zhou et al. [50], for the electrochemical determination of MG in fish by a Differential Pulse Voltammetry (DPV) method. The Ag/Cu-MOF material was synthesized by a one-step solvothermal synthesis from Copper(II) nitrate trihydrate, silver nitrate, and 1,3,5-benzenetricarboxylic acid through, and then it was modified on glassy carbon in order to be used as a voltammetric sensor for the detection of MG. The one-step direct synthesis was a simple and efficient. The obtained crystals showed a size distribution from tens to several hundreds of nanometers. The EDS and XPS analysis revealed an Ag to Cu elemental proportion of ~7.5%, with silver been homogeneously distributed at the entire particles.

Fish samples were homogenized after the addition of *p*-toluenesulfonic acid, hydroxylamine and acetonitrile followed by centrifugation (twice). The supernatants were collected together, dichloromethane was also added, and the solution was vortexed and centrifuged. Then, the organic phase was passed through a SCX SPE column. For the evaluation of the new developed method fish muscle samples were also analyzed with a commercial ELISA assay kit. The results showed no significant difference between the two methods suggesting the Ag/Cu-MOF modified electrode as a simple, high sensitive and accurate tool for MG determination in fish samples. Also, the detection limit of the proposed electrochemical sensor was estimated to be 2.2 nM [50].

3. Extraction of Metal Ions

The applications of MOFs for the extraction of metal ions from fish samples are summarized in Table 2. All fish and seafood samples were primarily digested with concentrated nitric acid for 4 h at 100 °C in Teflon beakers.

Table 2. Application of MOFs for the extraction of metal ions from fish samples.

Analyte	Metal/Organic Linker of MOF	Modification	Amount of Sorbent (mg)/Adsorption pH/Adsorption Time(min)	Desorption Time (min)/Type of Eluent	Matrix	Sample Preparation Technique	Detection Technique	Recovery (%)	LOD (ng mL^{-1})	Reusability	Ref.
Pd(II)	Cu/Trimesic acid	Fe$_3$O$_4$@Py	30/6.9/6	15.5/0.01 mol L^{-1} NaOH in 9.5 (w/v %) K$_2$SO$_4$	*Platycephalus indicus*	MSPE	FAAS	96.8–102.6	0.37	-	[80]
Hg(II)	Cu/Trimesic acid	SH@SiO$_2$	24/6/8	11/ 1.1 mol L^{-1} solution of thiourea	Fish, canned tuna	d-SPE	CV-AAS	91–105	0.02	-	[81]
	Cu/Trimesic acid	Fe$_3$O$_4$@4-(5)-imidazole-dithiocarboxylic acid	30/5.5/15	20/1 mol L^{-1} thiourea solution in 0.01 mol L^{-1} NaOH	*Platycephalus indicus*	MSPE	CVAAS	91	10	At least 12 times	[82]
	Zr/Benzoic acid and meso-tetrakis(4-Carboxyphenyl) porphyrin	-	2/5/-	-/ HCl (10% v/v)	Fish	PT-SPE	CVAAS	74.3–98.7	20 × 10^{-3}	At least 15 times	[83]
Cd(II), Pb(II)	Cu/Trimesic acid	Fe$_3$O$_4$@Py	30/6.3/14	16.5/0.8 mol L^{-1} EDTA in 0.01 mol L^{-1} NaOH	*Platycephalus indicus*	MSPE	FAAS	92.8–97.0	0.2–1.1	-	[84]
Cd(II) Pb(II) Ni(II)	Cu/Trimesic acid	Fe$_3$O$_4$@TAR	50/6.2/10	15.2/0.6 mol L^{-1} EDTA	Fish, shrimps	MSPE	FAAS	83–112	0.15–0.8	-	[85]
Cd(II), Pb(II), Ni(II), Zn(II)	Cu/Trimesic Acid	Fe$_3$O$_4$@DHz	25/6.4/13	20/ 0.01 mol L^{-1} NaOH in thiourea	*Platycephalus indicus*	MSPE	FAAS	88–92	0.12–1.2	-	[86]
Cd(II), Pb(II), Ni(II), Co(II)	Fe/Terephthalic Acid	Fe$_3$O$_4$@dipyridylamine	30/6.5/11	14/ 0.7 mol L^{-1} EDTA in 0.13 mol L^{-1} HNO$_3$	Fish liver, skin, muscle	MSPE	FAAS	88–108	0.13–0.75	-	[87]

3.1. Extraction of Palladium

Palladium (Pd) is a member of platinum group metals with various scientific and technological applications. Pd has been used in metallurgy, catalysts, electronic applications, capacitors, biomedical devices, and catalytic converters for car engines [80,88–91]. This element has no biological role and its compounds are considered toxic and carcinogenic [90,91]. Due to the increasing industrial applications of palladium, it can enter the aquatic environment and it is therefore a potential danger for humans and marine life. As a result, it is important to develop efficient analytical methods for the monitoring of Pd pollution in fish samples [89–91]. Pd(II) has been extracted from fish samples with a magnetic metal-organic framework derived from trimesic acid and copper nitrate trihydrate [80]. The MOF was modified with pyridine functionalized Fe_3O_4 (Fe_3O_4@Py) nanoparticles in order to increase selectivity towards palladium. The crystals of original $[Cu_3(BTC)_2(H_2O)_3]_n$ sample are octahedral with a smooth surface and have an average size of 10 mm (Figure 7a).

Figure 7. The SEM images of MOF (**a**) and magnetic MOF (**b–d**). Adapted with permission from [80]. Copyright (2012) Elsevier.

However, surface of the magnetic MOF tends to be rougher after immobilization by Fe_3O_4–Py (Figure 7b–d). The fish samples were initially digested with nitric acid. It was found that the optimum pH value for adsorption was 6.9. For the elution steps, the researchers used 0.01 mol L^{-1} sodium hydroxide in potassium sulfate to prevent sorbent decomposition that was observed in acidic environment. Satisfactory recovery values were observed, however no reusability data were provided. The developed method showed high sample clean-up as well as satisfactory recovery values (96.8–102.6%), however no sorbent reusability was reported.

3.2. Extraction of Mercury

Mercury is one of the most dangerous contaminants in the environment that threatens the human health and natural ecosystems. This element can enter the aquatic environment from a variety of sources including predominately mining and industrial production. Therefore, the accumulation of mercury in aquatic animals such as fish is unavoidable and through the dietary process, human health can be exposed to danger. Since, mercury is toxic even in low concentrations, its preconcentration from real samples is a necessary and demanding step for its determination [81,92,93].

Hg(II) has been extracted from fish samples with a porous thiol-functionalized metal-organic framework prepared from with trimesic acid and copper acetate monohydrate and functionalized with thiol-modified silica nanoparticles (SH@SiO$_2$) [81]. The SH@SiO$_2$ nanoparticles were prepared from silicon dioxide and (3-mercaptopropyl)-trimethoxysilane and were employed in order to increase the selectivity of the sorbent towards mercury ions. Compared to the pure MOF particles which showed an octahedral shaped particles of size 8 to 12 µm with smooth surfaces, the shape of the MOF particles in the case of the composite was no-so-well defined with rough surface covered with SiO$_2$ nanoparticles. No SiO$_2$ nanoparticles were detected separately and the authors reported the formation of an amorphous phase in between the MOF particles. Based on these observations and considering the XRD pattern of modified MOF (SH@SiO$_2$/Cu$_3$(BTC)$_2$), they concluded the formation of a nanocomposite rather than a physical mixture, with the molar ratio of sulfur to copper to be 3.9 mmol g^{-1}.

The prepared nanocomposite was employed for the dispersive solid-phase extraction of Hg ions from digested fish samples prior to their determination with Cold Vapor Atomic Absorption Spectrometry (CVAAS). Isolation of the sorbent was achieved with centrifugation and elution of the adsorbed analyte was performed with sodium hydroxide. It was found that this eluent provided satisfactory extraction recovery without decomposition of the sorbent. The maximum Hg(II) adsorption capacity by the nanocomposite under the optimum conditions was found 210 mg g-1, with the pseudo-second-order model to have the best fitting. However, no potential reusability of the sorbent was reported. It was indicated that although the preparation of sorbent was complicated, large quantities can be prepared at once [81].

Sohrabi prepared also an a HKUST-1 based magnetic MOF from trimesic acid and copper acetate that was modified with 4-(5)-imidazole-dithiocarboxylic acid functionalized Fe$_3$O$_4$ nanoparticles [82]. The modification of the synthesized MOF enhanced its selectivity towards mercury. The magnetic sorbent was used for the MSPE of mercury from fish and canned tuna samples prior to its determination with CVAAS. Compared to dispersive SPE (d-SPE), magnetic solid-phase extraction has the advantage of simple and rapid sorbent isolation with the implementation of an external magnetic field [88]. In this work, a solution of 0.01 mol L^{-1} thiourea was chosen for the elution of the adsorbed mercury ions and the sorbent was found to be reusable for up to 12 times, indicating satisfactory stability during adsorption and desorption steps.

Finally, mercury has been also extracted from fish samples with a mesoporous porphyrinic zirconium metal-organic framework (PCN-222/MOF-545) prior to CVAAS determination [83]. The sorbent was prepared from zirconyl chloride octahydrate, benzoic acid and meso-tetrakis(4-carboxyphenyl)porphyrin. Zirconium-based MOFs are known for their stability in aqueous environment. Following the strategy of use a porphyrin as linker compared to the mono-aromatic carboylic acids, the size of the pores/cages is increasing, resulting to enhanced mass transfer towards the active sites. Based on the theoretical calculations, the diameter of this MOF was estimated as 3.7 nm. The SEM micrographs revealed well-shaped hexagonal rod- and needle-like particles with diameter in the range from 300 to 800 nm, and length from 5 to 16 µm. For the pipette-tip extraction procedure, MOF was placed into a pipette-tip for the pipette-tip solid-phase extraction (PT-SPE) of Hg ions. In this work, only two milligrams of sorbent were required for the extraction of mercury. Moreover, the PCN-222/MOF-545 showed good stability under acidic conditions, since it was found to be reusable for at least 15 times after elution with HCl 10% v/v. Finally, the PT-SPE method was simple a rapid with a total extraction and desorption time span that was shorter than 7 min.

3.3. Multi-Element Extraction

Metal-organic frameworks have been used for the extraction of different metal ions i.e., Cd(II), Pb(II), Ni(II), Zn(II) and Co(II) from fish samples. Most of these ions are dangerous and toxic for human health. Since, those elements exist in real samples in low concentrations, their preconcentration is often considered mandatory [84–87].

Copper-(benzene-1,3,5-tricarboxylate) MOFs have been employed for the multi-element extraction of real samples after modification with Fe_3O_4 nanoparticles functionalized with chemical substances including pyridine (Py) [84], 4-(thiazolylazo) resorcinol (TAR) [85] and dithizone [86] in order to increase the extraction selectivity towards the target analytes. Moreover, a copper-(benzene-1,4-dicarboxylate) MOF, modified with Fe_3O_4 nanoparticles, functionalized with dipyridylamine has been also employed for the extraction and preconcentration of metal ions from fish samples [87]. Even though, the modified MOFs were found to provide satisfactory extraction recovery, enhancement factors and selectivity, no reusability data were reported indicating a limitation to their potential applications for the extraction of metal ions, since the Cu-based MOF are not so stable upon exposure to water. At this point we would like to mention that the utilization of the spend samples for other applications or either for the synthesis of alternative materials should gather more intense research attention, since by this strategy will close the reusability and the atom economy cycle.

Cu-BTC/HKUST-1 modified with pyridine functionalized Fe_3O_4 was used for the extraction of Cd(II) and Pb(II) ions from fish samples prior to FAAS determination [84]. The MOF material was stable under adsorption step (pH 6.3), however structure decomposition was noticed at the elution step with hydrochloric acid and nitric acid. Therefore, a solution of 0.01 mol L^{-1} sodium hydroxide in ethylenediaminetetraacetic acid (EDTA) was chosen as the optimum eluent. Ni(II), Cd(II), and Pb(II) ions were extracted from seafood (fish and shrimps) with a copper-(benzene-1,3,5-tricarboxylate) metal-organic framework that was modified by magnetic nanoparticles carrying covalently immobilized 4-(thiazolylazo) resorcinol (Fe_3O_4@TAR) [85]. Since this MOF is not stable at acidic environment, elution with EDTA was chosen. The above three ions plus Zn(II) ions were also extracted from fish samples with a copper-(benzene-1,3,5-tricarboxylate) MOF functionalized with dithizone-modified Fe_3O_4 nanoparticles (Fe_3O_4@DHz) prior to their determination with FAAS [86]. Elution of the adsorbed analytes was performed with a solution of 0.01 mol L^{-1} sodium hydroxide in thiourea to avoid any structure decomposition. Finally, in another work Cu-BTC modified with Fe_3O_4 dipyridylamine was employed to extract Cd(II), Pb(II), Co(II), and Ni(II) ions from fish samples prior to their determination by FAAS. Hydrochloric acid, nitric acid, sodium hydroxide, potassium sulfate, potassium chloride, thiourea and EDTA were evaluated as eluents and a solution of 0.7 mol L^{-1} EDTA in 0.13 mol L^{-1} nitric acid was chosen [87].

4. Extraction of Other Organic Compounds

Polycyclic aromatic hydrocarbons (PAHs) are a group of environmental contaminants consisting of two or more benzene rings fused in various arrangements. PAHs mainly come either from direct petroleum releasing or from incomplete combustion of organic materials and fossil fuel. They are persistent contaminants in environment and aquatic environment that can be transported over long distances and accumulate in living organisms, such as fish, due to their lipophilicity. By consumption of contaminated fish, the presence of PAHs is potential risk for human health, since they exhibit mutagenic, carcinogenic, teratogenic properties [94–96].

For the determination of PAHs in edible fish tissue samples, Hu et al. fabricated a hybrid magnetic MOF-5 with the chemical bonding approach and used it as adsorbent for the MSPE of PAHs prior to their determination with gas chromatography-mass spectrometry (GC-MS) [97]. For this purpose, MOF-5 was prepared from terephthalic acid and zinc acetate dihydrate in N,N'-dimethyl-formamide. Fish samples were initially ground and mixed with florisil and of n-hexane/methylene chloride (1:1, v/v). The mixture was shaken for in whirlpool bath and centrifuged, thrice. The combined extracts were dried and re-dissolved in n-hexane, followed by liquid-liquid extraction with sulfuric acid solution (60%, wt). After phase separation, the supernatant was collected and the magnetic sorbent was added for the MSPE procedure. Elution of the adsorbed analytes was performed with acetone. The MOF-5 sorbent was found to be stable for at least 100 extraction-desorption cycles

Polychlorinated biphenyls (PCBs) are a group of 209 chlorinated biphenyl rings with different physical-chemical properties and toxicity that depends on the number and the position of chlorine atoms [98]. PCBs are persistent organic pollutants that pose a great danger for human health and the environment because of their high toxicity and lipophilicity [99]. Fish accumulate polychlorinated biphenyls from the aquatic environment through their epithelial/dermal tissue or gills and by prey intake. PCBs can be transferred to humans via dietary intake of contaminated fish [99].

The research group of Lin and co-workers fabricated two different stir bar sorptive extraction (SBSE) bars for the extraction of polychlorinated biphenyls from fish tissue [100,101]. SBSE is an equilibrium extraction technique with large sorption phase volume that is known to provide good recovery and extraction capacity. Moreover, SBSE shows good reproducibility and low consumption of organic solvents [102]. For the first SBSE bar, a Fe_3O_4-MOF-5(Fe) material was used as a coating for a Nd-Fe-B permanent magnet. The bar was employed for the extraction of six PCBs from fish samples. Four different MOF materials (MIL-101(Cr), MOF-5(Zn), ZIF-8, and MOF-5(Fe)) were evaluated. The results showed that Fe_3O_4-MOF-5(Fe) (synthesized from terephthalic acid and ferric nitrate and modified with amine-functionalized Fe3O4 nanoparticles) provided the highest extraction efficiency. Prior to the SBSE procedure, fish samples were homogenized and extracted with n-hexane. The stir bar was found to be reusable for at least 60 times with recovery values more than 80% [100]. The second SBSE bar was based on the immobilization of aptamer in the surface of MOF-5. The immobilized aptamer exhibited selectivity towards two PCBs and the stir bar was fabricated by electro-deposition. The novel sorbent exhibited high surface area as well as high selectivity [101].

Low molecular weight alkylamines (including trimethylamine and triethylamine) were extracted from salmon samples with a modified zeolitic imidazolate framework (ZIF-8) coated on SPME Arrow fiber prior to their determination with GC-MS [103]. SPME Arrow is an interesting alternative to SPME and SBSE that combines the advantages of both techniques i.e the easy automation and flexibility of SPME with the larger sorption phase volumes of SBSE. At the same time, SPME Arrow avoids the drawbacks of both techniques including the limitation in automation of SBSE and the small sorption phase volumes as well as the low fiber robustness of conventional SPME technique [103–105].

Lan et al., fabricated a zeolitic imidazolate framework (A-ZIF-8) and utilized as a coating material with the assistance of poly(vinylchloride) (PVC) as adhesive. Subsequently, the pore size of ZIF-8 was modified by headspace exposure to hydrochloric acid to increase the extraction efficiency for amines. Salmon samples were treated with perchloric acid prior to the extraction procedure. The developed SPME Arrow fibers exhibited good repeatability, stability and batch-to-batch reproducibility and they were found to be reusable for at least 130 times [103].

Domoic acid, the primary amnesic shellfish poisoning toxin has been extracted from shellfish samples with a metal-organic framework magnetic nanocomposite prior to its determination by high performance liquid chromatography-tandem mass spectrometry (LC-MS/MS) [106]. For this purpose, Fe_3O_4@SiO_2 microspheres were synthesized through the solvothermal approach and were treated with glutaric acid anhydride for protection and to become carboxylate terminated. Subsequently, the Fe_3O_4@SiO_2 microspheres were mixed with terephthalic acid and zirconium(IV) chloride to form Fe_3O_4@SiO_2@UiO-66 core-shell microspheres. The carboxylic terminal groups acted as linkers for the growth of the framework, with the entire coated iron nanoparticles acting as seeds of the MOF growth (Figure 8). Similar behavior of graphitic carbon nitride nanospheres with carboxylic terminal surface groups acting as seeds of the UiO-66 or HKUST-1 growth were reported recently by Bandosz and co-workers [29,30,60,63]. The UiO-66/g-C_3N_4 nanocomposite showed higher adsorption capacity of CO_2 as well as catalytic detoxification of toxic chemical warfare agents, as for instance mustard gas. Figure 8 shows the synthetic procedure of Fe_3O_4@SiO_2@UiO-66 core-shell microspheres.

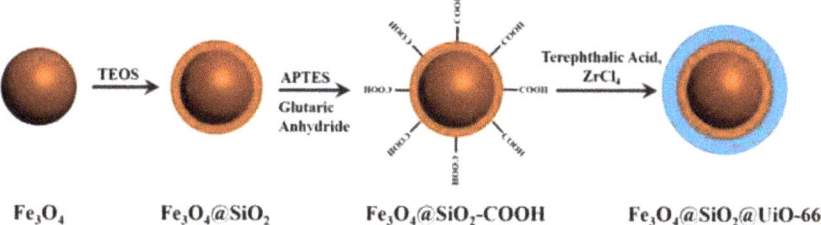

Figure 8. Synthetic procedure of Fe_3O_4@SiO_2@UiO-66 core-shell microspheres. Adapted with permission from [106]. Copyright (2015) Elsevier.

The XRD pattern of the Fe_3O_4@SiO_2@UiO-66 core-shell microspheres revealed to match perfectly with UiO-66 [107,108]. As revealed from the TEM micrographs (Figure 9), the Fe_3O_4 nanoparticles were successfully coated with a 10 nm in thickness silica layer.

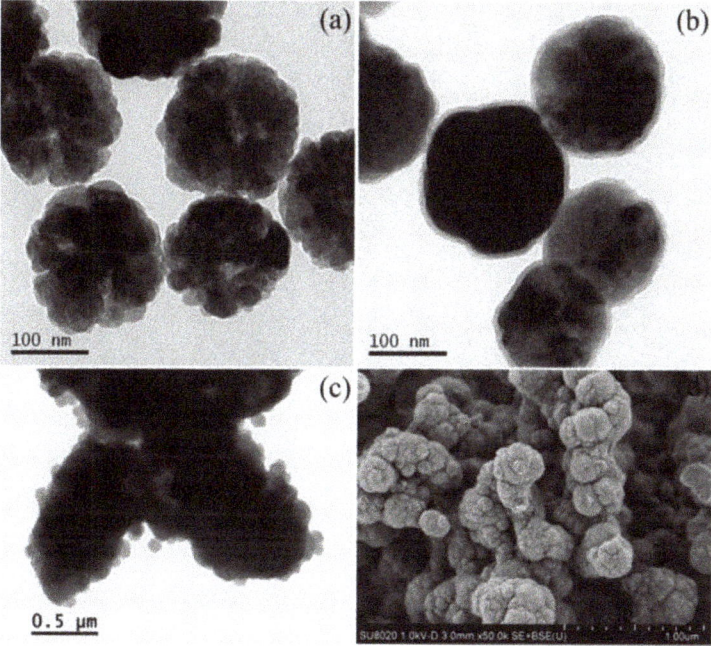

Figure 9. TEM images of Fe_3O_4 nanoparticles (**a**), after coating with SiO_2 (Fe_3O_4@SiO_2) (**b**), of the final nanocomposite Fe_3O_4@SiO_2@UiO-66 (**c**), and SEM of Fe_3O_4@SiO_2@UiO-66 (**d**). Adapted with permission from [106]. Copyright (2015) Elsevier.

For the composite material, the Fe_3O_4@SiO_2 microspheres were embedded inside the final framework, with the shape of the Fe_3O_4@SiO_2@UiO-66 to be spherical-shaped. The nanocomposite showed high porosity (816.3 m^2 g^{-1} surface area and 0.533 cm^3 g^{-1} total pore volume), with the pores to have predominately two sizes, 0.8 and 1.1 nm, characteristic as previously reported for the UiO-66. Prior to the MSPE procedure, shellfish samples were homogenized and extracted with methanol: water (1:1, v/v). Accordingly, one milligram of the magnetic sorbent was added for the extraction of domoic acid. Elution was performed with a mixture of acetonitrile:acetic acid (80:20, v/v), thrice. As a final step, the eluent was evaporated and re-dissolved in the mobile phase. Under optimum conditions, extraction recovery ranged between 93.1–107.3%, however no potential sorbent reusability was reported [106].

5. Extraction Mechanisms

The utilization of MOFs, as well as their composites, as adsorbents for a wide range of organic compounds or metals is a well explored field [109–111]. Although, the herein reported and discussed articles are predominately focused on the analytic part of application, and the interpretation of the involved mechanism is limited. Figure 10 collects the most commonly reported interactions/mechanisms between MOFs and organic compounds or metal/metalloids species. It is worth to mention that more than one mechanism are involved in many cases.

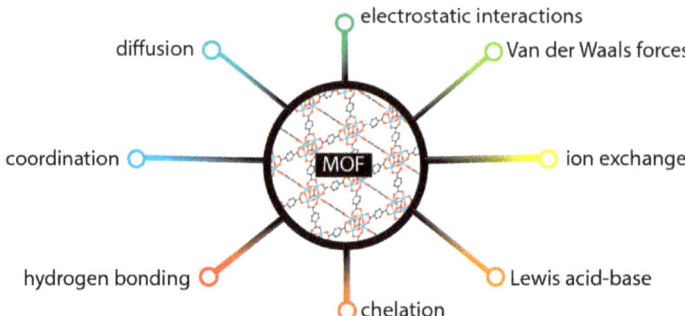

Figure 10. A schematic illustration of the interactions/mechanisms involved in the adsorption/extraction of organic compounds or metal ions by MOFs.

For the preconcentration/adsorption of metal ions, the commonest mechanism can be assigned to Lewis acid-base interactions [112]. Pre- or post-synthetic functionalization of the framework towards the incorporation of O-, N-, or/and S-containing functional groups, is presented as a successful approach. Alternative ways of interaction occur through coordination or chelation adsorption, and functionalization, predominately of the linkers, with specific groups such as thiol, hydroxyl, or amide showed as a prosperous method for metals, while addition of -NH_2 or -OH groups for organic compounds [113,114]. The presence of specific functional groups can positively influence also the physical-based adsorption like electrostatic interactions. The mass transfer phenomena/diffusion of the metal ions or the organic compounds thought the channels/pores towards the active sites is also an important aspect. The size and geometry of the pores is an important aspect, although the penetration of the adsorbate through the entrance/window of the porous framework is of a paramount importance [115]. A strategy to positively enhance the mass transfer is by design and synthesis of MOFs with bigger pores/cages, using larger in size linkers [116,117].

6. Conclusions

Metal-organic frameworks are crystalline porous materials consisting of metal ions or clusters coordinated to organic ligands forming one-, two-, or three-dimensional structures. Despite the fact that there have been only a few years since they first utilized in analytical chemistry, they have found a plethora of applications for the preparation and detection of a variety of organic and inorganic compounds in food samples. Metal-organic frameworks have been used for analytical purposes as alternative materials to conventional solid-phase extraction sorbents and they offer an interesting possibility by enriching the analytical toolbox for the easy pretreatment of fish muscle tissue and seafood samples. MOFs have also been utilized for the fabrication of electrochemical and fluorescent sensors for the selective and sensitive detection of organic compounds providing analytical techniques with very low limits of detection.

Compared with other SPE sorbents such as graphene-based nanomaterials, MOFs pose the advantage of significantly high surface area and hierarchically in structure pores/cages that result in high extraction efficiency as well as high pre-concentration factors. Therefore, MOFs have been used as adsorbents in various sample preparation techniques including SPE, SPME, d-SPE, MSPE, SBSE etc. By coupling MOFs with different micro-extraction techniques, the cost of the sample preparation and the consumption of organic solvents can be decreased and the simplicity of the pre-treatment step can be enhanced. MOFs can also be also applied for in-vivo experiments.

MOFs have been only recently utilized as sorbents for the extraction of organic compounds and metal ions from complex matrices such as fish and seafood samples. Most studies evaluate the application of functionalized MOF sorbents for the determination of specific analytes. The main advantage of metal-organic frameworks is their high total pore volume which results in high extraction recoveries. A wide variety of organic compounds including antibiotics, antimicrobial agents, polycyclic aromatic hydrocarbons, polychlorinated biphenyls etc. have been extracted with MOFs. Typically, the analytes were adsorbed from the samples and desorbed with the assistance of an organic solvents (e.g., methanol, acetonitrile etc.). Good extraction efficiency and reusability was reported in combination with low limits of detection and high enrichment factors.

Metal-organic frameworks have been also used for the extraction and preconcentration of metal ions such as palladium, mercury, cadmium, nickel, lead, cobalt etc. from fish muscle tissue and seafood samples. For inorganic analysis, adsorption was typically performed at intermediate pH value while desorption was performed with mild eluents including EDTA, sodium chloride, sodium hydroxide or thiourea. Acidic desorption was avoided since it was found to cause structure decomposition of the sorbent. However, even with mild eluents, poor reusability of MOFs for trace metals analysis has been reported. Moreover, a modification/functionalization step was required in order to enhance the selectivity of the MOFs towards the target analytes, resulting in complicated synthetic procedure for the preparation of the sorbent. Although, at the herein reported and discussed articles the explored materials as well as their performances are of a great interest, there is a luck of determine mechanistically the interactions and the features that play a key role. This is an aspect that we would like to trigger the research attention towards, since it will help to further explore different frameworks or/and to improve the performance of the already well performing ones.

In order to overcome the main drawbacks of various MOFs, like poor chemical stability in aquatic environments and in acidic/basic solutions as well as the difficulties in regeneration and reusability, various approaches have been proposed, including modifying the metal ions (i.e., by changing the oxidation state and/or by metal ion doping) or modifying the ligands (i.e., by inserting special functional groups) of the MOFs. Another strategy to overcome these drawbacks and to enhance simultaneously the adsorptive capability, is the formation of composite or hybrid novel materials. As thoroughly discussed herein, the main approaches can be summarized as coating of the MOFs' surface with nanoparticles, core-shell growth of the framework around the nanoparticles, encapsulation of nanoparticles inside the framework, growth of the MOF phase as individual particles in between/around the utilized nanoparticles, the incorporation of the nanoparticles simultaneously inside the matrix and on the surface of the frameworks, coating of the MOF particles with different phases (like aptamer or polymeric layer), or to coat fibers with a MOF phase.

Future perspectives can include more in-depth investigation of the use of MOFs as sorbents for the extraction of organic compounds and metal ions from fish and seafood samples, with a more intense emphasis on the determination of the involved mechanisms/interactions. Alternative or novel MOFs can be designed, synthesized, and examined for their extraction efficiency towards the desired analytes, while direct comparison for different ones will elevate the establishment of the most crucial features in each case. Furthermore, MOFs can be evaluated for their applicability after coupling with less studied sample preparation techniques such as PT-SPE and SBSE or on-line sample preparation techniques. Regarding the synthesis of MOFs, research has to be done in the field of exploring "greener" synthetic pathways, the synthesis of new generation biocompatible bio-MOFs, and application of scalable processes that could provide high quantities of MOFs, while the study of functionalization of the metal/metal clusters or of the linkers, and as a result the tuning of the pore sizes and the surface chemistry, will arise new advantageous perspectives.

Author Contributions: The authors have equally contributed to the manuscript. All authors have read and agreed to the published version of the manuscript.

Funding: The research work was supported by the Hellenic Foundation for Research and Innovation (HFRI) under the HFRI PhD Fellowship grant (Fellowship Number: 138).

Conflicts of Interest: The authors declare no conflict of interest.

References

1. Samanidou, V.F.; Evaggelopoulou, E.N. Analytical strategies to determine antibiotic residues in fish. *J. Sep. Sci.* **2007**, *30*, 2549–2569. [CrossRef] [PubMed]
2. Karanikolopoulos, G.; Gerakis, A.; Papadopoulou, K.; Mastrantoni, I. Determination of synthetic food colorants in fish products by an HPLC-DAD method. *Food Chem.* **2015**, *177*, 197–203. [CrossRef] [PubMed]
3. Evaggelopoulou, E.; Samanidou, V. Confirmatory development and validation of HPLC-DAD method for the determination of tetracyclines in gilthead seabream (*Sparus aurata*) muscle tissue. *J. Sep. Sci.* **2012**, *35*, 1372–1378. [CrossRef] [PubMed]
4. Samanidou, V.; Evaggelopoulou, E.; Guo, X.; Troetzmueller, M.; Lankmayr, E. Multi-residue determination of seven quinolones antibiotics in gilthead seabream using liquid chromatography–tandem mass spectrometry. *J. Chromatogr. A* **2008**, *1203*, 115–123. [CrossRef] [PubMed]
5. Lombardo-Agüí, M.; García-Campaña, A.; Cruces-Blanco, C.; Gámiz-Gracia, L. Determination of quinolones in fish by ultra-high performance liquid chromatography with fluorescence detection using QuEChERS as sample treatment. *Food Control* **2015**, *50*, 864–868. [CrossRef]
6. Evaggelopoulou, E.; Samanidou, V. Development and validation of an HPLC method for the determination of six penicillin and three amphenicol antibiotics in gilthead seabream (*Sparus aurata*) tissue according to the European Union Decision 2002/657/EC. *Food Chem.* **2013**, *136*, 1322–1329. [CrossRef]
7. Evaggelopoulou, E.; Samanidou, V.; Michaelidis, B.; Papadoyannis, I. Development and validation of an LC-DAD method for the routine analysis of residual quionolones in fish edible tissue and fish feed. Application to farmed gilthead sea bream following dietary administration. *J. Liq. Chromatog. Relat. Technol.* **2014**, *37*, 2142–2161. [CrossRef]
8. Vardali, S.; Samanidou, V.; Kotzamanis, Y. Development and validation of an ultra-performance liquid chromatography-quadrupole time of flight-mass spectrometry (in MSE mode) method for the quantitative determination of 20 antimicrobial residues in edible muscle tissue of European sea bass. *J. Chromatogr. A* **2018**, *1575*, 40–48. [CrossRef]
9. Kouroupakis, E.; Grigorakis, K.; Vardali, S.; Ilia, V.; Batjakas, I.; Kotzamanis, I. Evaluation of the fillet quality of wild-caught white sea bream (*Diplodus sargus* L.) and brown meagre (*Sciaena umbra* L.) captured from the Aegean Sea. *Mediterr. Mar. Sci.* **2019**, *20*, 373–379. [CrossRef]
10. Wang, P.; Zhang, Q.; Wang, Y.; Wang, T.; Li, X.; Ding, L.; Jiang, G. Evaluation of Soxhlet extraction, accelerated solvent extraction and microwave-assisted extraction for the determination of polychlorinated biphenyls and polybrominated diphenyl ethers in soil and fish samples. *Anal. Chim. Acta* **2010**, *663*, 43–48. [CrossRef]
11. Andersen, W.; Turnipseed, S.; Roybal, J. Quantitative and Confirmatory Analyses of Malachite Green and Leucomalachite Green Residues in Fish and Shrimp. *J. Agric. Food Chem.* **2006**, *54*, 4517–4523. [CrossRef]
12. Okpashi, V.; Ogugua, V.; Ubani, S.; Ujah, I.; Ozioko, J. Estimation of residual polycyclic aromatic hydrocarbons concentration in fish species: Implication in reciprocal corollary. *Cogent Environ. Sci.* **2017**, *3*, 1303979. [CrossRef]
13. Rambla-Alegre, M.; Peris-Vicente, J.; Esteve-Romero, J.; Carda-Broch, S. Analysis of selected veterinary antibiotics in fish by micellar liquid chromatography with fluorescence detection and validation in accordance with regulation 2002/657/EC. *Food Chem.* **2010**, *123*, 1294–1302. [CrossRef]
14. Roudaut, B.; Yorke, J. High-performance liquid chromatographic method with fluorescence detection for the screening and quantification of oxolinic acid, flumequine and sarafloxacin in fish. *J. Chromatogr. B* **2002**, *780*, 481–485. [CrossRef]
15. Manousi, N.; Raber, G.; Papadoyannis, I. Recent Advances in Microextraction Techniques of Antipsychotics in Biological Fluids Prior to Liquid Chromatography Analysis. *Separations* **2017**, *4*, 18. [CrossRef]
16. Kabir, A.; Locatelli, M.; Ulusoy, H. Recent Trends in Microextraction Techniques Employed in Analytical and Bioanalytical Sample Preparation. *Separations* **2017**, *4*, 36. [CrossRef]

17. Filippou, O.; Bitas, D.; Samanidou, V. Green approaches in sample preparation of bioanalytical samples prior to chromatographic analysis. *J. Chromatogr. B* **2017**, *1043*, 44–62. [CrossRef]
18. Seyhan Bozkurt, S.; Erdogan, D.; Antep, M.; Tuzmen, N.; Merdivan, M. Use of ionic liquid based chitosan as sorbent for preconcentration of fluoroquinolones in milk, egg, fish, bovine, and chicken meat samples by solid phase extraction prior to HPLC determination. *J. Liq. Chromatogr. Relat. Technol.* **2016**, *39*, 21–29. [CrossRef]
19. Chen, L.; Zhou, T.; Zhang, Y.; Lu, Y. Rapid determination of trace sulfonamides in fish by graphene-based SPE coupled with UPLC/MS/MS. *Anal. Methods* **2013**, *5*, 4363. [CrossRef]
20. Mehdinia, A.; Ramezani, M.; Jabbari, A. Preconcentration and determination of lead ions in fish and mollusk tissues by nanocomposite of Fe_3O_4@graphene oxide@polyimide as a solid phase extraction sorbent. *Food Chem.* **2017**, *237*, 1112–1117. [CrossRef]
21. Es'haghi, Z.; Khooni, M.; Heidari, T. Determination of brilliant green from fish pond water using carbon nanotube assisted pseudo-stir bar solid/liquid microextraction combined with UV–vis spectroscopy–diode array detection. *Spectrochim. Acta A* **2011**, *79*, 603–607. [CrossRef]
22. Sun, X.; Wang, J.; Li, Y.; Yang, J.; Jin, J.; Shah, S.; Chen, J. Novel dummy molecularly imprinted polymers for matrix solid-phase dispersion extraction of eight fluoroquinolones from fish samples. *J. Chromatogr. A* **2014**, *1359*, 1–7. [CrossRef]
23. Manousi, N.; Zachariadis, G.; Deliyanni, E.; Samanidou, V. Applications of Metal-Organic Frameworks in Food Sample Preparation. *Molecules* **2018**, *23*, 2896. [CrossRef]
24. James, S.L. Metal-organic frameworks. *Chem. Soc. Rev.* **2003**, *32*, 276–288. [CrossRef]
25. Zhou, H.; Long, J.R.; Yaghi, O.M. Introduction to Metal–Organic Frameworks. *Chem. Rev.* **2012**, *112*, 673–674.
26. Li, H.; Libo, L.; Lin, R.; Zhou, W.; Zhang, Z.; Xiang, S.; Chen, B. Porous metal-organic frameworks for gas storage and separation: Status and challenge. *EnergyChem* **2019**, *1*, 100006. [CrossRef]
27. Kampouraki, Z.C.; Giannakoudakis, D.A.; Nair, V.; Hosseini-Bandegharaei, A.; Colmenares, J.C.; Deliyanni, E. Metal Organic Frameworks as Desulfurization Adsorbents of DBT and 4,6-DMDBT from Fuels. *Molecules* **2019**, *24*, 4525. [CrossRef]
28. Koo, W.; Jang, J.; Kim, I. Metal-Organic Frameworks for Chemiresistive Sensors. *Chem* **2019**, *5*, 1938–1963. [CrossRef]
29. Giannakoudakis, D.A.; Bandosz, T.J. Defectous UiO-66 MOF Nanocomposites as Reactive Media of Superior Protection against Toxic Vapors. *ACS Appl. Mater. Interfaces* **2019**. [CrossRef]
30. Giannakoudakis, D.A.; Bandosz, T.J. Building MOF Nanocomposites with Oxidized Graphitic Carbon Nitride Nanospheres: The Effect of Framework Geometry on the Structural Heterogeneity. *Molecules* **2019**, *24*, 4529. [CrossRef]
31. Rojas, S.; Baati, T.; Njim, L.; Manchego, L.; Neffati, F.; Abdeljeli, N.; Saguem, S.; Serre, C.; Najjar, M.F.; Zakhama, A.; et al. Metal–Organic Frameworks as Efficient Oral Detoxifying Agents. *J. Am. Chem. Soc.* **2018**, *140*, 9581–9586. [CrossRef] [PubMed]
32. Li, D.; Hai-Qun, X.; Long, J.; Hai-Long, J. Metal-organic frameworks for catalysis: State of the art, challenges, and opportunities. *EnergyChem* **2019**, *1*, 100005. [CrossRef]
33. Cai, W.; Chu, C.C.; Liu, G.; Wáng, Y.X. Metal-Organic Framework-Based Nanomedicine Platforms for Drug Delivery and Molecular Imaging. *Small* **2015**, *11*, 4806–4822. [CrossRef]
34. Wen, Y.; Chen, L.; Li, J.; Liu, D.; Chen, L. Recent advances in solid-phase sorbents for sample preparation prior to chromatographic analysis. *Trends Anal. Chem.* **2014**, *59*, 26–41. [CrossRef]
35. Pettinari, C.; Marchetti, F.; Mosca, N.; Tosi, G.; Drozdov, A. Application of metal–organic frameworks. *Polym. Int.* **2017**, *66*, 731–744. [CrossRef]
36. Furukawa, H.; Cordova, K.E.; O'Keeffe, M.; Yaghi, O.M. The chemistry and applications of metal-organic framework. *Science* **2013**, *341*, 1230444. [CrossRef]
37. Salarian, M.; Ghanbarpour, A.; Behbahani, M.; Bagheri, S.; Bagheri, A. A metal-organic framework sustained by a nanosized Ag12 cuboctahedral node for solid-phase extraction of ultra traces of lead(II) ions. *Microchim. Acta* **2014**, *181*, 999–1007. [CrossRef]
38. Lan, H.; Pan, D.; Sun, Y.; Guo, Y.; Wu, Z. Thin metal-organic frameworks coatings by cathodic electrodeposition for solid-phase microextraction and analysis of trace exogenous estrogens in milk. *Anal. Chim. Acta* **2016**, *937*, 53–60. [CrossRef]

39. Kalantari, H.; Manoochehri, M. A nanocomposite consisting of MIL-101(Cr) and functionalized magnetite nanoparticles for extraction and determination of selenium(IV) and selenium(VI). *Microchim. Acta* **2018**, *185*, 196. [CrossRef]
40. Moghaddam, Z.; Kaykhaii, M.; Khajeh, M.; Oveisi, A. Synthesis of UiO-66-OH zirconium metal-organic framework and its application for selective extraction and trace determination of thorium in water samples by spectrophotometry. *Spectrochim. Acta A* **2018**, *194*, 76–82. [CrossRef]
41. Fan, L.; Yan, X. Evaluation of isostructural metal–organic frameworks coated capillary columns for the gas chromatographic separation of alkane isomers. *Talanta* **2012**, *99*, 944–950. [CrossRef] [PubMed]
42. Yan, Z.; Zheng, J.; Chen, J.; Tong, P.; Lu, M.; Lin, Z.; Zhang, L. Preparation and evaluation of silica-UIO-66 composite as liquid chromatographic stationary phase for fast and efficient separation. *J. Chromatogr. A* **2014**, *1336*, 45–53. [CrossRef] [PubMed]
43. Geng, Z.; Song, Q.; Yu, B.; Cong, H. Using ZIF-8 as stationary phase for capillary electrophoresis separation of proteins. *Talanta* **2018**, *188*, 493–498. [CrossRef] [PubMed]
44. Fei, Z.; Zhang, M.; Zhang, J.; Yuan, L. Chiral metal–organic framework used as stationary phases for capillary electrochromatography. *Anal. Chim. Acta* **2014**, *830*, 49–55. [CrossRef]
45. Liu, Q.; Ning, D.; Li, W.J.; Du, X.M.; Wang, Q.; Li, Y.; Ruan, W.J. Metal–organic framework-based fluorescent sensing of tetracycline-type antibiotics applicable to environmental and food analysis. *Analyst* **2019**, *144*, 1916–1922. [CrossRef]
46. Liu, S.; Bai, J.; Huo, Y.; Ning, B.; Peng, Y.; Li, S.; Han, D.; Kang, W.; Gao, Z. A zirconium-porphyrin MOF-based ratiometric fluorescent biosensor for rapid and ultrasensitive detection of chloramphenicol. *IJBSBE* **2020**, *149*, 111801. [CrossRef]
47. Yu, L.; Chen, H.; Yue, J.; Chen, X.; Sun, M.; Tan, H.; Asiri, A.M.; Alamry, K.A.; Wang, X.; Wang, S. Metal-Organic Framework Enhances Aggregation-Induced Fluorescence of Chlortetracycline and the Application for Detection. *Anal. Chem.* **2019**, *91*, 5913–5921. [CrossRef]
48. Yu, L.; Chen, H.; Ji, Y.; Chen, X.; Sun, M.; Hou, J.; Alamry, K.A.; Marwani, H.M.; Wang, X.; Wang, S. Europium metal-organic framework for selective and sensitive detection of doxycycline based on fluorescence enhancement. *Talanta* **2020**, *207*, 120297. [CrossRef]
49. Huang, S.; Gan, N.; Zhang, X.; Wu, Y.; Shao, Y.; Jiang, Z.; Wang, Q. Portable fluoride-selective electrode as signal transducer for sensitive and selective detection of trace antibiotics in complex samples. *IJBSBE* **2019**, *128*, 113–121. [CrossRef]
50. Zhou, Y.; Li, X.; Pan, Z.; Ye, B.; Xu, M. Determination of Malachite Green in Fish by a Modified MOF-Based Electrochemical Sensor. *Food Anal. Methods* **2019**, *12*, 1246–1254. [CrossRef]
51. Tchounwou, P.B.; Yedjou, C.G.; Patlolla, A.K.; Sutton, D.J. Heavy metals toxicity and the environment. *EXS* **2012**, *101*, 133–164. [PubMed]
52. Rashed, M. Monitoring of environmental heavy metals in fish from Nasser Lake. *Environ. Int.* **2001**, *27*, 27–33. [CrossRef]
53. Ghanemi, K.; Nikpour, Y.; Omidvar, O.; Maryamabadi, A. Sulfur-nanoparticle-based method for separation and preconcentration of some heavy metals in marine samples prior to flame atomic absorption spectrometry determination. *Talanta* **2011**, *85*, 763–769. [CrossRef] [PubMed]
54. Thongsaw, A.; Sananmuang, R.; Udnan, Y.; Ross, G.; Chaiyasith, W. Speciation of mercury in water and freshwater fish samples using two-step hollow fiber liquid phase microextraction with electrothermal atomic absorption spectrometry. *Spectrochim. Acta B* **2019**, *152*, 102–108. [CrossRef]
55. Guo, T.; Baasner, J. On-line microwave sample pretreatment for the determination of mercury in blood by flow injection cold vapor atomic absorption spectrometry. *Talanta* **1993**, *40*, 1927–1936. [CrossRef]
56. Fallah, A.; Saei-Dehkordi, S.; Nematollahi, A.; Jafari, T. Comparative study of heavy metal and trace element accumulation in edible tissues of farmed and wild rainbow trout (*Oncorhynchus mykiss*) using ICP-OES technique. *Microchem. J.* **2011**, *98*, 275–279. [CrossRef]
57. Zhu, S.; Chen, B.; He, M.; Huang, T.; Hu, B. Speciation of mercury in water and fish samples by HPLC-ICP-MS after magnetic solid phase extraction. *Talanta* **2017**, *171*, 213–219. [CrossRef]
58. Rocío-Bautista, P.; Taima-Mancera, T.; Pasán, J.; Pino, V. Metal-Organic Frameworks in Green Analytical Chemistry. *Separations* **2019**, *6*, 33. [CrossRef]
59. Wang, Q.; Bai, J.; Lu, Z.; Pan, Y.; You, X. Finely tuning MOFs towards high-performance post-combustion CO_2 capture materials. *Chem. Commun.* **2016**, *52*, 443–452. [CrossRef]

60. Giannakoudakis, D.A.; Seredych, M.; Rodríguez-Castellón, E.; Bandosz, T.J. Mesoporous Graphitic Carbon Nitride-Based Nanospheres as Visible-Light Active Chemical Warfare Agents Decontaminant. *ChemNanoMat* **2016**, *2*, 268–272. [CrossRef]
61. Giannakoudakis, D.A.; Travlou, N.A.; Secor, J.; Bandosz, T.J. Oxidized g-C 3 N 4 Nanospheres as Catalytically Photoactive Linkers in MOF/g-C$_3$N$_4$ Composite of Hierarchical Pore Structure. *Small* **2017**, *13*, 1601758. [CrossRef] [PubMed]
62. Li, N.; Xu, J.; Feng, R.; Hu, T.L.; Bu, X.H. Governing metal-organic frameworks towards high stability. *Chem. Commun.* **2016**, *52*, 8501–8513. [CrossRef] [PubMed]
63. Giannakoudakis, D.A.; Barczak, M.; Florent, M.; Bandosz, T.J. Analysis of interactions of mustard gas surrogate vapors with porous carbon textiles. *Chem. Eng. J.* **2019**, *362*, 758–766. [CrossRef]
64. Giannakoudakis, D.A.; Hu, Y.; Florent, M.; Bandosz, T.J. Smart textiles of MOF/g-C$_3$N$_4$ nanospheres for the rapid detection/detoxification of chemical warfare agents. *Nanoscale Horiz.* **2017**, *2*, 356–364. [CrossRef]
65. Florent, M.; Giannakoudakis, D.A.; Wallace, R.; Bandosz, T.J. Carbon Textiles Modified with Copper-Based Reactive Adsorbents as Efficient Media for Detoxification of Chemical Warfare Agents. *ACS Appl. Mater. Interfaces* **2017**, *9*, 26965–26973. [CrossRef] [PubMed]
66. Giannakoudakis, D.A.; Barczak, M.; Pearsall, F.; O'Brien, S.; Bandosz, T.J. Composite porous carbon textile with deposited barium titanate nanospheres as wearable protection medium against toxic vapors. *Chem. Eng. J.* **2020**, *384*, 123280. [CrossRef]
67. Gu, Z.; Yang, C.; Chang, N.; Yan, X. Metal–Organic Frameworks for Analytical Chemistry: From Sample Collection to Chromatographic Separation. *Acc. Chem. Res.* **2012**, *45*, 734–745. [CrossRef]
68. Rocío-Bautista, P.; Termopoli, V. Metal–Organic Frameworks in Solid-Phase Extraction Procedures for Environmental and Food Analyses. *Chromatographia* **2019**, *82*, 1191–1205. [CrossRef]
69. Maya, F.; Palomino Cabello, C.; Frizzarin, R.; Estela, J.; Turnes Palomino, G.; Cerdà, V. Magnetic solid-phase extraction using metal-organic frameworks (MOFs) and their derived carbons. *Trac Trends Anal. Chem.* **2017**, *90*, 142–152. [CrossRef]
70. Rocío-Bautista, P.; Pacheco-Fernández, I.; Pasán, J.; Pino, V. Are metal-organic frameworks able to provide a new generation of solid-phase microextraction coatings? – A review. *Anal. Chim. Acta* **2016**, *939*, 26–41. [CrossRef]
71. Mondal, S.; Xu, J.; Chen, G.; Huang, S.; Huang, C.; Yin, L.; Ouyang, G. Solid-phase microextraction of antibiotics from fish muscle by using MIL-101(Cr)NH$_2$-polyacrylonitrile fiber and their identification by liquid chromatography-tandem mass spectrometry. *Anal. Chim. Acta* **2019**, *1047*, 62–70. [CrossRef] [PubMed]
72. Xia, L.; Liu, L.; Lv, X.; Qu, F.; Li, G.; You, J. Towards the determination of sulfonamides in meat samples: A magnetic and mesoporous metal-organic framework as an efficient sorbent for magnetic solid phase extraction combined with high-performance liquid chromatography. *J. Chromatogr. A* **2017**, *1500*, 24–31. [CrossRef] [PubMed]
73. Wang, Y.; Tong, Y.; Xu, X.; Zhang, L. Metal-organic framework-derived three-dimensional porous graphitic octahedron carbon cages-encapsulated copper nanoparticles hybrids as highly efficient enrichment material for simultaneous determination of four fluoroquinolones. *J. Chromatogr. A* **2018**, *1533*, 1–9. [CrossRef] [PubMed]
74. Mohammadnejad, M.; Hajiashrafi, T.; Rashnavadi, R. Highly efficient determination of malachite green in aquatic product using Tb-organic framework as sorbent. *J. Porous Mat.* **2018**, *25*, 1771–1781. [CrossRef]
75. Zhou, Z.; Fu, Y.; Qin, Q.; Lu, X.; Shia, X.; Zhao, C.; Xu, G. Synthesis of magnetic mesoporous metal-organic framework-5 for the effective enrichment of malachite green and crystal violet in fish samples. *J. Chromatogr. A* **2018**, *1560*, 19–25. [CrossRef]
76. Wei, X.; Wang, Y.; Chen, J.; Xu, P.; Xu, W.; Ni, R.; Meng, J.; Zhou, Y. Poly (deep eutectic solvent)-functionalized magnetic metal-organic framework composites coupled with solid-phase extraction for the selective separation of cationic dyes. *Anal. Chim. Acta* **2019**, *1056*, 47–61. [CrossRef]
77. Culp, S.J.; Beland, F.A. Malachite green: A toxicological review. *J. Am. Coll. Toxicol.* **1996**, *15*, 219–238. [CrossRef]
78. Wang, Y.; Liao, K.; Huang, X.; Yuan, D. Simultaneous determination of malachite green, crystal violet and their leuco-metabolites in aquaculture water samples using monolithic fiber-based solid-phase microextraction coupled with high performance liquid chromatography. *Anal. Methods* **2015**, *7*, 8138–8145. [CrossRef]

79. Fallah, A.A.; Barani, A. Determination of malachite green residues in farmed rainbow trout in Iran. *Food Control.* **2014**, *40*, 100–105. [CrossRef]
80. Bagheri, A.; Taghizadeh, M.; Behbahani, M.; Akbar Asgharinezhad, A.; Salarian, M.; Dehghani, A.; Ebrahimzadeh, H.; Amini, M. Synthesis and characterization of magnetic metal-organic framework (MOF) as a novel sorbent, and its optimization by experimental design methodology for determination of palladium in environmental samples. *Talanta* **2012**, *99*, 132–139. [CrossRef]
81. Sohrabi, M. Preconcentration of mercury(II) using a thiol-functionalized metal-organic framework nanocomposite as a sorbent. *Microchim. Acta* **2013**, *181*, 435–444. [CrossRef]
82. Tadjarodi, A.; Abbaszadeh, A. A magnetic nanocomposite prepared from chelator-modified magnetite (Fe_3O_4) and HKUST-1 (MOF-199) for separation and preconcentration of mercury(II). *Microchim. Acta* **2016**, *183*, 1391–1399. [CrossRef]
83. Rezaei Kahkha, M.; Daliran, S.; Oveisi, A.; Kaykhaii, M.; Sepehri, Z. The Mesoporous Porphyrinic Zirconium Metal-Organic Framework for Pipette-Tip Solid-Phase Extraction of Mercury from Fish Samples Followed by Cold Vapor Atomic Absorption Spectrometric Determination. *Food Anal. Methods* **2017**, *10*, 2175–2184. [CrossRef]
84. Sohrabi, M.; Matbouie, Z.; Asgharinezhad, A.; Dehghani, A. Solid-phase extraction of Cd(II) and Pb(II) using a magnetic metal-organic framework, and their determination by FAAS. *Microchim. Acta* **2013**, *180*, 589–597. [CrossRef]
85. Ghorbani-Kalhor, E. A metal-organic framework nanocomposite made from functionalized magnetite nanoparticles and HKUST-1 (MOF-199) for preconcentration of Cd(II), Pb(II), and Ni(II). *Microchim. Acta* **2016**, *183*, 2639–2647. [CrossRef]
86. Taghizadeh, M.; Asgharinezhad, A.; Pooladi, M.; Barzin, M.; Abbaszadeh, A.; Tadjarodi, A. A novel magnetic metal-organic framework nanocomposite for extraction and preconcentration of heavy metal ions, and its optimization via experimental design methodology. *Microchim. Acta* **2013**, *180*, 1073–1084. [CrossRef]
87. Babazadeh, M.; Hosseinzadeh Khanmiri, R.; Abolhasani, J.; Ghorbani-Kalhor, E.; Hassanpour, A. Synthesis and Application of a Novel Functionalized Magnetic Metal–Organic Framework Sorbent for Determination of Heavy Metal Ions in Fish Samples. *Bull. Chem. Soc. Jpn.* **2015**, *88*, 871–879. [CrossRef]
88. Giakisikli, G.; Anthemidis, A.N. Magnetic materials as sorbents for metal/metalloid preconcentration and/or separation. A review. *Anal. Chim. Acta* **2013**, *789*, 1–16. [CrossRef]
89. Balcerzak, M. Sample Digestion Methods for the Determination of Traces of Precious Metals by Spectrometric Techniques. *Anal. Sci.* **2002**, *18*, 737–750. [CrossRef]
90. Liang, P.; Zhao, E.; Li, F. Dispersive liquid–liquid microextraction preconcentration of palladium in water samples and determination by graphite furnace atomic absorption spectrometry. *Talanta* **2009**, *77*, 1854–1857. [CrossRef]
91. Jamali, M.; Assadi, Y.; Kozani, R.R. Determination of Trace Amounts of Palladium in Water Samples by Graphite Furnace Atomic Absorption Spectrometry after Dispersive Liquid-Liquid Microextraction. *J. Chem.* **2013**, *2013*, 1–6. [CrossRef]
92. Zachariadis, G.; Kapsimali, D. Effect of sample matrix on sensitivity of mercury and methylmercury quantitation in human urine, saliva, and serum using GC-MS. *J. Sep. Sci.* **2008**, *31*, 3884–3893. [CrossRef] [PubMed]
93. Anthemidis, A.; Zachariadis, G.A.; Stratis, J.A. Development of a sequential injection system for trace mercury determination by cold vapour atomic absorption spectrometry utilizing an integrated gas-liquid separator/reactor. *Talanta* **2004**, *64*, 1053–1057. [CrossRef] [PubMed]
94. Wang, D.; Yu, Y.; Zhang, X.; Zhang, S.; Pang, Y.; Zhang, X.; Yu, Z.; Wu, M.; Fu, J. Polycyclic aromatic hydrocarbons and organochlorine pesticides in fish from Taihu Lake: Their levels, sources, and biomagnification. *Ecotox. Environ. Saf.* **2012**, *82*, 63–70. [CrossRef]
95. Takeuchi, I.; Miyoshi, N.; Mizukawa, K.; Takada, H.; Ikemoto, T.; Omori, K.; Tsuchiya, K. Biomagnification profiles of polycyclic aromatic hydrocarbon d 13C alkylphenols and polychlorinated biphenyls in Tokyo Bay elucidated by d 15N and isotope ratios as guides to trophic webstructure. *Mar. Pollut. Bull.* **2009**, *58*, 663–671. [CrossRef]
96. Tuvikene, A. Responses of fish to polycyclic aromatic hydrocarbons (PAHs). *Ann. Zool. Fennici.* **1995**, *32*, 295–309.

97. Hu, Y.; Huang, Z.; Liao, J.; Li, G. Chemical Bonding Approach for Fabrication of Hybrid Magnetic Metal–Organic Framework-5: High Efficient Adsorbents for Magnetic Enrichment of Trace Analytes. *Anal. Chem.* **2013**, *85*, 6885–6893. [CrossRef]
98. Shang, X.; Dong, G.; Zhang, H.; Zhang, L.; Yu, X.; Li, J.; Wang, X.; Yue, B.; Zhao, Y.; Wu, Y. Polybrominated diphenyl ethers (PBDEs) and indicator polychlorinated biphenyls (PCBs) in various marine fish from Zhoushan fishery, China. *Food Control.* **2016**, *67*, 240–246. [CrossRef]
99. Ranjbar Jafarabadi, A.; Riyahi Bakhtiari, A.; Mitra, S.; Maisano, M.; Cappello, T.; Jadot, C. First polychlorinated biphenyls (PCBs) monitoring in seawater, surface sediments and marine fish communities of the Persian Gulf: Distribution, levels, congener profile and health risk assessment. *Environ. Pollut.* **2019**, *253*, 78–88. [CrossRef]
100. Lin, S.; Gan, N.; Qiao, L.; Zhang, J.; Cao, Y.; Chen, Y. Magnetic metal-organic frameworks coated stir bar sorptive extraction coupled with GC-MS for determination of polychlorinated biphenyls in fish samples. *Talanta* **2015**, *144*, 1139–1145. [CrossRef]
101. Lin, S.; Gan, N.; Zhang, J.; Qiao, L.; Chen, Y.; Cao, Y. Aptamer-functionalized stir bar sorptive extraction coupled with gas chromatography–mass spectrometry for selective enrichment and determination of polychlorinated biphenyls in fish samples. *Talanta* **2016**, *149*, 266–274. [CrossRef]
102. Nazyropoulou, C.; Samanidou, V. Stir bar sorptive extraction applied to the analysis of biological fluids. *Bioanalysis* **2015**, *7*, 2241–2250. [CrossRef] [PubMed]
103. Lan, H.; Rönkkö, T.; Parshintsev, J.; Hartonen, K.; Gan, N.; Sakeye, M.; Sarfraz, J.; Riekolla, M. Modified zeolitic imidazolate framework-8 as solid-phase microextraction Arrow coating for sampling of amines in wastewater and food samples followed by gas chromatography-mass spectrometry. *J. Chromatogr. A* **2017**, *1486*, 76–85. [CrossRef] [PubMed]
104. Kremser, A.; Jochmann, M.; Schmidt, T. PAL SPME Arrow—Evaluation of a novel solid-phase microextraction device for freely dissolved PAHs in water. *Anal. Bioanal. Chem.* **2015**, *408*, 943–952. [CrossRef]
105. Eckert, K.; Carter, D.; Perrault, K. Sampling Dynamics for Volatile Organic Compounds Using Headspace Solid-Phase Microextraction Arrow for Microbiological Samples. *Separations* **2018**, *5*, 45. [CrossRef]
106. Zhang, W.; Yan, Z.; Gao, J.; Tong, P.; Liu, W.; Zhang, L. Metal-organic framework UiO-66 modified magnetite@silica core-shell magnetic microspheres for magnetic solid-phase extraction of domoic acid from shellfish sample. *J. Chromatogr. A* **2015**, *1400*, 10–18. [CrossRef] [PubMed]
107. González-Rodríguez, G.; Taima-Mancera, I.; Lago, A.B.; Ayala, J.H.; Pasán, J.; Pino, V. Mixed functionalization of organic ligands in UiO-66: A tool to design metal-organic frameworks for tailored microextraction. *Molecules* **2019**, *24*, 3656. [CrossRef] [PubMed]
108. Bůžek, D.; Demel, J.; Lang, K. Zirconium Metal–Organic Framework UiO-66: Stability in an Aqueous Environment and Its Relevance for Organophosphate Degradation. *Inorg. Chem.* **2018**. [CrossRef]
109. Kumar, V.; Kumar, S.; Kim, K.H.; Tsang, D.C.W.; Lee, S.S. Metal organic frameworks as potent treatment media for odorants and volatiles in air. *Environ. Res.* **2019**, *168*, 336–356. [CrossRef]
110. Kumar, P.; Vellingiri, K.; Kim, K.H.; Brown, R.J.C.; Manos, M.J. Modern progress in metal-organic frameworks and their composites for diverse applications. *Microporous Mesoporous Mater.* **2017**, *253*, 251–265. [CrossRef]
111. Vikrant, K.; Tsang, D.C.W.; Raza, N.; Giri, B.S.; Kukkar, D.; Kim, K.H. Potential Utility of Metal-Organic Framework-Based Platform for Sensing Pesticides. *ACS Appl. Mater. Interfaces* **2018**, *10*, 8797–8817. [CrossRef]
112. Vu, T.A.; Le, G.H.; Dao, C.D.; Dang, L.Q.; Nguyen, K.T.; Nguyen, Q.K.; Dang, P.T.; Tran, H.T.K.; Duong, Q.T.; Nguyen, T.V.; et al. Arsenic removal from aqueous solutions by adsorption using novel MIL-53(Fe) as a highly efficient adsorbent. *RSC Adv.* **2015**, *5*, 5261–5268. [CrossRef]
113. Audu, C.O.; Nguyen, H.G.T.; Chang, C.; Katz, M.J.; Mao, L.; Farha, O.K.; Hupp, J.T.; Nguyen, S.T. The dual capture of AsV and AsIII by UiO-66 and analogues. *Chem. Sci.* **2016**, *7*, 6492–6498. [CrossRef]
114. Fang, Q.-R.; Yuan, D.-Q.; Sculley, J.; Li, J.-R.; Han, Z.-B.; Zhou, H.-C. Functional mesoporous metal-organic frameworks for the capture of heavy metal ions and size-selective catalysis. *Inorg. Chem.* **2010**, *49*, 11637–11642. [CrossRef]
115. Jian, M.P.; Liu, B.; Zhang, G.S.; Liu, R.P.; Zhang, X.W. Adsorptive removal of arsenic from aqueous solution by zeolitic imidazolate framework-8 (ZIF-8) nanoparticles. *Colloids Surf. A Physicochem. Eng. Asp.* **2014**, *465*, 67–76. [CrossRef]
116. Manousi, N.; Giannakoudakis, D.A.; Rosenberg, E.; Zachariadis, G.A. Extraction of Metal Ions with Metal–Organic Frameworks. *Molecules* **2019**, *24*, 4605. [CrossRef]

117. Giliopoulos, D.; Zamboulis, A.; Giannakoudakis, D.A.; Bikiaris, D.; Triantafyllidis, K. Polymer/Metal Organic Framework (MOF) Nanocomposites for Biomedical Applications. *Molecules* **2020**, *25*, 185. [CrossRef]

© 2020 by the authors. Licensee MDPI, Basel, Switzerland. This article is an open access article distributed under the terms and conditions of the Creative Commons Attribution (CC BY) license (http://creativecommons.org/licenses/by/4.0/).

Review

Polymer/Metal Organic Framework (MOF) Nanocomposites for Biomedical Applications

Dimitrios Giliopoulos [1,*], Alexandra Zamboulis [2], Dimitrios Giannakoudakis [1], Dimitrios Bikiaris [2,*] and Konstantinos Triantafyllidis [1,*]

1. Laboratory of Chemical and Environmental Technology, Department of Chemistry, Aristotle University of Thessaloniki, GR-54124 Thessaloniki, Greece; dagchem@gmail.com
2. Laboratory of Polymer Chemistry and Technology, Department of Chemistry, Aristotle University of Thessaloniki, GR-54124 Thessaloniki, Greece; azampouli@chem.auth.gr
* Correspondence: dgiliopo@chem.auth.gr (D.G.); dbic@chem.auth.gr (D.B.); ktrianta@chem.auth.gr (K.T.); Tel.: +30-23-1099-7730 (D.G. & K.T.); +30-23-1099-7812 (D.B.)

Academic Editor: Roman Dembinski
Received: 18 November 2019; Accepted: 28 December 2019; Published: 1 January 2020

Abstract: The utilization of polymer/metal organic framework (MOF) nanocomposites in various biomedical applications has been widely studied due to their unique properties that arise from MOFs or hybrid composite systems. This review focuses on the types of polymer/MOF nanocomposites used in drug delivery and imaging applications. Initially, a comprehensive introduction to the synthesis and structure of MOFs and bio-MOFs is presented. Subsequently, the properties and the performance of polymer/MOF nanocomposites used in these applications are examined, in relation to the approach applied for their synthesis: (i) non-covalent attachment, (ii) covalent attachment, (iii) polymer coordination to metal ions, (iv) MOF encapsulation in polymers, and (v) other strategies. A critical comparison and discussion of the effectiveness of polymer/MOF nanocomposites regarding their synthesis methods and their structural characteristics is presented.

Keywords: metal organic framework; polymer nanocomposites; drug delivery; magnetic resonance imaging

---

1. Introduction

The homogeneous dispersion of inorganic, organic, or hybrid nanoscale components inside a polymeric matrix results in materials with physically and/or chemically distinct phases that are called polymer nanocomposites. Polymer nanocomposites have unique or improved properties when compared to pristine polymers or conventional composites and these properties can easily be tuned by controlling the type or the concentration of the additives, selecting specific production methods, and functionalizing the surface of the additives, etc. [1–8]. Due to the superior properties and the diversity of products, polymer nanocomposites are used in a variety of applications in most industrial and research fields. Among them, biomedicine has greatly benefited from the progress in nanocomposite materials regarding the advances that have been made in the areas of diagnosis, monitoring, and therapy. Some of the biomedical applications of polymer nanocomposites may include drug or gene delivery, skin regeneration, soft-tissue engineering, bone or joint replacement, bioimaging, biosensors, dental or antimicrobial applications, and many other [9–11].

Many types of nanostructured materials have been used in combination with biocompatible polymers to produce nanocomposites for biomedical applications such as clays, carbon nanotubes, graphene, metal oxides, porous nanomaterials, magnetic nanoparticles, and others. As part of more complex systems for biomedical applications, nanostructured materials may exhibit various functions. For example, they can reinforce the polymer matrix or offer some new property, they can interact with a substrate or a substance when it would be impossible for the polymer, and they can control the transport phenomena through the polymer matrix, etc. [10,12–16].

MOFs are a class of crystalline materials possessing structures formed from the coordination of metal ions to multidentate organic groups. The main characteristics of MOFs are the high degree of porosity and the tunable architecture of the structure by selecting appropriate metal ions and linkers. Furthermore, MOFs can have their surface further modified, thereby increasing their functionality. These characteristics make MOFs ideal candidates for biomedical applications like drug delivery and magnetic resonance imaging (MRI) [17–19]. As it concerns drug delivery, the high surface areas and large pore sizes of MOFs are favorable for the encapsulation of high drug loadings [20], while the high structural and functional flexibility of MOFs allow their adaption to the shape, size, and functionality of the drug molecules [20,21]. On the other hand, regarding imaging applications, MOFs can be modified with chemical groups and uniquely affect the delivery of contrast imaging agents [22]. Moreover, MOFs have the advantage of acting simultaneously as MRI contrast agents and drug carriers, serving both purposes of diagnosis and therapy [23]. As can be understood, the use of MOFs in biomedical applications offers serious advantages to scientists in the fields of diagnosis, monitoring, and therapy. As a result, numerous studies over the last years have focused on the combined use of MOFs and biocompatible polymers, aiming at the development of more sophisticated systems that would be more effective than previous products while ensuring a higher quality of life for patients.

In this review, we examine the various types of polymer/MOF nanocomposites used in biomedical applications, and more specifically in drug delivery and imaging. Although there have been many reviews covering various aspects of the use of MOFs in biomedical applications, no work at the present has reviewed the composite materials of polymer matrix and drug loaded MOF additives in biomedical applications. More specifically, we focused on the different approaches followed to produce the composites and discuss the findings regarding the behavior of the composites in each application.

2. Metal Organic Frameworks

Metal organic frameworks, also known as porous coordination networks (PCNs) or porous coordination polymers (PCPs), are in general highly porous 1-, 2-, or 3-dimensional extended organic-inorganic coordination structures [24,25]. Their network is composed of metal centers (ions, clusters of ions, or better multinuclear complexes) linked by di- or polydentate organic bridges called linkers (Figure 1a,b). Even though coordination chemistry between metal ions and organic linkers to form coordination polymers (like Werner complexes or Prussian blue compounds) has a prolonged history [26], Hoskins and Robson were the first to suggest in 1989 of the potential synthesis of solid porous polymeric materials based on coordination bonds [27]. The introduction in the literature of the terminology 'metal organic framework' occurred in 1995 from Yaghi and Li, who reported the hydrothermal synthesis of a "zeolite-like" crystalline structure by the polymeric coordination of copper with 4,4'-bipyridine and nitrate ions [28]. It took some years in order for this class of new supramolecular materials to become a mainstream topic of research, with the most influential reports published in 1999 for two 3-D frameworks that still act as benchmark representatives [29]: HKUST-1 by Chui et al. [30] and MOF-5 by Li et al. [31] (Figure 1c,d). The former one, known also as MOF-199, took its name from the place of synthesis (Hong Kong University of Science and Technology) and is built up from a paddlewheel shaped $Cu_2(CO_2)_4$ metal cluster/subunit (called the secondary building unit, SBU) consisting of a dimer of Cu^{2+} ions, where each copper ion has been coordinated with four benzene-1,3,5-tricarboxylic acid (BTC) as a tritopic linker. 3-D illustrations of the structure of the polymeric framework, as reported in the original article, can be seen in Figure 1c,d. MOF-5 (known also as IRMOF-1) has an octahedral multinuclear complex/SBU, $Zn_4O(CO_2)_6$, in which an O^{2-} ion is tetrahedrally linked with four Zn^{2+} ions, and each zinc ion is coordinated with three oxygens from three different 1,4-benzenedicarboxylate (terephthalate, BDC) linkers, resulting in a cubic framework (Figure 1e).

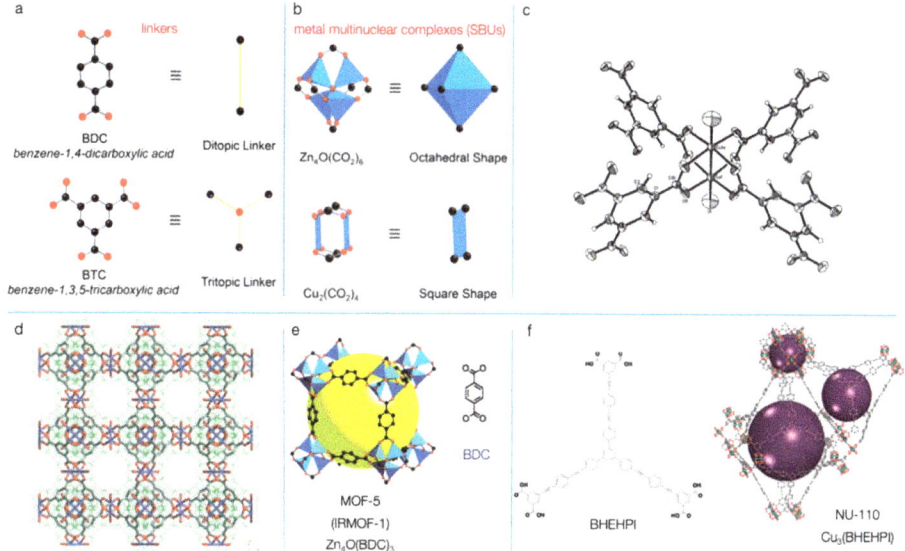

Figure 1. (a) The linkers [32]; (b) the metal clusters/multinuclear complexes (Secondary Building Units, SBUs) of HKUST-1 and MOF-5 (color assignment: black for C, red for O, blue for Cu squares and Zn polyhedrals; H atoms are omitted) [32]; (c) the dicopper(II) tetracarboxylate building block of HKUST-1 [30]; (d) the polymeric framework of HKUST-1 (viewed down the direction) [30]; (e) the single crystal structure of MOF-5 (the yellow spheres represent the maximum volume of the biggest cavity) [33]; (f) the chemical structure of the ligand and the different cages of the NU-110 framework [34].

As S. Kaskel mentions in his book [35], all of the known MOFs up until 2002 could be summarized within a book chapter. Nowadays, there are more than ten thousand 3-D registered MOFs in the Cambridge Structural Database and more than twenty thousand hypothetical or real known MOFs [36]. The research effort continues to be extensive, with many new MOFs commercially available. Due to the high surface area, porosity, and tailorable size of the pores/cages as a result of the diversity of combination of metal and linkers, MOFs have garnered an enormous boost in attention in the last decades for a wide range of potential applications such as adsorption, gas storage, purification, separation, chemical sensing, and even for selective catalytic processes against toxic compounds [24,25,37–41]. The pores/cavities are created as free spaces, cages, or voids inside the structure. The reported surface area values in the initial article for HKUST-1, calculated based on N_2 adsorption/desorption tests, were 692.2 $m^2\ g^{-1}$ using the Brunauer–Emmett–Teller (BET) equation and 917.6 $m^2\ g^{-1}$ based on the Langmuir approach, while the single-point total pore volume was 0.333 $cm^3\ g^{-1}$ [30]. In the case of MOF-5, the authors reported (based on liquid nitrogen vapor sorption test) an estimated Langmuir surface area of 2900 $m^2\ g^{-1}$ (2320 $m^2\ g^{-1}$ based on the BET method) and a pore volume (based Dubinin–Raduskhvich equation) up to 1.04 $cm^3\ g^{-1}$ [31].

Great effort has been given to achieve higher porosity, predominately by increasing the size of the linkers, leading to significantly higher structural feature values when compared to commonly used activated carbons and zeolites [42–46]. A characteristic example is Cu_3(BHEHPI) or NU-110 (NU stands for Northwestern University in Chicago, USA), which has the highest reported surface area and total pore volume up to now [34,44]. This copper based MOF (Figure 1f) was reported by O. Farha, J. Hupp, and co-workers in 2012 [31], where a hexacarboxylate macromolecule was used as a ligand (BHEHPI– stands for 5,5′,5″-((((benzene-1,3,5-triyltris(benzene-4,1-diyl)) tris(ethyne-2,1-diyl))-tris(benzene-4,1-diyl)) tris(ethyne-2,1-diyl)) triisophthalate). The reported BET surface area by N_2 sorption experiments was 7140 $m^2\ g^{-1}$ and the total pore volume was 4.4 $cm^3\ g^{-1}$, values that are the highest experimentally obtained up today. Interestingly, the obtained nitrogen isotherm was closer to type-IV rather than to type-I and revealed multiple sizes of pores, a fact that is consistent with the different types of illustrated cages in Figure 1f. The authors also showed that in general, the theoretical surface of the MOFs could reach up to 14,600 m^2 g^{-1} [34].

An important factor that should be taken into consideration in the design and synthesis of MOFs for application in aquatic environments is that their stability depends on the strength of coordination between the metal and linker [47]. The reason behind this instability is the ability of water to interact with the metal ions/clusters competitively to the linkers, leading to the collapse of the framework. There are also various other factors that play a crucial role in the stability of the MOFs, with the most important being crystallinity, hydrophobicity, and the extent of the defectous sites [43]. Additionally, the temperature and pH should also be considered. In general, the hard/soft acid/base (HSAB) principles can predict the level of metal/linker coordination strength [48,49]. Hard acidic metal ions (like Zr^{4+}, Cr^{3+}, Al^{3+}, and Fe^{3+}) combined with carboxylate-based linkers acting as hard bases result in frameworks with a significant water resistivity/stability. Stability against water is also due to the coordination between weak acidic metal ions (like Cu^{2+}, Mn^2, Zn^{2+}, Ag^+, and Ni^{2+}) and linkers with a weak basic character (like pyrazolates, triazolates, and imidazolates). Combining strong acidic metal ions with weak basic linkers, and vice versa, results in a vulnerability to water frameworks.

2.1. Biological Metal Organic Frameworks (BioMOFs)

In the last decade, a novel and attractive sub-class of MOFs, the biological metal organic framework (BioMOFs), has had an augmented degree of interest, giving rise to new opportunities for their utilization in a plethora of biological and medical applications. Although there is no specific definition for these new generation biocompatible materials, in order for an inorganic–organic framework to be classified as a BioMOF, it should either consist of at least a biomolecule or have a direct application across medicine and biology. With the exception of biocompatibility, the other two are features of utmost importance with regard to BioMOFs design are to possess the appropriate size of pores/cages and to be able to selectively and strongly retain the targeted therapeutic/drug. The latter aspect is known as host–guest chemistry, which is critical for supramolecular recognition features. The most important fields of BioMOF utilization can be summarized as adsorption/encapsulation, the protection and delivery of molecular therapeutics (drug delivery), enantioseparation, magnetic resonance imaging (MRI), photothermal therapy, biomimetic catalysis, biobanking, biosensing, and cell and various manipulations, etc. [50,51].

Prior to the appearance of the BioMOF, drug delivery methods were based on two routes. In the first and "organic route", a biocompatible host (such as polymers or dendritic macromolecules) was used as the host. Even though it is possible to encapsulate a wide range of therapeutics via the organic route, the controlled release is challenging due to there being no well-defined porosity or a homogeneous distribution of the drug inside the host matrix [52–54]. For the second and "inorganic route", a mesoporous inorganic substance (like silicate or zeolite) acts as the host through grafting of the pore's walls, leading to a lowering of the porosity and the therapeutic-loading capacity [55,56]. In 2006, the innovative work of Horcajada et al. [53] introduced a "hybrid route", in which a MOF structure was utilized as the host. They synthesized two cubic zeotypic MOFs, abbreviated as MIL-100 and MIL-101 (MIL, Materials Institute Lavoisier). MIL-100 and MIL-101 were built from trimers of chromium octahedras and di-(1,3,5-benzene tricarboxylic acid, BTC) or tri-carboxylic acid (1,4-benzenedicarboxylic acid, BDC), respectively (Figure 2). MIL-100 showed pore/cage sizes between 25–29 Å and a specific surface area of 3340 m^2 g^{-1}, while the respective values for MIL-101 were reported as 29–34 Å and 5510 m^2 g^{-1}. The material showed a remarkably great capacity toward ibuprofen, reaching a loading of 1.4 g per one gram in the case of MIL-101 [53]. Even though this study was criticized due to the known toxicity of Cr, it opened the road for many other MOFs to be designed and tested as hosts for controllable drug delivery. Interestingly, in 2010, the same team showed that analogue structured MOFs could be obtained based on Fe in aqueous or ethanolic solutions, even by avoiding the use of other organic solvents and chromium [21]. The low toxicity of these BioMOFs was demonstrated by in vivo rat and in vitro cell studies. The nanoscaled Fe-MIL-100 showed a 31.9% loading per weight for the antitumoral drug, busulfan, a value five-fold higher than that of the existing busulfan delivery platforms and with a similar cytotoxic activity as the free drug. Additionally, loading with the anti-HIV agent (AZT-TP) was revealed as promising for the "in vitro inhibition of virus replication".

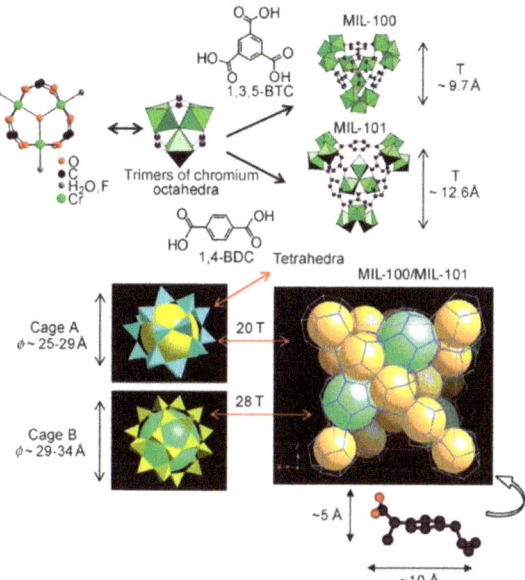

Figure 2. Schematic 3-D representation of the tetrahedra (T) consisting of trimers of chromium octahedra and 1,3,5-benzene tricarboxylic acid (BTC) or 1,4-benzene dicarboxylic acid (BDC) in MIL-101 and MIL-100, respectively (**top**) and a schematic 3-D illustration of the zeotype-architecture MIL-100 and MIL-101 (**bottom**) [53].

2.2. Metal Organic Frameworks (MOFs) for Biomedical Applications

Initially, many of the already known MOFs were examined for potential bio-applications. However, the modern strategy toward the exploration of novel BioMOFs is the usage of biological molecules as ligands. Even though some biomolecules have been successfully utilized as organic linkers, their complicated chemistry (like molecular symmetry, geometry, flexibility etc.) has hindered the possibilities of obtaining crystalline frameworks with the desired properties. Among the most intensively studied biomolecules are nucleobases, amino acids, peptides and proteins, porphyrins, metalloporphyrins, and cyclodextrin [21,50,57,58]. More details can be found in the very recent comprehensive review by Cai et al. [50].

Extensive efforts have been given to alternative approaches for the utilization of biomolecules in the MOF matrix. The main concept is to use the biomolecules in addition to conventional linkers, or use combinations of biomolecules and common linkers. An example is the utilization of a symmetric auxiliary molecule in order to compensate the limited symmetry of the biomolecule. ZnBTCA (where BTC stands for benzene-1,3,5-tricarboxyl and A for adenine) is a characteristic paradigm of the utilization of nucleobase moieties, as reported by Cai and co-workers in 2015 [59]. Adeninate moieties were periodically introduced into the framework, providing sufficient and available Watsin–Click faces (Figure 3a). The kinetic and thermodynamic studies revealed unusual hysteresis of the interaction of the Watsin–Click faces with the amino groups of the guest. It was also reported that the combination of adenine and thymine conferred a pronounced adaptive recognition/response.

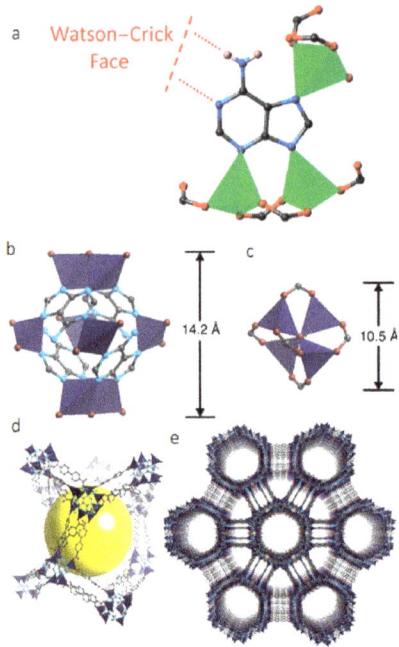

Figure 3. (**a**) Open Watson–Crick sites and the coordination environment of adenine in ZnBTCA [59]. (**b**,**c**) A comparative illustration of the structure and size of the building units in bio-MOF-100 and the basic zinc-carboxylate building [57]. (**d**,**e**) The 3-D crystal structure of bio-MOF-100 where the cavities (yellow sphere) and the large channels can be seen (Zn^{2+}: green or dark blue tetrahedra, C: grey spheres, O: red spheres, N: blue spheres, H: omitted for clarity) [57].

Another alternative approach is based on the use of asymmetric biomolecules for the formation of a metal–biomolecule cluster as a secondary building unit. In 2012, An et al. reported that zinc-adeninate SBU can be interconnected with a relatively short dicarboxylate linker (biphenyldicarboxylate, BPDC), forming an exclusively mesoporous bioMOF, bio-MOF-100 [57]. This material showed a pioneering high surface area (4300 m^2 g^{-1}) and total pore volume (4.3 cm^3 g^{-1}) as well as very low crystal density (~0.3 g cm^{-3}). The structure and the zinc-anadinate SBU as well as an illustration of the three-dimensional structure with large cavities can be seen in Figure 3b–e. Other strategies involve the use of low symmetry small biomolecules in order to form cyclic oligomers or post-synthetic covalently attaching biomolecules on the existing MOFs, or encapsulating biomolecules inside the pores by permeation or diffusion [50].

3. Polymer/MOF Nanocomposites

Polymer/MOF nanocomposites have attracted wide attention because they combine both the advantages of highly porous MOFs and flexible polymer materials. The combination of MOF with polymers has been reported in a variety of contexts. For mixed-matrix membranes, polymers are often co-blended with MOFs. In composite materials, MOF particles are cross-linked through polymer chains, where some repeating units in the polymer chain act as ligands of the MOF structure. In biomedical applications, MOF nanoparticles are coated with a polymer layer to form core-shell-like architectures. The ideal coating should: (i) be selectively attached on the external surface, avoiding intrusion inside the porous structure; (ii) display suitable stability under physiological conditions; (iii) not interfere with the entrapped drugs; (iv) be obtained in a single step (or few steps), under mild conditions, and

(v) enhance the MOF performances for bio-applications by improving their colloidal stability, retarding their degradation, prolonging blood circulation (stealth), and allowing targeting, etc. [60,61].

The polymer coating is generally set up by post-synthetic modification. The strategies developed to coat MOF nanoparticles can be divided in non-covalent and covalent approaches. Non-covalent approaches lie principally on electrostatic interactions or hydrogen bonds. Covalent approaches can be divided in "grafting to" and "grafting from" methods. "Grafting to" involves the reaction of end-functionalized polymers with functional groups located on the MOF, the coordinatively unsaturated metal sites or groups on the ligands, while "grafting from" involves polymerization from active sites on the MOF.

3.1. Non-Covalent Attachment

Liu et al. investigated the non-covalent surface modification of iron(III) carboxylate nano-MOFs with copolymers bearing a fluorescence probe [62]. MIL-101-NH_2(Fe) bears on its surface positive charges, hydrophobic channels, and open metal sites. It was rationalized that by bearing ionizable carboxylic acid groups, fluorescein (F) would bind to MIL-101-NH_2(Fe) due to a synergy of electrostatic and hydrophobic interactions. Copolymers comprising of poly(oligoethylene glycol monomethyl ether methacrylate) (pOEGMA) and different amounts of poly(2-aminoethyl methacrylate) (pAEMA) conjugated to fluorescein were prepared (Figure 4A) and a very strong binding affinity to MIL-101-NH_2(Fe) nanoparticles was observed. Interestingly, it was observed that the binding of the copolymers to MIL-101-NH_2(Fe) was non-sheddable. In other words, when the free polymers in solution were completely removed, the bound polymers remained bound on the nanoMOFs, instead of partially diffusing into solution (Figure 4B, step 5). It was shown that the surface polymers significantly slowed the degradation of the MIL-101-NH_2(Fe) nanoparticles, most likely because the diffusion of water in the MOF particles was restricted. Finally, as the degradation of MIL-101-NH_2(Fe) took place, the amount of polymer adsorbed on the nanoMOFs remained constant, suggesting that it bound to newly formed sites during the degradation of the MOF structure (Figure 4B, step 6).

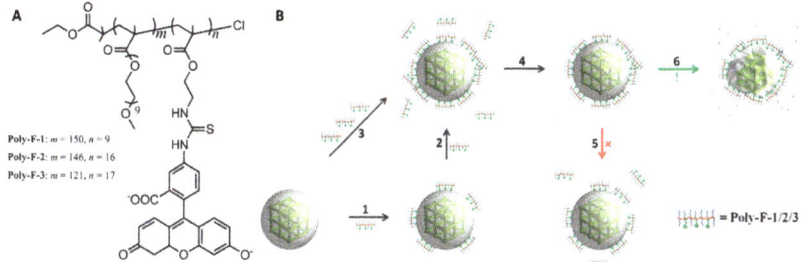

Figure 4. (**A**) Structure of pOEGMA/pAEMA copolymer–fluorescein conjugates (the segment of free AEMA units was omitted for clarification). (**B**) Diagram illustrating the binding/assembly of polymers onto the surface of MIL-101-NH_2(Fe): (1–3) different concentrations of polymers incubated; (4) free polymers removed by centrifugation; (5) dissociation of polymer from the surface (not observed); (6) degradation of MIL-101-NH_2(Fe) and redistribution of surface polymers [62].

Azizi Vahed et al. reported the preparation of a novel MOF: MIL-100-metformin(Fe), an antihyperglycemic agent used for the treatment of type II diabetes that also presents anti-cancer properties [63]. In the MOF structure, the metformin molecules are believed to coordinate iron ions, but without bridging two different ions. As they are prone to hydrolysis in aqueous media, MIL-100-Metformin(Fe) nanoparticles were coated to increase their stability. Sodium alginate was chosen to bring about a pH-controlled behavior and formed a complex structure with MIL-100-Metformin(Fe) stabilized through hydrogen bonds. The coated nanoparticles were characterized by Fourier-transform infrared spectroscopy (FT-IR), thermogravimetric analysis (TGA), and x-ray diffraction (XRD) (indicating the

MOF crystallinity is retained). The MIL-100-Metformin(Fe) nanoparticles were further loaded with metformin by incubation in a metformin solution, resulting in a total 42% metformin content. The release of metformin was studied at two different pHs and monitored through ultraviolet–visible (UV–Vis) spectroscopy. At pH 1.5 (stomach acidity), the release of metformin was almost negligible (10% in 8 h). In contrast, at pH 8 (intestinal pH), release was much more important (87% within 8 h) and furthermore, no initial burst release was observed. The pH-sensitive behavior was attributed to the carboxylic acid groups of sodium alginate. At low pH, they are protonated and neutral. In basic pH, carboxylate ions are formed, which repel each other due to their negative charge, the polymer expands, and cargo molecules can diffuse out of the MOF nanoparticles.

The same group extended the use of sodium alginate to the coating of ZIF-8, a Zn-based MOF [64]. The coating process was carried out in situ by ball-milling zinc acetate and 2-methylimidazole with sodium alginate. Sodium alginate is believed to coat the particles through interactions between its carboxylate groups and the Lewis acid sites of the framework of ZIF-8 or the functional groups of the linkers. Successful coating was confirmed by infrared spectroscopy (IR), while the similar XRD patterns of the coated and uncoated particles proved that the crystalline structure of ZIF-8 was preserved. Uncoated ZIF-8 particles were loaded with metformin and coated with sodium alginate by immersion. Similar to the release of metformin from the alginate-coated MIL-100(Fe), a negligible release was observed at pH 1.5, while the release was much more important at pH 8.

Combining covalent modifications and non-covalent interactions, Wang et al. reported an interesting smart drug delivery device based on a polymer-coated MOF (TTMOF) bearing stimuli responsive features [65]. Post-synthetic modification of MIL-101-NH$_2$(Fe) MOF nanoparticles afforded azide-functionalized MIL-101-N$_3$(Fe), which were subsequently loaded with doxorubicin (DOX). Then, the nanoparticles were modified with β-cyclodextrins (β-CD) by a strain-promoted [3 + 2] azide-alkyne cycloaddition reaction between the azide groups of the MOF particles and the triple bond of the β-CD derivatives. Finally, polyethylene glycol (PEG) chains, functionalized with an adamantane group and a lysine-arginine-glycine-asparagine-serine peptide (K(ad)RGDS) for targeting purposes, were attached to the particles through host–guest interactions between the β-CD and the adamantane group of the PEG chains (Figure 5A). The stimuli responsive behavior was implemented through the benzoic-imine bond, which linked the PEG chains to the targeting peptide and a disulfide bond between the β-CD and the MOF nanoparticles.

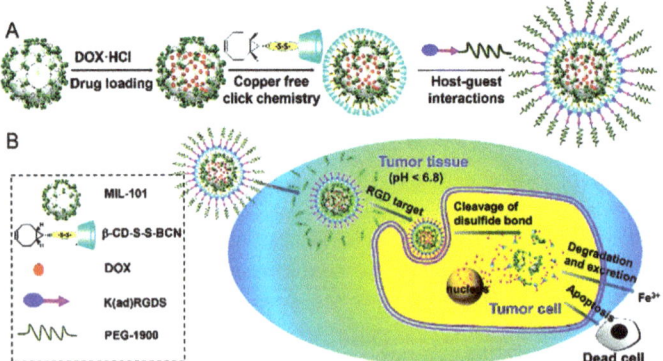

Figure 5. Schematic illustration of (**A**) the drug loading and post-synthetic modification procedure and (**B**) the tumor targeting drug delivery and cancer therapy procedure of the multifunctional MOF based drug delivery system [65].

β-CD were attached to the MOF nanoparticles via a disulfide bond and they blocked the pores, preventing drug release. Indeed, in vitro, less than 15% of the drug was released after a 5-day incubation in phosphate-buffered saline (PBS). However, in the presence of dithiothreitol (DTT), a reducing agent, up to 78% of DOX could be released. This was attributed to the reduction and cleavage of the disulfide bond, resulting in the removal of the β-CD, thus freeing the pore entrances and releasing DOX. In contrast to blood and extracellular fluids, inside the cells, the concentration of glutathione, a biological reducing agent, was 100–1000 times higher, ensuring the rapid cleavage of the S–S bond and the selective, intracellular release of DOX (Figure 5B). Due to the targeting RGD peptide, negligible cellular uptake was observed for non-cancerous cells, but cancerous HeLa cells internalized the nanoparticles; furthermore, uptake was more important at pH 5.0 than at pH 7.4. This is due to the benzoic-imine bond, which linked the PEG chains to the targeting peptide. The benzoic-imine bond was stable at neutral pH. As a result, the targeting peptide was shielded by the PEG chains and the cellular internalization was lower. Under slightly acidic conditions, the benzoic–imine bond was cleaved, the PEG chains were removed, and the targeting peptide was exposed: the outcome was an increased cellular uptake. Finally, the in vivo antitumor efficacy of these nanoparticles was investigated with hepatoma H22 tumor bearing mice (the H22 tumor is integrin positive). Both free doxorubicin and TTMOF nanoparticles exhibited an important tumor growth inhibition, however, side effects, monitored through body weight fluctuations, were considerably lower for TTMOFs.

3.2. Covalent Attachment

3.2.1. "Grafting to" Approaches

Zhao et al. reported the successful functionalization of a copper MOF bearing alkynyl functionalized ligands with azide-modified PEG chains via a copper-catalyzed click reaction [66]. Likewise, based on click chemistry but also employing coordination modulation, Lázaro et al. reported on the covalent functionalization of zirconium MOFs through a click modulation strategy. Initially, appropriately functionalized monodentate ligands are introduced in the MOF synthesis, along with bidentate ligands. In the second step, the polymer is installed directly or indirectly on the modulator by a click reaction. Zirconium MOF UiO-66 nanoparticles coated with polyethyleneglycol (PEG) [67], poly(L-lactide) (PLLA), poly(N-isopropylacrylamide) (PNIPAM), and heparin [68] were prepared. UiO-66-L1 was synthesized in the presence of modulator L1 bearing an azide moiety, N_3. Then, employing a copper(I)-catalyzed azide–alkyne cycloaddition (CuAAC), PEG and PLLA polymer chains functionalized with a complementary propargylic moiety, –C≡CH, were covalently bonded to the modulator to produce UiO-66-L1-polymer particles. A slightly different process was adopted for PNIPAM. Starting from UiO-66-L1 via a surface–ligand exchange, modulator L2 bearing a propargylic moiety, –C≡CH, was introduced, followed by click chemistry with an azide-modified PNIPAM polymer.

The attachment of the polymers on the MOF nanoparticles was evidenced through IR, TGA, and mass spectrometry (MS). Powder x-ray diffraction (PXRD) confirmed that the crystallinity of the MOF nanoparticles had not been altered. Scanning electron microscopy (SEM) images showed particles with a more rounded shape and a larger size after the addition of the polymer chains. N_2 uptake experiments showed that the surface area of the polymer-MOFs had decreased. Dynamic light scattering (DLS) measurements showed that the particles did not aggregate in PBS at pH 7.4. Finally, the polymer–MOFs showed a slower degradation compared to UiO-66-L1/L2. The drug delivery potential of the coated MOF nanoparticles was investigated with calcein as a model drug, and dichloroacetic acid (DCA). The drug was added during the synthesis of the UiO-66-L1/L2 MOFs. It was shown that the drug-loaded nanoparticles were successfully internalized, the endocytosis process depended on the coating, and induced significant cell death. Furthermore, the PEGylated particles showed a pH-responsive behavior, as a faster calcein release was observed at a pH 5.5 compared to 7.4 (Figure 6) [67]. Although some cytotoxicity issues need to be improved, these polymer-coated, DCA-loaded MOFs show promising therapeutic potential.

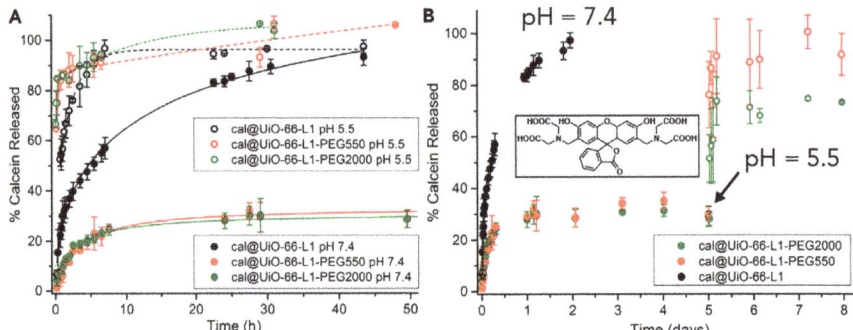

Figure 6. pH-responsive release of calcein from PEGylated UiO-66. (**A**) Calcein-release profiles from UiO-66-L1, UiO-66-L1-PEG550, and UiO-66-L1-PEG2000 in PBS (pH 7.4 and 5.5). (**B**) pH-responsive release of calcein from the PEGylated MOFs. Inset: chemical structure of calcein. Error bars denote standard deviations from triplicate experiments [67].

This strategy was further extended to Zr-fumarate MOFs (Zr-fum): a p-azidomethyl benzoic acid modulator (L1) was introduced in Zr-fum through surface ligand exchange and the azide group of L1 was subsequently used to covalently attach propargyl-terminated PEG chains to the outer surface of the MOFs [69]. Colloidal stability was slightly improved upon PEGylation, and degradation in phosphate buffer saline at pH 7.4 was initially slowed down (induction period) before degrading at a similar rate to non-coated MOF nanoparticles, possibly due to the detachment of the PEG corona. DCA/Zr-fum-L1-PEG exhibited some cytotoxicity toward healthy cells at high concentrations; however, according to the authors, DCA/Zr-fum-L1-PEG had a higher therapeutic efficiency than DCA/UiO-66-L1-PEG.

The modulation strategy was likewise employed by Rijnaarts et al. to introduce PEG chains in MIL-88A MOF particles [70]. Small amounts (0.1–5%) of fumaric acid, an ordinary multivalent ligand in MIL-88A, were replaced by a monovalent PEGylated derivative of succinic acid that acted as a capping ligand, while maintaining a 1:1 stoichiometry between the binding groups. It was shown that the size of the PEG-MIL-88A particles depended on the PEG length and concentration [71]. XRD experiments confirmed that the crystalline structure of MIL-88A was preserved after the insertion of the PEGylated ligand. Elemental analysis evidenced that the PEGylated ligands did not considerably penetrate the bulk of the crystals. The Brunauer–Emett–Teller (BET) surface area decreased probably because the PEG chains blocked the access to the MOF porosity. PEGylated MIL-88A was loaded with sulforhodamine B by counterion exchange. It was observed that encapsulation in the PEG-functionalized particles was more important than in the uncoated particles. This phenomenon was attributed to the higher surface area of the coated particles due to their smaller size/volume.

He et al. reported nanodevices that would simultaneously co-deliver a photosensitizer necessary for photodynamic therapy and a hypoxia-activated prodrug to implement a combined photodynamic and hypoxia-activated therapy (Figure 7) [72]. The outer surface of the zirconium terephthalate UiO-66 nanoparticles was functionalized with photochlor (HPPH), the photosensitizer, and azide groups, N_3, by using monocarboxyl photochlor and p-azidomethylbenzoic acid as modulators during the synthesis of UiO-66 nanoparticles. Then, the nanoparticles were loaded with banoxantrone (AQ4N), the hypoxia-activated prodrug. Finally, to improve the stability of the nanodevices, PEG chains were introduced through a simple copper-free click reaction between the azide groups of the p-azidomethylbenzoic acid and alkyne-terminated PEG, DBCO-PEG.

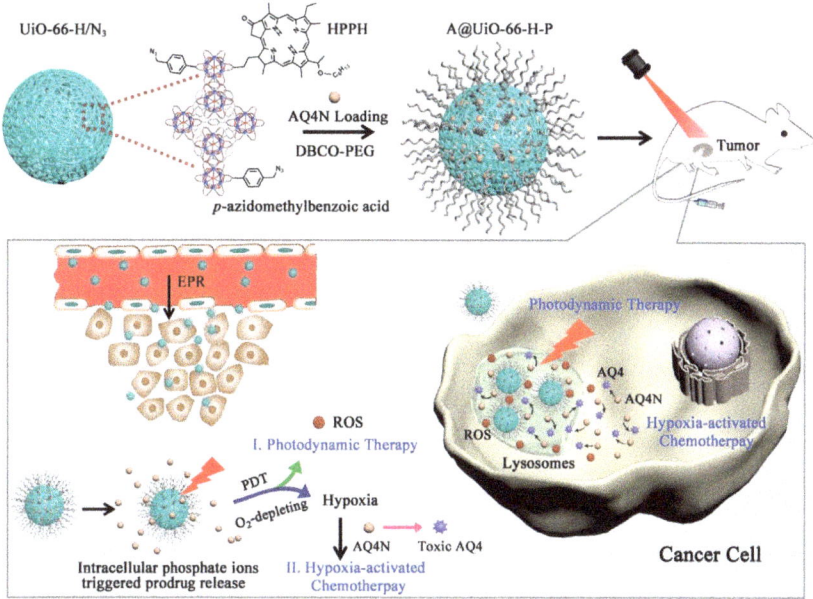

Figure 7. Synthetic procedure of A/UiO-66-H-P nanoparticles and mechanism of photodynamic therapy and hypoxia-activated cascade chemotherapy [72].

The resultant nanoparticles (A/UiO-66-H-P) were duly characterized and their increased stability in saline solution and low concentration PBS solutions was demonstrated. The nanoparticles efficiently produced reactive oxygen species (ROS), $^1O_2$, under laser irradiation. It was further shown that the PEGylation had a beneficial effect on the generation rate of $^1O_2$. In vitro studies demonstrated that the capacity of A/UiO-66-H-P to produce ROS was preserved after the cell internalization of the nanoparticles. The prodrug release studies evidenced a phosphate-controlled release as the release of AQ4N is slow at low PBS concentrations but fast at higher PBS concentrations. The nanoparticles exhibited good biocompatibility and important cellular uptake. In vitro studies with U87MG cells showed that A/UiO-66-H-P inhibited cell growth while in vivo studies showed that A/UiO-66-H-P combined with laser irradiation outperformed any other control therapy.

Zimpel et al. investigated the covalent modification of MOF nanoparticles by exploiting the unsaturated functional groups of the organic linker [73]. This approach allowed for a selective external functionalization, preserving the porous scaffold of the MOF nanoparticles. More explicitly, MIL-100(Fe) nanoparticles were modified by two amino-terminated polymers by coupling the carboxylic acid groups of trimesic acid with the amino groups of the polymers in a carbodiimide-mediated reaction. The two polymers used were an amino-terminated polyethylene glycol and Stp10-C, an oligo-amino-amide bearing a thiol group, which can be further used for the attachment of a fluorescent probe or additional functionalization. XRD and transmission electron microscopy (TEM) confirmed that the crystalline structure of MIL-100(Fe) was retained, the colloidal stability (in water and 10% fetal bovine serum) was significantly increased, a slight decrease of the BET surface area was observed (attributed to the mass increase rather than the loss of porosity); FT-IR and TGA analysis further confirmed the successful attachment of the polymers; and fluorescence correlation spectroscopy (FCS) and DLS measurement showed that the hydrodynamic radius of the particles was 135 ± 45 nm. ^1H nuclear magnetic resonance spectroscopy (NMR), complemented by some other observations, evidenced the covalent nature of the bond between trimesic acid and the polymers. All of these elements indicated a successful polymer coating of MIL-100(Fe) particles, although the functionalization degree of the nanoparticles was estimated to be rather low. This was attributed to the limited amount of free carboxylic acid groups on the external surface of the MOF nanoparticles. As MIL-100(Fe) is active in magnetic resonance, the relaxivities of the coated nanoparticles were calculated. Albeit having a lower activity than uncoated MIL-100(Fe), visualization of the polymer-coated nanoparticles by magnetic resonance imaging was possible. Finally, Stp10-C-coated MIL-100(Fe) particles were functionalized on the free thiol group of Stp10-C chains with a fluorescent probe, cyanine 5 (Cy5). MIL-100(Fe)/Stp10-C*Cy5 were successfully internalized by murine neuroblastoma N2A cells (as revealed by fluorescence microscopy) and well tolerated.

Marqués et al. used the GraftFast process to covalently coat iron and aluminum trimesate MOF with PEG derived polymeric chains [74]. The process is based on the iron mediated reduction of aryldiazonium salts that generates aryl radicals. The aryl radicals have two roles. First, they react directly with the MIL-100(Fe) particle surface to form a polyphenylene sublayer. Second, they act as initiators for the radical polymerization of acryl-PEG (PEG chains functionalized with acryl moieties). Oligomer chains were formed and, in turn, they reacted with the polyphenylene sublayer to yield the grafted coating layer. This coating occurred in a single step, very quickly, and in aqueous solutions. The PEG coating increased the colloidal stability of the MIL-100(Fe) nanoparticles and slowed down the degradation of the MOF particles without affecting their low cytotoxicity. The coating was found to be rather stable in different aqueous media and under ultrasound sonication. The porosity of the nanoparticles was not affected, and caffeine and tritium-labelled gemcitabine were successfully loaded in the PEG-coated nanoparticles. Finally, it was demonstrated that the PEG coating prevented recognition and removal by macrophages. The GraftFast process was similarly applied to ZIF-8 nanoparticles, with the sole difference being that ascorbic acid was used instead of iron for the reduction of the aryldiazonium salt, thus avoiding the presence of iron-based impurities in the zinc MOFs. As with the MIL-100(Fe) nanoparticles, the colloidal and water stability of ZIF-8 were both increased after coating [75].

In their work, Cai et al. described the coating of Fe-soc-MOF nanocrystals (constructed from oxygen-centered iron carboxylate trimermolecular building blocks and 3,3′,5,5′-azobenzenetetracarboxylic acid as a linker) by polypyrrole (Ppy), resulting in the formation of a core-shell structure [76]. The Fe-soc-MOF nanocrystals were initially modified with functionalized PEG chains bearing a thiol and a carboxylic acid moiety. The PEG chains were attached to the oleic acid, stabilizing the Fe-soc-MOF nanocrystals through their thiol-terminated end via a UV-induced thiol-ene reaction. The resulting nanoparticles were dissolved in a polyvinyl alcohol aqueous solution in the presence of pyrrole monomers. Once pyrrole adsorbed on the surface of the particles, an oxidant was added to initiate an oxidation polymerization to finally obtain Fe-soc-MOF core-shell nanoparticles with a thick Ppy layer (Fe-soc-MOF/PPy). The crystallinity of the Fe-soc-MOF core remained intact after the modifications. However, the Fe-soc-MOF/PPy nanoparticles were almost nonporous because Ppy occupied the porosity of Fe-soc-MOF. Fe-soc-MOF/PPy nanoparticles were characterized by UV–Vis near infrared (NIR) absorption, FT-IR, and TGA, and were found to be stable in PBS solution (37 °C), even after repeated irradiation at 808 nm. It was shown that Fe-soc-MOF/PPy could be used as a T2 contrast agent for T2-weighted magnetic resonance imaging. The nanoparticles (in aqueous dispersions) exhibited photothermal properties, and after a 10-min irradiation at 808 nm, temperatures ranging from 40.6 to 72.4 °C were recorded, depending on the concentration. The photothermal conversion efficiency of thee Fe-soc-MOF/PPy nanoparticles was significantly lower when compared to the pure PPy particles, perhaps due to the small amount of PPy they contained. In vitro, Fe-soc-MOF/PPy nanoparticles had a low toxicity and efficiently inhibited the growth of breast cancer cells (murine breast cancer 4T1 cell line). In vivo studies demonstrated that Fe-soc-MOF/PPy could efficiently convert laser irradiation in thermal energy, and as a result, suppress tumor growth.

Li et al. reported the preparation of a hybrid polymer-MOF architecture for enzyme immobilization [77]. UiO-66-NH$_2$ was chosen as the backbone of the structure and post-synthetically, the amino groups were modified into propargylic moieties, –C≡CH. Azide-terminated poly(tert-butyl methacrylate), prepared via atom transfer radical polymerization, was clicked on the alkyne functions of the UiO-66-NH-CH$_2$-C≡CH nanoparticles through a copper-catalyzed alkyne-azide cycloaddition. Finally, the tert-butyl protecting groups were removed to yield poly(methacrylic acid) (PMMA)-modified UiO-66-NH$_2$ nanoparticles, UiO-66-NH$_2$/PMMA. Pectinase was immobilized on UiO-66-NH$_2$/PMMA (UiO-66-NH$_2$/PMMA/pect) through electrostatic attractions between the carboxylic acid groups of PMMA and the amino groups of pectinase. FT-IR and ^{1}H NMR were used to confirm the success of the various modifications. PXRD analysis of all the MOF-containing nanoparticles demonstrated that the crystalline structure of UiO-66-NH$_2$ was maintained throughout all the post-synthetic modifications carried out. Additionally, no modifications were observed in the PXRD pattern of UiO-66-NH$_2$/PMMA/pect after exposition to a citrate buffer, indicating an increased structural stability in aqueous environments. The colloidal stability after PMMA coating was also increased. The BET surface area decreased considerably, especially the microporosity. This was attributed to the PMMA chains and pectinase molecules covering the pores of UiO-66-NH$_2$. Compared to free pectinase, pectinase immobilized on UiO-66-NH$_2$/PMMA exhibited an increased stability in acidic and basic media, a good catalytic activity in a wider range of temperatures, and a clearly enhanced long-term stability. Furthermore, UiO-66-NH$_2$/PMMA/pect maintained 80% of its catalytic activity after eight continuous recycling cycles.

Nagata and his coworkers capitalized on the thermoresponsive behavior of poly(N-isopropylacrylamide) (PNIPAM) to develop a polymer–MOF device for controlled release [78]. PNIPAM-NHS chains were covalently grafted on amino-functionalized UiO-66 crystals to afford UiO-66/PNIPAM. The size of the UiO-66/PNIPAM crystals was around 200 nm, the crystals had an octahedral shape, and their crystalline structure was similar to UiO-66. Based on ^{1}H NMR, the modification percentage of the organic ligands grafted by PNIPAM was calculated to be 11%. PNIPAM diffused only slowly in the pores of UiO-66-NH$_2$ because of its size; thus, grafting occurred predominantly on the outer surface. The cloud point of PNIPAM was 32 °C. Below 32 °C, PNIPAM is dissolved in water; above 32 °C, it aggregates. Therefore, the pores of UiO-66/PNIPAM were expected to be accessible below 32 °C, but blocked above 32 °C. The temperature-dependent release of guest molecules was investigated using resorufin, caffeine, and procainamide. At 25 °C (coil conformation), the cargo molecules were released within four days. At 40 °C (globule conformation), less than 20% of the cargo molecules were released, even after seven days. After the complete release of guest molecules, UiO-66/PNIPAM could be reloaded, still exhibiting a very similar release behavior to the initial one. Finally, controlled, on-off, stepwise release was demonstrated by switching the temperature between 25 °C and 40 °C every 20 min.

Chen et al. recently reported a polyacrylamide hydrogel coating of UiO-68 zirconium MOF nanoparticles [79]. The polyacrylamide hydrogel was cross-linked through DNA sequences recognizing adenosine triphosphate (ATP). In the presence of ATP, overexpressed in cancer cells, the cross-links dissociated via the formation of ATP complexes and the hydrogel became more permeable, allowing the release of the MOF load. An oligonucleotide was bonded to the linkers of the nanoMOF via a triazine linker, the nucleic acid sequence was hybridized with a complementary one that could interact with the polymer chains of the hydrogel in order to bind the hydrogel to the MOF. The hydrogel coating did not affect the crystallinity of the MOF nanoparticles. The UiO-68 nanoparticles were loaded with Rhodamine 6G fluorophore and doxorubicin before installing the hydrogel layer and ATP-triggered release of the loading was demonstrated.

3.2.2. "Grafting from" Approaches

The typical strategy for covalent modification via the "grafting from" approach is the introduction of bromoisobutyrate moieties on the MOFs and the subsequent polymerization via atom transfer radical polymerization (ATRP), with the MOF particles acting as initiators. For example, Xie et al. employed this strategy to modify the external surface of UiO-66-NH$_2$ [80]. The amine groups of the UiO-66-NH$_2$ nanoparticles were coupled to α-bromoisobutyryl bromide and the modified MOF nanoparticles were used as multifunctional initiators for the polymerization of poly(ethylene glycol) methyl ether methacrylate (PEGMA). X-ray photoelectron spectroscopy (XPS), XRD, DLS, and SEM confirmed the successful synthesis of the polymer-coated MOF nanoparticles and the retention of the MOF crystallinity and porosity. Interestingly, the modified nanoparticles presented a reversible pH-switchable dispersity in water: clear solutions were obtained at pH 9 and cloudy suspensions were observed at pH 4–7, depending on the length of the PEG chains. This behavior was attributed to the interactions of the PEG chains with the –COOH groups of the superficial partially uncoordinated MOF ligands.

Similarly, Liu et al. grafted copolymers of 2-(2-methoxyethoxy)ethyl methacrylate (MEO$_2$MA) and oligo(ethylene glycol) methacrylate (OEGMA) on MIL-101(Al)-NH$_2$ [81] where the crystalline structure of MIL-101(Al)-NH$_2$ was preserved. Due to the polymer grafting, the polymer-coated MOF particles exhibited a reversible and temperature-dependent hydrophilic/hydrophobic transition. The lower critical solution temperature (LCST) of poly(MEO$_2$MA-co-OEGMA) was reported to be 39 °C (molar composition: 90% MEO$_2$MA and 10% OEGMA). Above the LCST, the copolymer is hydrophobic and insoluble in water; below the LCST, it is hydrophilic and soluble. As a result, stable dispersions of the polymer-coated MIL-101(Al) were observed below 35 °C, and complete precipitation was observed above 45 °C.

Dong et al. reported the synthesis of a dendritic catiomer based on the functionalization of UiO-66-NH$_2$ with poly(glycidyl methacrylate) chains [82]. After the polymerization of glycidyl methacrylate, the ring-opening of the epoxide rings with ethanolamine afforded UiO-PGMA-EA bearing a secondary amine and a hydroxyl group. FT-IR and XPS demonstrated the successful synthesis of UiO-PGMA-EA; XRD confirmed the preservation of the structure of UiO-66-NH$_2$; and TGA was used to evaluate the amount of grafted polymer. Unlike UiO-66, UiO-PGMA-EA did not aggregate, and due to the hydroxyl groups, exhibited reduced protein adsorption. UiO-PGMA-EA was found to form stable complexes with pDNa due to the abundant amino groups on the PGMA-EA chains, and had high transfection efficiencies, therefore exhibiting potential as a gene carrier for gene therapy [83]. UiO-PGMA-EA was successfully used for the complexation and delivery of mRNA as well as having better performances than the linear PGMA-EA or available commercial products [82].

Likewise, Chen et al. developed an elaborate drug delivery platform founded on the functionalization of zirconium MOF nanoparticles with poly(glycidyl methacrylate) (PGMA) [84]. UiO-PGMA was synthesized as described previously [77], and further reaction of the glycidyl groups with ethylenediamine afforded UiO-PGEDA with two additional amine groups, positively charged at physiological pH. The polymer-coated MOF was loaded with aggregates of doxorubicin (DOX) with a tetraphenylene derivative bearing four –COOH groups (TPE). The DOX-TPE aggregates were used to monitor the DOX release, taking advantage of the fluorescence resonance energy transfer between TPE and DOX. The aggregates were not loaded in the MOF cavities, but were complexed in the polymer layer due to electrostatic attractions. Finally, cucurbit[7]uril (CB[7]) was bound to the residual positively charged amino groups in the polymer layer. CB[7] was used in order to prevent the membrane cell destabilization by the positively charged amino groups of the PGEDA chains, and to regulate the release of DOX. At pH 7, CB[7] and the positively charged amino groups were tightly bound, preventing any DOX leakage. It was demonstrated that at pH 5.0 (endosomal pH), the CB[7] disassembled, allowing DOX release. The empty drug delivery devices showed low cytotoxicity while the DOX-TPE loaded ones had a higher cytotoxicity than free DOX.

Grafting polymers on MOFs is often linked to a decrease in porosity, either because the entrance of the pores is hampered by the polymer chains or because the polymer chains extend into the MOF structure, filling the pores. In response to this drawback, McDonald et al. reported a core-shell MOF architecture with polymer chains grafted on the outer shell, which preserved the core porosity [85]. A shell of IRMOF-3 (zinc ions and 2-aminobenzenedicarboxylate ligand) was grown on a core of MOF-5 (zinc ions and benzenedicarboxylate ligand). Post-synthetic modification of the amino groups of IRMOF-3 with 2-bromoisobutyric anhydride afforded the formation of the initiator for the subsequent polymerization step. Finally, copper mediated atom transfer radical polymerization (ATRP) was carried out using methyl methacrylate as a monomer to yield poly(methyl methacrylate)/IRMOF-3/MOF-5 (PMMA/IRMOF-3/MOF-5). PXRD demonstrated that the crystalline structure of MOF-5 was well preserved throughout all of the modifications. Furthermore, the surface area of PMMA/IRMOF-3/MOF-5 was measured to be 2857 $m^2\ g^{-1}$ and 2289 $m^2\ g^{-1}$, depending on the polymerization duration (5 min and 1 h), showing that the porosity of MOF-5 was intact and accessible. Finally, Raman mapping of the PMMA/IRMOF-3/MOF-5 cross sections showed that the polymer chains are localized on the exterior and within the IRMOF-3 shell.

An interesting functionalization method, lent from the field of mixed matrix membranes, was described by Molavi et al., who modified UiO-66-NH_2 nanoparticles in order to introduce vinyl moieties and, subsequently grew PMMA chains directly from the surface of the MOF particles [86]. The grafting was confirmed by FT-IR and NMR; PXRD showed that the crystalline structure of UiO-66-NH_2 was retained; thermal stability was assessed with TGA; and the BET surface area was significantly lower when compared to the vinyl-modified UiO-66-NH_2 due to the thick and dense PMMA layer that blocked the entrances of the pores. Upon the formation of the polymeric shell, the stability in the PMMA solutions dramatically increased.

Hou et al. used UV-polymerization to graft polymer brushes on MOF particles [87]. The advantage of UV-photoinduced polymerization is that it can be selectively applied to the external surface of MOF particles, thus preserving the porosity of the MOF structure. A suspension of IRMOF-3 particles and methyl methacrylate, styrene, or 2-isopropenyl-2-oxazoline was subjected to UV light. According to the authors, surface radicals are formed on the MOFs by the abstraction of hydrogen atoms and, in turn, these surface radicals initiate a free-radical polymerization. FT-IR was used to evidence the formation of the polymer chains on the MOF particles. PXRD demonstrated that the crystalline structure of IRMOF-3 was not affected by the polymer grafting and TEM images showed the formation of a polymer shell around the nanoparticles. Unlike uncoated IRMOF-3, PMMA-IRMOF retained its crystalline structure even after exposure to air for three days. Besides improving the stability in air, the grafted polymer brushes also prevented the aggregation of the nanoparticles in solution. This method was successfully extended to MOF-5, UiO-66, UiO-66-NH_2, ZIF-8, MIL-125(Ti), and $[Cu(BTCA)_{0.5}(H_2O)_3] \cdot 2H_2O$ (BTCA = 1,2,3,4-butanetetracarboxylic acid) particles.

3.3. Polymer Coordination to Metal Ions

The superficial metal ions of MOFs in MOF nanoparticles are coordinatively unsaturated, therefore, they can be exploited for the coordination of polymers bearing functional groups with a high affinity to metal ions such as amines or phosphate groups.

Gadolinium nanoparticles have attracted a great deal of attention due to their potential as magnetic resonance imaging (MRI) contrast agents, biosensors, and in drug delivery applications. Rowe et al. reported the covalent modification of gadolinium nanoparticles with interesting, from a biomedical point of view, polymers [88]. The polymers were synthesized via reversible addition-fragmentation chain transfer (RAFT) polymerization, which allows for good control over molecular weights and thus, a low polydispersity index. The RAFT agent used in this work was S-1-dodecyl S'-(α,α-dimethylacetic acid) trithiocarbonate, DATC, and it afforded polymer chains terminated with a trithiocarbonate group. Aminolysis of this group generated a free thiol, which was used to covalently graft the polymers to the Gd MOF (gadolinium 1,4-benzenedicarboxylate) nanoparticles by complexation of the thiolates to the Gd^{3+} ions at the surface of the Gd MOF nanoparticles. The studied polymers were poly[N-(2-hydroxypropyl)methacrylamide] (PHPMA), polystyrene (PS), PNIPAM, poly(2-(dimethylamino) ethyl acrylate) (PDMAEA), poly(((poly)ethylene glycol methyl ether) acrylate) (PPEGMEA), and poly(acrylic acid) (PAA) homopolymers. Successful modification of the nanoparticles was confirmed by FT-IR. TEM images showed that the polymers formed a uniform coating on the surface of the Gd MOF nanoparticles. The thickness of the coating depended on the molecular weight of the coating polymer—increased molecular weight afforded increased coating thickness—and could be tuned by varying the polymerization parameters. The polymer coating had good stability and remained intact after several months in aqueous and organic media at room or physiological temperatures. Calculations of the polymer grafting density determined that the coated polymers were in the "brush" regime and a decrease in the grafting density with increased molecular weight of the grafted polymer was observed. Nonetheless, the grafted density values were rather high. The relaxation properties of the unmodified and polymer modified Gd nanoparticles were determined by in vitro MRI and compared with two clinically employed MRI contrast agents: gadopentetate dimeglumine (Magnevist) and gadobenate dimeglumine (Multihance). The Gd MOF nanoparticles modified with hydrophilic polymers exhibited much higher relaxivities compared to the unmodified Gd MOF nanoparticles and Magnevist and Multihance, which is advantageous for their use as clinical positive contrast agents. The relaxivity values tended to increase with increasing polymer molecular weight. In contrast, PS-modified Gd MOF nanoparticles had low longitudinal relaxivity values, attributed to the low water retention due to the hydrophobic nature of PS, and were unsuitable for use as positive contrast agents.

In parallel, aiming at the preparation of novel theragnostic nanodevices, this work was extended to a more elaborate, multifunctional, biocompatible copolymer [89]. Gadolinium 1,4-benzenedicarboxylate nanoparticles were coated with copolymers composed of N-isopropylacrylamide, N-acryloxysuccinimide (NAOS), and fluorescein O-methacrylate (FMA). FMA was employed for fluorescence tagging of the nanoparticles and NAOS was introduced for further modifications to tailor the nanoparticles for specific applications. Indeed, the polymer backbone was functionalized with a targeting ligand (H-glycine-arginine-glycine-aspartate-serine-NH_2 peptide (GRGDS-NH_2)) or a therapeutic agent (methotrexate (MTX), an antineoplastic drug). The polymer coating enhanced the stability of the Gd MOF nanoparticles and growth inhibition studies established an increase in the cell viability in the presence of the coated nanoparticles. The relaxivity properties of the Gd MOF polymer-coated nanoparticles were studied and it was determined that they could produce clinically exploitable results at lower Gd^{3+} concentrations than the contrast agents currently in use, therefore combining fluorescence imaging and magnetic resonance imaging. The targeting potential of these nanoparticles was successfully demonstrated by introducing the GRGDS-NH_2 targeting pentapeptide on the succinimide groups of the copolymer. Finally, when MTX was attached to the succinimide groups of the copolymer, the MTX-containing polymer-modified Gd MOF nanoparticles inhibited the growth of FITZ-HSA tumor cells, in vitro, likewise free MTX.

Horcajada et al. engineered the surfaces of iron(III) carboxylate MOF nanoparticles through the coordination of amine-bearing polymers to the iron(III) ions of the MOF [21]. Modifications were performed during the synthesis of the nanoparticles or post-synthetically. More specifically, adding alpha monomethoxy-omega-amino poly(ethyleneglycol) (CH_3-O-PEG-NH_2) during the synthesis of MIL-88A and MIL-89 afforded the corresponding PEGylated nanoparticles and, similarly, chitosan modified MIL-88A particles were prepared using chitosan grafted with lauryl side chains. MIL-88A was also modified post-synthetically with PEG chains, and MIL-100 was post-synthetically modified with PEG chains and dextran-fluorescein-biotin. MIL-88 and MIL-100 PEGylated nanoparticles were loaded with azidothymidine triphosphate by impregnation and submitted to HIV-activity tests. It was observed that the polymer coating prevented the aggregation of pure MOF nanoparticles in aqueous media without affecting the therapeutic results. When evaluated as MRI contrast agents, the PEGylated nanoparticles showed slightly higher transversal relaxivities than the non-PEGylated ones.

Heparin is a sulfated glycosaminoglycan (polysaccharide) best known for its anticoagulant properties. Heparin coating is expected to confer hydrophilic properties to the surface of the MOF nanoparticles to improve their colloidal stability and to increase blood circulation times, as a result of a lower uptake by macrophages [61]. Bellido et al. studied the surface modification of iron(III) trimesate, MIL-100(Fe), with heparin. Surface modification of MIL-100(Fe) nanoparticles with heparin to afford MIL-100(Fe)/hep was carried out through a simple impregnation of the nanoparticles using a heparin water/ethanol solution. Fast grafting kinetics were observed (around 85% of heparin in solution was associated to MIL-100(Fe) within 4 min), indicating a high affinity of heparin for the MIL-100(Fe) outer surface. Experimental evidence pointed out a coordination of the sulfate groups of heparin to the coordinatively unsaturated iron atoms of MIL-100(Fe). Heparin chains extend partially from the surface in a dense "brush". The crystalline structure of MIL-100(Fe) was not altered by the heparin coating and no significant changes were observed in the BET surface area, indicating that heparin was only grafted on the outer surface of the nanoparticles and did not block access to the pores of MIL-100(Fe). The robustness of the heparin coating was important in water and cell culture medium, however, this stability was challenged in PBS (serum conditions), most likely because the phosphate groups contained in PBS are stronger complexing groups and replace the sulfate groups of heparin in the iron(III) coordination sphere. The colloidal stability of MIL-100(Fe) nanoparticles was improved, but their hydrolytic degradation was not affected by the heparin coating. These results are in contrast with those reported by Lázaro et al., according to whom, the UiO-66 heparin-coated nanoparticles showed undesirable degradation kinetics and a lower colloidal stability compared to UiO-66 particles [68].

Agostoni et al. reported on the coordination of phosphorylated β-cyclodextrins to the iron ions of MIL-100(Fe) nanoparticles via the coordination of the phosphate groups to the coordinatively unsaturated iron ions of the MOF [60]. It was demonstrated that the phosphorylated β-CD remained at the external surface of the MOF nanoparticles, crystallinity and porosity were preserved, and that the β-CD-coating was very stable in both the phosphate buffer solution and cell culture media, in contrast to the PEG-modified MIL-100(Fe) nanoparticles. Additionally, the authors reported the PEG coating of MOF nanoparticles by the coordination of β-CD-P:Ad-PEG supramolecular assemblies to the external surface of MIL-100(Fe), where β-CD-P:Ad-PEG stands for the host–guest complexes formed by adamantyl-modified PEG chains (Ad-PEG) with phosphorylated β-CD (β-CD-P).

In line with this work, Aykaç et al. reported on the coating of MIL-100(Fe) nanoMOFs with polymeric β-CD derivatives [90]. For this purpose, phosphorylated derivatives of epichlorohydrin crosslinked β-CD polymer were used (polyCD). Modification was carried out by impregnation with aqueous solutions containing the phosphorylated derivative to afford MIL-100/polyCD. The polymers were adsorbed onto the nanoparticles in less than one hour, demonstrating the strong affinity of the nanoMOFs for the phosphorylated polyCD. Isothermal titration calorimetry showed the strong interactions between the phosphate groups of the polymeric CD and the iron trimers in MIL-100(Fe). The number of grafted phosphate groups on polyCD did not seem to affect the interactions with the nanoMOFS, suggesting that only some of the grafted group are accessible to the iron trimers. The nanoMOF particles preserved their morphology, crystalline porous structure, and BET surface area after coating. Rhodamine polyCD was used to assess the stability of the polymeric shell. The more phosphorylated polymeric β-cyclodextrin showed a better stability, with less than 30% detachment after a 24 h incubation in PBS. In order to study the drug loading/releasing capacities of MIL-100/polyCD particles, MIL-100(Fe) nanoparticles were impregnated with 3′-azidothymidine triphosphate before being coated. After one day, a 13% slower releaser was observed for the MIL-100/polyCD nanoparticles when compared to the uncoated ones.

Li et al. described ZIF-8 MOF nanoparticles coated with hyaluronic acid for the pH dependent release of curcumin (CCM), an anticancer drug. CCM-loaded ZIF-8 nanoparticles were synthesized in a one-step process and further embedded in hyaluronic acid with pendant imidazole moieties via the complexation of the imidazole units to the zinc ions of ZIF-8 [91]. According to the XRD measurements, the crystalline structure of ZIF-8 was not affected during drug loading or polymer complexation, the dispersity in aqueous solutions was clearly improved, and TEM imaging, ζ-potential measurement, and ^{1}H spectroscopy confirmed the successful preparation of CCM/ZIF-8/HA nanoparticles. In vitro curcumin release was studied at pH 5.5 (tumor tissues pH) and 7.4 (pH of healthy tissues). At pH 7.4, the release of curcumin was slow, and the cumulative release of curcumin was 25% after one week. At pH 5.5, 80% of the loaded curcumin was released within four days. At pH 5.5, the pendant imidazole groups of the hyaluronic acid polymer became protonated and thus were no longer coordinated to the zinc ions. As a result, the protective polymer coating was removed from the ZIF-8 nanoparticles and the degradation of ZIF-8 was faster, resulting in a faster curcumin release. Cytotoxicity was lower for ZIF-8/HA compared to non-coated ZIF-8. Hyaluronic acid binds to the CD44 receptor that is overexpressed by many growing tumor cells. Therefore, compared to free curcumin, an improved cellular uptake was expected, and indeed observed, for CCM/ZIF-8/HA. Furthermore, CCM/ZIF-8/HA induced significant cell death when incubated with HeLa cells.

MIL-100(Fe) nanoparticles were coated by chitosan simply by mixing suspensions of the two materials [92]. The coating was restricted to the outer surface of the MOF nanoparticles and the crystallinity and porosity of the MIL-100(Fe) particles was preserved. X-ray absorption near edge structure spectra of chitosan, MIL-100(Fe) nanoparticles, and chitosan-coated MIL-100(Fe) nanoparticles revealed that the hydroxyl groups of chitosan interacted with the superficial iron ions of the MOF nanoparticles. The chemical stability of MIL-100(Fe) nanoparticles in different physiologic media was increased upon chitosan-coating, however, a general faster aggregation was observed, attributed to the bioadhesive properties of chitosan. The cytotoxicity of the coated nanoMOF was low and the chitosan coating seemed to reduce the inflammatory response and increase the cellular uptake of MIL-100(Fe) nanoparticles.

3.4. Encapsulation of MOF into Polymers

Filippousi and her coworkers reported a non-covalent microencapsulation method for the fabrication of MOF/polymer drug delivery devices [93]. UiO-66 and UiO-67 nanocrystals were loaded with either cisplatin (hydrophilic) or taxol (hydrophobic). Drug-loaded MOF were then encapsulated in a poly(ε-caprolactone)–tocopheryl polyethylene-glycol-succinate (PCL–TPGS) copolymer using a solid/oil/water emulsion method. PXRD measurements demonstrated that the crystalline structure of the MOFs was preserved, while the drugs were in an amorphous form. Annular dark field scanning transmission electron microscopy (ADF-STEM) measurements confirmed that the drug-loaded MOF particles were located inside the polymeric microparticles. In vitro drug release studies displayed an enhanced drug-release profile and a smaller initial burst effect when compared to the corresponding drug-loaded MOFs. Finally, as assessed by cell viability data, the polymer coated drug-loaded MOFs nanoparticles were not cytotoxic, even at high concentrations (tested cell lines: U-87 MG (human glioblastoma grade IV; astrocytoma) and HSC-3 (human oral squamous carcinoma)).

He et al. reported an interesting and generalizable strategy for the polymer coating of MOFs (Figure 8) [94]. The MOF nanoparticles were initially embedded in an inter-chain hydrogen bond self-assembled network of a random copolymer. The random copolymer (RCP) contained carboxylic acid groups, as the hydrogen bond directed assembly and the bromoisobutyrate functional groups (Figure 8A). The latter acted as initiators for the atom transfer radical polymerization of a mixture of monomer and cross-linker (e.g., styrene and 1,4-butanediol diacrylate), affording a uniform layer of cross-linked polymer around the MOF nanoparticles. It was demonstrated that polymerization occurred only on the surface of the MOF nanoparticles, the crystallinity and porosity of the nanoparticles were retained, and that the coating was chemically stable, preserving the MOF structure under acidic and basic conditions. The thickness of the polymer coating could be tuned by polymerization time and monomer concentration and it was shown that the wettability of the polymer-coated MOF nanoparticles could be manipulated through the choice of polymer coating. Furthermore, a second polymer layer could be added through a second polymerization step to introduce new functionalities. To demonstrate its wide scope, this surface modification method was applied to UiO-66, ZIF-8, ZIF-67, MIL-96, and MIL-101(Cr) MOF nanoparticles and besides styrene, 2-hydroxyethyl acrylate, n-butyl acrylate, 1H,1H,2H,2H-perfluorodecyl methacrylate, and a mixture of benzyl methacrylate and 1H,1H,2H,2H-perfluorodecyl methacrylate (1/1).

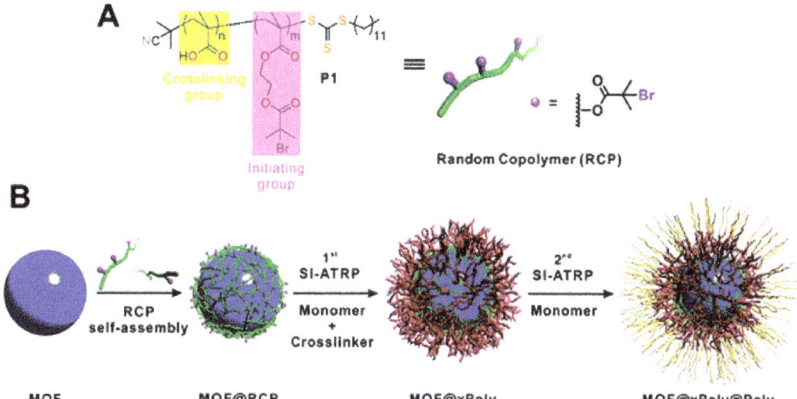

Figure 8. (**A**) Molecular structure of the random copolymer macroinitiator, P1. (**B**) Schematic illustration of typical experimental procedures for growing polymer shells on a MOF particle [94].

Márquez et al. reported on the fabrication of polymer–MOF patches for cutaneous applications [95]. MIL-100(Fe) caffeine (CAF)-loaded MOF nanoparticles were encapsulated in gelatin from pig skin (GEL) or low molecular weight polyvinyl alcohol (PVA) through a three-step press-molding process: the components of the patches were milled, mixed, and pressed into a wafer. It was found that the patches had a non-bioadhesive character. In release studies, caffeine was progressively released within 48 h from the patches. It is noteworthy that the burst release for the GEL_MIL-100_CAF was only 15%. This was attributed to the lower hydrophilicity and higher strength of GEL compared to PVA, which might slow down the diffusion of caffeine, and the possible coordination of the iron ions of MIL-100 with the amino or carboxylic acid groups of gelatin, which resulted in a more compact network. Overall, polymer-MOF particles exhibited a better release than pure polymer patches, showing the beneficial role of MOF encapsulation. This methodology was successfully extended to ibuprofen.

Liu et al. reported on the modification of hafnium MOF nanoparticles (Hf^{4+} with tetrakis (4-carboxyphenyl) porphyrin) by PEG-grafted poly(maleicanhydride-alt-1-octadecene) [96]. The resulting nanoparticles, thanks to the PEG chains, could undergo dispersion in aqueous media and exhibited rather long blood circulation (blood circulation half-life ca. 3.3 h). The nanoparticles were used for combined radiotherapy and photodynamic therapy and were found to successfully inhibit tumor growth in vivo.

Cai et al. reported on the use of iron MOF nanoparticles (MIL-100) for photothermal therapy (PTT) [97]. The particles were loaded with indocyanine green (ICG), a photo-responsive organic dye, with an absorption maximum in the NIR, and used for clinical applications. In order to increase the binding affinity selectively for the surface of cancer cells, the nanoparticles were non-covalently conjugated to hyaluronic acid (HA), a natural biopolymer, which was found to mediate the targeting recognition of CD44 over-expressing cancer cells. The MIL-100 coated nanoparticles, MIL-100(Fe)/HA, were obtained simply by mixing the MOF nanoparticles with an aqueous solution of hyaluronic acid, and the ICG molecules were then loaded in the coated nanoparticles to obtain MIL-100(Fe)/HA/ICG nanoparticles (Figure 9). The nanoparticles showed a good colloidal stability in water, and the crystalline structure of MIL-100(Fe) was retained. FT-IR confirmed the successful conjugation with hyaluronic acid. UV–Vis measurements showed that ICG was successfully loaded in the particles and the loading was evaluated around 43 wt %. After encapsulation in the MOF, the ICG absorption and emission maxima were slightly red shifted. When compared to free ICG, MIL-100(Fe)/HA/ICG showed enhanced photostability and photothermal efficiency. Furthermore, it was demonstrated that the MIL-100(Fe)/HA/ICG nanoparticles could be used as MRI contrast agents. According to in vitro studies, MIL-100(Fe)/HA/ICG showed a low cell cytotoxicity, were internalized by CD44-positive

MCF-7 cancer cells, and could efficiently kill cells when irradiated at 808 nm (near IR wavelength). When used to treat tumor-bearing mice, PTT was performed 48 h after the injection of the nanoparticles. The temperatures of tumors gradually increased and reached 52 °C after 10 min of irradiation, complete ablation of the tumor was observed after 14 days, and an 80% survival rate was recorded after 20 days.

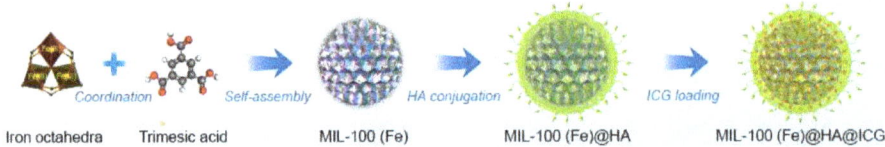

Figure 9. Schematic representation of the synthesis procedure, HA conjugation and ICG loading of MIL-100(Fe) NPs [97].

Different research groups reported polymer coatings by the in-situ polymerization of adsorbed monomers on MOF particles. For example, Wang et al. coated UiO-66 nanoparticles with polyaniline (PAN) for applications in photothermal therapy [98]. Aniline is initially adsorbed on the negatively charged surface of UiO-66 particles because of electrostatic interactions, and then polymerized in situ with the addition of ammonium persulfate as an oxidizing agent. The synthesized UiO-66/PAN nanoparticles were around 100 nm. XRD characterization showed that the crystalline structure of UiO-66 was preserved. The UV–Vis spectrum of UiO-66/PAN nanoparticles dispersed in water showed an absorption around 800 nm, typical of PAN in the form of its emeraldine salt. FT-IR and elemental analysis confirmed the formation of polyaniline on the surface of the UiO-66 nanoparticles. The photothermal performance of UiO-66/PAN were evaluated in vitro and in vivo. In vivo studies were carried out on mice bearing subcutaneous colon cancer xenografts. Upon laser irradiation, the local tumor temperature was between 42 and 45 °C, and the tumor showed complete regression after 10 days.

Wu et al. used polydopamine to aggregate isolated ZIF-8 nanocrystals loaded with glucose oxidase into micrometer-sized particles to facilitate the repeated use of the enzyme [99]. Dopamine was incubated with the enzyme-loaded ZIF-8 at pH 8.5, and polymerized during the incubation. The crystallinity of the MOF was unchanged by the loading and the coating, and though lower activities were observed, higher stability and excellent reusability were demonstrated. Using the same principle, Feng et al. reported on the preparation of MOF nanoparticles for chemo-photothermal therapy [100]. MOF nanoparticles (ZIF-8, UiO-66 and MIL-101) were loaded with doxorubicin (DOX) before being coated with polydopamine, as mentioned earlier. The polydopamine coating was further functionalized with targeting molecules: sgc-8 aptamer and/or folic acid. pH-dependent DOX release was observed, as the ZIF-8 structure is stable at neutral pH but degrades at acidic pH. The coated particles exhibited photothermal activity, attributed to the polydopamine coating. Finally, it was demonstrated that in vivo chemo-photothermal treatment by sgc-8 aptamer-PDA-DOX/ZIF-8 and near infra-red illumination resulted in tumor elimination, without noticeable systemic toxicity and favorable biocompatibility. Yang et al. applied a similar method for the surface functionalization of MOF, aiming to increase the hydrophobicity and water stability [101]. This time, it was based on the easy polymerization of free-base dopamine at room temperature under a mild oxygen atmosphere, affording a polydopamine layer on the MOF. The polydopamine coating was further fluorinated with 1H,1H,2H,2H-perfluorodecanethiol, affording very hydrophobic surfaces. When modified according to this two-step process, HKUST-1 retained its crystalline structure and its stability in water, acidic, or basic media was dramatically increased. Although some amount of dopamine could have potentially penetrated the MOF pores and bound to the Cu ions, in situ diffuse reflectance infrared Fourier transform spectroscopy (in situ DRIFTs) demonstrated that many open metal sites remained after the modification. The polymer loading was optimized in order to retain a high surface area. This modification was successfully extended to ZIF-67, ZIF-8, UiO-66, Mg-MOF-74, MIL-100 (Fe), and Cu-TDPAT.

Likewise, Castells-Gil et al. exploited the easy copper-catalyzed polymerization of catechol derivatives to coat the copper MOF HKUST to improve its moisture tolerance [102]. According to the authors, a ligand exchange took place initially, with the catechols partially replacing the ligands of HKUST on the outer surface of the particles. Then, copper-catalyzed polymerization of catechols was initiated, resulting in the formation of a superficial, tightly bound, thick layer of polymer. Modification with fluorinated 4-undecylcatechol afforded a robust coverage that retained 92% of the MOF porosity and allowed incubation in water for at least one week without significant decrease in crystallinity. The properties of the MOF could be tuned by using appropriately functionalized catechols.

Another strategy to preserve the integrity of the MOF structure and porosity is the coating of MOF particles with microporous organic polymers (MOP). For example, Chun et al. coated UiO-66(Zr)-NH_2 particles by a microporous organic network assembled from tetra(4-ethynylphenyl)methane and 1,4-diiodobenzene or 4,4′-diiodobiphenyl [103]. The thickness of the coating was controlled by the number of organic building blocks. PXRD demonstrated that the crystalline structure of the UiO-66(Zr)-NH_2 particles was retained during coating, and coated particles retained their structure when exposed to water. Decrease in the measured surface area was attributed to a partial inclusion of the organic microporous network in the pores of the MOF.

MOF/MOP hybrid nanoparticles (UNP) were prepared by growing a boron-dipyrromethene (BODIPYs)–imine based polymer (by the reaction of 1,3,5-tris(4-aminophenyl)benzene (TAPB) with dialdehyde-substituted BODIPYs (CHO-BDP)) on the surface of pre-synthesized amine-modified UiO-66 MOF seeds (UiO-66 with 50 mol% amino groups), where the amino groups of modified UiO-66 were used as grafting points of the polymer on the MOF particles [104]. The successful formation of the imine polymer was confirmed by FT-IR and solid state ^{13}C NMR spectroscopy, while XRD measurements demonstrated that the crystalline structure of modified UiO-66 was preserved. N_2 sorption experiments evidenced the coexistence of microporous and mesoporous pores within the hybrid nanoparticles, with a calculated BET surface area lying between that of amine-modified UiO-66 MOF and that of pure BDP-imine MOPs. From the view of targeting biomedical applications, folic acid was grafted on the MOP coating by condensation of the amine groups of folic acid with the residual unreacted aldehyde groups of CHO-BDP on the hybrid nanoparticles. No significant cell cytotoxicity was observed for the MOF/MOP hybrid nanoparticles, with or without folic acid. All particles could pass across the cell membrane in the cytoplasm, though cellular uptake was much higher for the folic acid-modified nanoparticles.

Akin to this work, Wang et al. reported the coating of UiO-66 nanocrystals with a cyanine-containing organic polymer built on the surface of UiO-66 particles via a multicomponent Passerini reaction of carboxyl cyanine, o-nitrobenzaldehyde, and 1,6-diisocyanatohexane [105]. Although XRD patterns suggested that the crystalline structure was not affected by the coating, the decreased BET surface area and the increased pore size implied that some structural defects might have occurred on the surface of the UiO-66 crystals. The particles exhibited a good photothermal response and low toxicity, and in vivo tests demonstrated that upon laser irradiation, UiO-66/CyP successfully killed cancerous cells and inhibited tumor growth.

3.5. Other Strategies

Finally, to conclude this survey on the modification of MOF particles for biomedical applications, a couple of other interesting strategies will be mentioned. Ostermann et al., aiming at the construction of hierarchical nanostructures, obtained nanofibers of ZIF-8/PVP by electrospinning a PVP/ZIF-8 dispersion [106]. The diameter of the fibers could be adjusted by the polymer concentration. Homogeneous distribution of the nanoparticles inside the fibers was evidenced by microscopy (SEM and TEM) and adsorption measurements showed that the MOF nanoparticles were accessible. The method was further extended to polystyrene and polyethylene oxide.

To address water stability issues, Gamage et al. reported the preparation of MOF-5 composites with polystyrene (PS) [107]. Polystyrene was grafted onto MOF-5 crystals by heating in neat pure styrene at 65 °C. The crystallinity of MOF-5 in the MOF-5-PS composites was not altered by the polymerization of PS. However, the BET surface areas calculated by N_2 sorption experiments were lower when compared to MOF-5, suggesting that PS chains partially block the pores of MOF-5. Indeed, according to the Raman mapping image and the white-light image, polystyrene is uniformly distributed throughout the MOF-5 crystal. The hydrolytic stability of the MOF-5-PS composites was significantly improved. Dye-adsorption studies showed a decrease of polarity in the pores when PS was grafted. This modification methodology was further extended to other MOFs (IRMOF-3, MOF-177, and HKUST-1) and functionalized styrene monomers (4-bromostyrene). Zhang et al. developed a strategy for the coating of MOF particles with a thin polydimethysiloxane (PDMS) coating [108] and applied it to three different MOFs: MOF-5 with $Zn_4O(COO)_6$ clusters; HKUST-1 with paddle wheel $Cu_2(COO)_4$ centers; and $[Zn(bdc)(ted)_{0.5}]\cdot 2DMF\cdot 0.2H_2O$ with $Zn_2(COO)_4N_2$ clusters. The MOF particles were heated in the presence of PDMS, and thermal degradation of PDMS generated volatile silicone oligomers that adsorbed on the MOF and cross-linked to form a PDMS coating. For all of the studied MOFs, retention of the crystalline structure and the porosity was observed after coating and accessibility of the active sites. Water stability was indeed increased as the hydrophobic PDMS coating prevented water molecules from interacting with the metal ions.

Pastore et al. reported a different approach that is worth mentioning, although not exploitable for drug delivery application [109]. In this work, poly(amic acid) was conjugated to MOF crystals using post-synthetic ligand exchange (Figure 10). The difference is that the ligand moieties were already incorporated into the poly(amic acid) backbone before the ligand exchange. This strategy was successfully applied to three different polymers, namely MOF-5, ZIF-8 and UiO-66. The crystallinity of the MOF particles was retained and the porosity was preserved, depending on the size of the MOF crystals. It is noteworthy that for micro-crystalline MOF-5, a 99.8% retention of porosity was calculated.

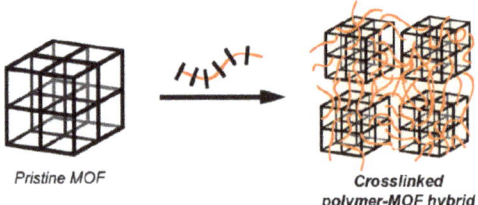

Figure 10. Schematic of direct integration through post-synthetic ligand exchange to form a crosslinked polymer-MOF network with preserved porosity [109].

4. Concluding Remarks

In summary, we can say that a variety of strategies have been set up for the modification of MOF by polymers to exploit electrostatic interactions and complexation to the metal ions of the MOF structure of covalent grafting. When a MOF is combined with a polymer, its colloidal stability is enhanced without loss of crystallinity. However, a recurrent issue is the decrease of porosity due to the polymer obstructing the entrance to the pores, or the penetration of the polymer chains inside the MOF cavities. Besides increasing the stability, the polymer coating offers the possibility of adding targeting functionalities or introducing a stimuli-responsive release, allowing for the preparation of improved drug delivery or imaging devices.

Up until now, the developed technologies for the use of polymer/MOF and BioMOF nanocomposites are limited due to the lack of knowledge surrounding the interactions of these materials with the human body. In our opinion, the future research efforts should focus, among other things, on the (bio)stability of these systems in physiological conditions and the long-term impact of such compounds on living organisms. Furthermore, since the use of therapeutic proteins is a very promising route in the development of specific drugs, focus should be given to the production of appropriate systems for the delivery of proteins without disrupting their bioavailability and activity. Other critical aspects include biocompatibility issues, batch-to-batch repeatability, pharmacokinetics/pharmacodynamics, dose response, clinical applications, etc.

In conclusion, polymer/MOF nanocomposites constitute a next generation class of multifunctional devices for biomedical applications that can greatly contribute to the improvement of personalized medicine and healthcare in general.

Author Contributions: Conceptualization D.G. (Dimitrios Giliopoulos), D.B. and K.T.; literature survey, organization and critical analysis of data D.G. (Dimitrios Giliopoulos), D.G. (Dimitrios Giannakoudakis) and A.Z.; initial writing D.G. (Dimitrios Giliopoulos), D.G. (Dimitrios Giannakoudakis), A.Z.; revisions and proof-reading D.G. (Dimitrios Giliopoulos), D.G. (Dimitrios Giannakoudakis), A.Z., D.B. and K.T.; funding acquisition, D.G. (Dimitrios Giliopoulos), D.B. and K.T. All authors have read and agreed to the published version of the manuscript.

Funding: This research was implemented through the IKY scholarships program and co-financed by the European Union (European Social Fund, ESF) and Greek national funds through the action entitled "Reinforcement of Postdoctoral Researchers" (MIS: 5001552), and in the framework of the Operational Program "Human Resources Development Program, Education and Lifelong Learning" of the National Strategic Reference Framework (NSRF) 2014–2020.

Conflicts of Interest: The authors declare no conflict of interest.

References

1. Azizi Samir, M.A.S.; Alloin, F.; Dufresne, A. Review of recent research into cellulosic whiskers, their properties and their application in nanocomposite field. *Biomacromolecules* **2005**, *6*, 612–626. [CrossRef] [PubMed]
2. Fu, S.Y.; Feng, X.Q.; Lauke, B.; Mai, Y.W. Effects of particle size, particle/matrix interface adhesion and particle loading on mechanical properties of particulate-polymer composites. *Compos. Part B Eng.* **2008**, *39*, 933–961. [CrossRef]
3. Hussain, F.; Hojjati, M.; Okamoto, M.; Gorga, R.E. Review article: Polymer-matrix nanocomposites, processing, manufacturing, and application: An overview. *J. Compos. Mater.* **2006**, *40*, 1511–1575. [CrossRef]
4. Moniruzzaman, M.; Winey, K.I. Polymer nanocomposites containing carbon nanotubes. *Macromolecules* **2006**, *39*, 5194–5205. [CrossRef]
5. Ramanathan, T.; Abdala, A.A.; Stankovich, S.; Dikin, D.A.; Herrera-Alonso, M.; Piner, R.D.; Adamson, D.H.; Schniepp, H.C.; Chen, X.; Ruoff, R.S.; et al. Functionalized graphene sheets for polymer nanocomposites. *Nat. Nanotechnol.* **2008**, *3*, 327–331. [CrossRef]
6. Sinha Ray, S.; Okamoto, M. Polymer/layered silicate nanocomposites: A review from preparation to processing. *Prog. Polym. Sci.* **2003**, *28*, 1539–1641. [CrossRef]
7. Spitalsky, Z.; Tasis, D.; Papagelis, K.; Galiotis, C. Carbon nanotube-polymer composites: Chemistry, processing, mechanical and electrical properties. *Prog. Polym. Sci.* **2010**, *35*, 357–401. [CrossRef]
8. Zou, H.; Wu, S.; Shen, J. Polymer/Silica Nanocomposites: Preparation, characterization, properties, and applications. *Chem. Rev.* **2008**, *108*, 3893–3957. [CrossRef]
9. Armentano, I.; Dottori, M.; Fortunati, E.; Mattioli, S.; Kenny, J.M. Biodegradable polymer matrix nanocomposites for tissue engineering: A review. *Polym. Degrad. Stab.* **2010**, *95*, 2126–2146. [CrossRef]
10. Paul, D.R.; Robeson, L.M. Polymer nanotechnology: Nanocomposites. *Polymer* **2008**, *49*, 3187–3204. [CrossRef]
11. Reddy, M.M.; Vivekanandhan, S.; Misra, M.; Bhatia, S.K.; Mohanty, A.K. Biobased plastics and bionanocomposites: Current status and future opportunities. *Prog. Polym. Sci.* **2013**, *38*, 1653–1689. [CrossRef]
12. Bianco, A.; Kostarelos, K.; Partidos, C.D.; Prato, M. Biomedical applications of functionalised carbon nanotubes. *Chem. Commun.* **2005**, *5*, 571–577. [CrossRef] [PubMed]
13. Chimene, D.; Alge, D.L.; Gaharwar, A.K. Two-Dimensional nanomaterials for biomedical applications: Emerging trends and future prospects. *Adv. Mater.* **2015**, *27*, 7261–7284. [CrossRef] [PubMed]
14. Li, Z.; Barnes, J.C.; Bosoy, A.; Stoddart, J.F.; Zink, J.I. Mesoporous silica nanoparticles in biomedical applications. *Chem. Soc. Rev.* **2012**, *41*, 2590–2605. [CrossRef]
15. Liu, Z.; Tabakman, S.; Welsher, K.; Dai, H. Carbon nanotubes in biology and medicine: In vitro and in vivo detection, imaging and drug delivery. *Nano Res.* **2009**, *2*, 85–120. [CrossRef]

16. Vallet-Regí, M.; Balas, F.; Arcos, D. Mesoporous materials for drug delivery. *Angew. Chem. Int. Ed.* **2007**, *46*, 7548–7558. [CrossRef]
17. McKinlay, A.C.; Morris, R.E.; Horcajada, P.; Ferey, G.; Gref, R.; Couvreur, P.; Serre, C. BioMOFs: Metal–organic frameworks for biological and medical applications. *Angew. Chem. Int. Ed. Engl.* **2010**, *49*, 6260–6266. [CrossRef]
18. Simon-Yarza, T.; Mielcarek, A.; Couvreur, P.; Serre, C. Nanoparticles of metal–organic frameworks: On the road to in vivo efficacy in biomedicine. *Adv. Mater.* **2018**, *30*, 1707365. [CrossRef]
19. Wu, M.-X.; Yang, Y.-W. Metal–Organic Framework (MOF)-based drug/cargo delivery and cancer therapy. *Adv. Mater.* **2017**, *29*, 1606134. [CrossRef]
20. An, J.; Geib, S.J.; Rosi, N.L. Cation-Triggered drug release from a porous Zinc–adeninate metal–organic framework. *J. Am. Chem. Soc.* **2009**, *131*, 8376–8377. [CrossRef]
21. Horcajada, P.; Chalati, T.; Serre, C.; Gillet, B.; Sebrie, C.; Baati, T.; Eubank, J.F.; Heurtaux, D.; Clayette, P.; Kreuz, C.; et al. Porous metal–organic-framework nanoscale carriers as a potential platform for drug delivery and imaging. *Nat. Mater.* **2010**, *9*, 172–178. [CrossRef] [PubMed]
22. Beg, S.; Rahman, M.; Jain, A.; Saini, S.; Midoux, P.; Pichon, C.; Ahmad, F.J.; Akhter, S. Nanoporous metal organic frameworks as hybrid polymer–metal composites for drug delivery and biomedical applications. *Drug Discov. Today* **2017**, *22*, 625–637. [CrossRef] [PubMed]
23. Chowdhury, M.A. Metal–organic-frameworks for biomedical applications in drug delivery, and as MRI contrast agents. *J. Biomed. Mater. Res. Part A* **2017**, *105*, 1184–1194. [CrossRef] [PubMed]
24. Meek, S.T.; Greathouse, J.A.; Allendorf, M.D. Metal–organic frameworks: A rapidly growing class of versatile nanoporous materials. *Adv. Mater.* **2011**, *23*, 249–267. [CrossRef] [PubMed]
25. Zhou, H.-C.; Long, J.R.; Yaghi, O.M. Introduction to metal–organic frameworks. *Chem. Rev.* **2012**, *112*, 673–674. [CrossRef] [PubMed]
26. Naka, K. Metal Organic Framework (MOF). In *Encyclopedia of Polymeric Nanomaterials*; Kobayashi, S., Müllen, K., Eds.; Springer: Berlin/Heidelberg, Germany, 2015; pp. 1233–1238. [CrossRef]
27. Hoskins, B.F.; Robson, R. Infinite polymeric frameworks consisting of three dimensionally linked rod-like segments. *J. Am. Chem. Soc.* **1989**, *111*, 5962–5964. [CrossRef]
28. Yaghi, O.M.; Li, H. Hydrothermal synthesis of a Metal–organic framework containing large rectangular channels. *J. Am. Chem. Soc.* **1995**, *117*, 10401–10402. [CrossRef]
29. Sun, Y.; Zhou, H.C. Recent progress in the synthesis of Metal–organic frameworks. *Sci. Technol. Adv. Mater.* **2015**, *16*, 054202. [CrossRef]
30. Chui, S.S.; Lo, S.M.; Charmant, J.P.; Orpen, A.G.; Williams, I.D. A chemically functionalizable nanoporous material. *Science* **1999**, *283*, 1148–1150. [CrossRef]
31. Li, H.; Eddaoudi, M.; O'Keeffe, M.; Yaghi, O.M. Design and synthesis of an exceptionally stable and highly porous Metal–organic framework. *Nature* **1999**, *402*, 276–279. [CrossRef]
32. Alshammari, A.; Jiang, Z.; Cordova, K.E. Metal organic frameworks as emerging photocatalysts. In *Semiconductor Photocatalysis: Materials, Mechanisms and Applications*; Cao, W., Ed.; InTech: London, UK, 2016; pp. 302–341.
33. Eddaoudi, M.; Kim, J.; Rosi, N.; Vodak, D.; Wachter, J.; O'Keeffe, M.; Yaghi, O.M. Systematic design of pore size and functionality in isoreticular MOFs and their application in methane storage. *Science* **2002**, *295*, 469–472. [CrossRef] [PubMed]
34. Farha, O.K.; Eryazici, I.; Jeong, N.C.; Hauser, B.G.; Wilmer, C.E.; Sarjeant, A.A.; Snurr, R.Q.; Nguyen, S.T.; Yazaydin, A.O.; Hupp, J.T. Metal–organic framework materials with ultrahigh surface areas: Is the sky the limit? *J. Am. Chem. Soc.* **2012**, *134*, 15016–15021. [CrossRef] [PubMed]
35. Burrows, A.D. The Chemistry of Metal–Organic Frameworks. Synthesis, Characterization, and Applications, 2 Volumes. Edited by Stefan Kaskel. *Angew. Chem. Int. Ed.* **2017**, *56*, 1449. [CrossRef]
36. McGuire, C.V.; Forgan, R.S. The surface chemistry of metal–organic frameworks. *Chem. Commun.* **2015**, *51*, 5199–5217. [CrossRef]
37. Furukawa, H.; Cordova, K.E.; O'Keeffe, M.; Yaghi, O.M. The chemistry and applications of Metal–organic frameworks. *Science* **2013**, *341*, 1230444. [CrossRef]
38. Giannakoudakis, D.A.; Hu, Y.; Florent, M.; Bandosz, T.J. Smart textiles of MOF/g-$_3$N$_4$ nanospheres for the rapid detection/detoxification of chemical warfare agents. *Nanoscale Horiz.* **2017**, *2*, 356–364. [CrossRef]

39. Giannakoudakis, D.A.; Travlou, N.A.; Secor, J.; Bandosz, T.J. Oxidized g-$_3$N$_4$ nanospheres as catalytically photoactive linkers in MOF/G-C$_3$N$_4$ composite of hierarchical pore structure. *Small* **2017**, *13*, 1601758. [CrossRef]
40. Petit, C.; Bandosz, T.J. Engineering the surface of a new class of adsorbents: Metal–organic framework/graphite oxide composites. *J. Colloid Interface Sci.* **2015**, *447*, 139–151. [CrossRef]
41. Van Assche, T.R.C.; Duerinck, T.; Gutiérrez Sevillano, J.J.; Calero, S.; Baron, G.V.; Denayer, J.F.M. High adsorption capacities and two-step adsorption of polar adsorbates on copper–benzene-1,3,5-tricarboxylate metal–organic framework. *J. Phys. Chem. C* **2013**, *117*, 18100–18111. [CrossRef]
42. Bandosz, T.J. (Ed.) *Activated Carbon Surfaces in Environmental Remediation*; Elsevier: Oxford, UK, 2006; Volume 7, pp. 1–571.
43. Florent, M.; Giannakoudakis, D.A.; Bandosz, T.J. Mustard gas surrogate interactions with modified porous carbon fabrics: Effect of oxidative treatment. *Langmuir* **2017**, *33*, 11475–11483. [CrossRef]
44. Kyzas, G.Z.; Fu, J.; Lazaridis, N.K.; Bikiaris, D.N.; Matis, K.A. New approaches on the removal of pharmaceuticals from wastewaters with adsorbent materials. *J. Mol. Liq.* **2015**, *209*, 87–93. [CrossRef]
45. Landers, J.; Gor, G.Y.; Neimark, A.V. Density functional theory methods for characterization of porous materials. *Colloids Surf. A Physicochem. Eng. Asp.* **2013**, *437*, 3–32. [CrossRef]
46. Stout, S.C.; Larsen, S.C.; Grassian, V.H. Adsorption, desorption and thermal oxidation of 2-CEES on nanocrystalline zeolites. *Microporous Mesoporous Mater.* **2007**, *100*, 77–86. [CrossRef]
47. Feng, M.; Zhang, P.; Zhou, H.-C.; Sharma, V.K. Water-stable Metal–organic frameworks for aqueous removal of heavy metals and radionuclides: A review. *Chemosphere* **2018**, *209*, 783–800. [CrossRef] [PubMed]
48. Alfarra, A.; Frackowiak, E.; Béguin, F. The HSAB concept as a means to interpret the adsorption of metal ions onto activated carbons. *Appl. Surf. Sci.* **2004**, *228*, 84–92. [CrossRef]
49. Dhakshinamoorthy, A.; Asiri, A.M.; Garcia, H. 2D metal–organic frameworks as multifunctional materials in heterogeneous catalysis and electro/photocatalysis. *Adv. Mater.* **2019**, *31*, e1900617. [CrossRef]
50. Cai, H.; Huang, Y.-L.; Li, D. Biological metal–organic frameworks: Structures, host–guest chemistry and bio-applications. *Coord. Chem. Rev.* **2019**, *378*, 207–221. [CrossRef]
51. Jiao, L.; Seow, J.Y.R.; Skinner, W.S.; Wang, Z.U.; Jiang, H.-L. Metal–organic frameworks: Structures and functional applications. *Mater. Today* **2019**, *27*, 43–68. [CrossRef]
52. Freiberg, S.; Zhu, X.X. Polymer microspheres for controlled drug release. *Int. J. Pharm.* **2004**, *282*, 1–18. [CrossRef]
53. Horcajada, P.; Serre, C.; Vallet-Regí, M.; Sebban, M.; Taulelle, F.; Férey, G. Metal–Organic frameworks as efficient materials for drug delivery. *Angew. Chem. Int. Ed.* **2006**, *45*, 5974–5978. [CrossRef]
54. Soppimath, K.S.; Aminabhavi, T.M.; Kulkarni, A.R.; Rudzinski, W.E. Biodegradable polymeric nanoparticles as drug delivery devices. *J. Control. Release* **2001**, *70*, 1–20. [CrossRef]
55. Muñoz, B.; Rámila, A.; Pérez-Pariente, J.; Díaz, I.; Vallet-Regí, M. MCM-41 organic modification as drug delivery rate regulator. *Chem. Mater.* **2003**, *15*, 500–503. [CrossRef]
56. Vallet-Regi, M.; Rámila, A.; del Real, R.P.; Pérez-Pariente, J. A new property of MCM-41: Drug delivery system. *Chem. Mater.* **2001**, *13*, 308–311. [CrossRef]
57. An, J.; Farha, O.K.; Hupp, J.T.; Pohl, E.; Yeh, J.I.; Rosi, N.L. Metal-adeninate vertices for the construction of an exceptionally porous Metal–organic framework. *Nat. Commun.* **2012**, *3*, 604. [CrossRef] [PubMed]
58. Jurow, M.; Varotto, A.; Manichev, V.; Travlou, N.A.; Giannakoudakis, D.A.; Drain, C.M. Self-organized nanostructured materials of alkylated phthalocyanines and underivatized C$_{60}$ on ITO. *RSC Adv.* **2013**, *3*, 21360–21364. [CrossRef]
59. Cai, H.; Li, M.; Lin, X.R.; Chen, W.; Chen, G.H.; Huang, X.C.; Li, D. Spatial, hysteretic, and adaptive host-guest chemistry in a Metal–organic framework with open watson-crick sites. *Angew. Chem. Int. Ed.* **2015**, *54*, 10454–10459. [CrossRef]
60. Agostoni, V.; Horcajada, P.; Noiray, M.; Malanga, M.; Aykaç, A.; Jicsinszky, L.; Vargas-Berenguel, A.; Semiramoth, N.; Daoud-Mahammed, S.; Nicolas, V.; et al. A "green" strategy to construct non-covalent, stable and bioactive coatings on porous MOF nanoparticles. *Sci. Rep.* **2015**, *5*, 7925. [CrossRef]
61. Bellido, E.; Hidalgo, T.; Lozano, M.V.; Guillevic, M.; Simon-Vazquez, R.; Santander-Ortega, M.J.; Gonzalez-Fernandez, A.; Serre, C.; Alonso, M.J.; Horcajada, P. Heparin-engineered mesoporous iron Metal–organic framework nanoparticles: Toward stealth drug nanocarriers. *Adv. Healthc. Mater.* **2015**, *4*, 1246–1257. [CrossRef]

62. Liu, S.; Zhai, L.; Li, C.; Li, Y.; Guo, X.; Zhao, Y.; Wu, C. Exploring and exploiting dynamic noncovalent chemistry for effective surface modification of nanoscale Metal–organic frameworks. *ACS Appl. Mater. Interfaces* **2014**, *6*, 5404–5412. [CrossRef]
63. Azizi Vahed, T.; Naimi-Jamal, M.R.; Panahi, L. (Fe)MIL-100-Met@alginate: A hybrid polymer–MOF for enhancement of metformin's bioavailability and pH-controlled release. *New J. Chem.* **2018**, *42*, 11137–11146. [CrossRef]
64. Azizi Vahed, T.; Naimi-Jamal, M.R.; Panahi, L. Alginate-coated ZIF-8 Metal–organic framework as a green and bioactive platform for controlled drug release. *J. Drug Deliv. Sci. Technol.* **2019**, *49*, 570–576. [CrossRef]
65. Wang, X.-G.; Dong, Z.-Y.; Cheng, H.; Wan, S.-S.; Chen, W.-H.; Zou, M.-Z.; Huo, J.-W.; Deng, H.-X.; Zhang, X.-Z. A multifunctional metal–organic framework based tumor targeting drug delivery system for cancer therapy. *Nanoscale* **2015**, *7*, 16061–16070. [CrossRef] [PubMed]
66. Zhao, D.; Tan, S.; Yuan, D.; Lu, W.; Rezenom, Y.H.; Jiang, H.; Wang, L.Q.; Zhou, H.C. Surface functionalization of porous coordination nanocages via click chemistry and their application in drug delivery. *Adv. Mater.* **2011**, *23*, 90–93. [CrossRef] [PubMed]
67. Abánades Lázaro, I.; Haddad, S.; Sacca, S.; Orellana-Tavra, C.; Fairen-Jimenez, D.; Forgan, R.S. Selective surface PEGylation of UiO-66 nanoparticles for enhanced stability, cell uptake, and pH-responsive drug delivery. *Chem* **2017**, *2*, 561–578. [CrossRef] [PubMed]
68. Abánades Lázaro, I.; Haddad, S.; Rodrigo-Muñoz, J.M.; Orellana-Tavra, C.; del Pozo, V.; Fairen-Jimenez, D.; Forgan, R.S. mechanistic investigation into the selective anticancer cytotoxicity and immune system response of surface-functionalized, dichloroacetate-loaded, UiO-66 nanoparticles. *ACS Appl. Mater. Interfaces* **2018**, *10*, 5255–5268. [CrossRef]
69. Abánades Lázaro, I.; Haddad, S.; Rodrigo-Muñoz, J.M.; Marshall, R.J.; Sastre, B.; del Pozo, V.; Fairen-Jimenez, D.; Forgan, R.S. Surface-Functionalization of Zr-fumarate MOF for selective cytotoxicity and immune system compatibility in nanoscale drug delivery. *ACS Appl. Mater. Interfaces* **2018**, *10*, 31146–31157. [CrossRef]
70. Rijnaarts, T.; Mejia-Ariza, R.; Egberink, R.J.M.; van Roosmalen, W.; Huskens, J. Metal–Organic Frameworks (MOFs) as multivalent materials: Size control and surface functionalization by monovalent capping ligands. *Chem. Eur. J.* **2015**, *21*, 10296–10301. [CrossRef]
71. Mejia-Ariza, R.; Huskens, J. The effect of PEG length on the size and guest uptake of PEG-capped MIL-88A particles. *J. Mater. Chem. B* **2016**, *4*, 1108–1115. [CrossRef]
72. He, Z.; Dai, Y.; Li, X.; Guo, D.; Liu, Y.; Huang, X.; Jiang, J.; Wang, S.; Zhu, G.; Zhang, F.; et al. Hybrid nanomedicine fabricated from photosensitizer-terminated Metal–organic framework nanoparticles for photodynamic therapy and hypoxia-activated cascade chemotherapy. *Small* **2019**, *15*, e1804131. [CrossRef]
73. Zimpel, A.; Preiß, T.; Röder, R.; Engelke, H.; Ingrisch, M.; Peller, M.; Rädler, J.O.; Wagner, E.; Bein, T.; Lächelt, U.; et al. Imparting functionality to MOF nanoparticles by external surface selective covalent attachment of polymers. *Chem. Mater.* **2016**, *28*, 3318–3326. [CrossRef]
74. Giménez-Marqués, M.; Bellido, E.; Berthelot, T.; Simón-Yarza, T.; Hidalgo, T.; Simón-Vázquez, R.; González-Fernández, Á.; Avila, J.; Asensio, M.C.; Gref, R.; et al. GraftFast surface engineering to improve MOF nanoparticles furtiveness. *Small* **2018**, *14*, 1801900. [CrossRef]
75. Benzaqui, M.; Semino, R.; Carn, F.; Tavares, S.R.; Menguy, N.; Giménez-Marqués, M.; Bellido, E.; Horcajada, P.; Berthelot, T.; Kuzminova, A.I.; et al. Covalent and selective grafting of polyethylene glycol brushes at the surface of ZIF-8 for the processing of membranes for pervaporation. *ACS Sustain. Chem. Eng.* **2019**, *7*, 6629–6639. [CrossRef]
76. Cai, X.; Deng, X.; Xie, Z.; Shi, Y.; Pang, M.; Lin, J. Controllable synthesis of highly monodispersed nanoscale Fe-soc-MOF and the construction of Fe-soc-MOF@polypyrrole core-shell nanohybrids for cancer therapy. *Chem. Eng. J.* **2019**, *358*, 369–378. [CrossRef]
77. Li, Y.; Liu, J.; Zhang, K.; Lei, L.; Lei, Z. UiO-66-NH_2@PMAA: A hybrid Polymer–MOFs architecture for pectinase immobilization. *Ind. Eng. Chem. Res.* **2018**, *57*, 559–567. [CrossRef]
78. Nagata, S.; Kokado, K.; Sada, K. Metal–Organic framework tethering PNIPAM for ON–OFF controlled release in solution. *Chem. Commun.* **2015**, *51*, 8614–8617. [CrossRef] [PubMed]
79. Chen, W.-H.; Liao, W.-C.; Sohn, Y.S.; Fadeev, M.; Cecconello, A.; Nechushtai, R.; Willner, I. Stimuli-Responsive nucleic acid-based polyacrylamide hydrogel-coated metal–organic framework nanoparticles for controlled drug release. *Adv. Funct. Mater.* **2018**, *28*, 1705137. [CrossRef]

80. Xie, K.; Fu, Q.; He, Y.; Kim, J.; Goh, S.J.; Nam, E.; Qiao, G.G.; Webley, P.A. Synthesis of well dispersed polymer grafted metal–organic framework nanoparticles. *Chem. Commun.* **2015**, *51*, 15566–15569. [CrossRef]
81. Liu, H.; Zhu, H.; Zhu, S. Reversibly dispersible/collectable Metal–organic frameworks prepared by grafting thermally responsive and switchable polymers. *Macromol. Mater. Eng.* **2015**, *300*, 191–197. [CrossRef]
82. Sun, P.; Li, Z.; Wang, J.; Gao, H.; Yang, X.; Wu, S.; Liu, D.; Chen, Q. Transcellular delivery of messenger RNA payloads by a cationic supramolecular MOF platform. *Chem. Commun.* **2018**, *54*, 11304–11307. [CrossRef]
83. Dong, S.; Chen, Q.; Li, W.; Jiang, Z.; Ma, J.; Gao, H. A dendritic catiomer with an MOF motif for the construction of safe and efficient gene delivery systems. *J. Mater. Chem. B* **2017**, *5*, 8322–8329. [CrossRef]
84. Chen, S.; Chen, Q.; Dong, S.; Ma, J.; Yang, Y.-W.; Chen, L.; Gao, H. Polymer brush decorated MOF nanoparticles loaded with AIEgen, anticancer drug, and supramolecular glue for regulating and in situ observing DOX release. *Macromol. Biosci.* **2018**, *18*, 1800317. [CrossRef] [PubMed]
85. McDonald, K.A.; Feldblyum, J.I.; Koh, K.; Wong-Foy, A.G.; Matzger, A.J. Polymer@MOF@MOF: "grafting from" atom transfer radical polymerization for the synthesis of hybrid porous solids. *Chem. Commun.* **2015**, *51*, 11994–11996. [CrossRef] [PubMed]
86. Molavi, H.; Shojaei, A.; Mousavi, S.A. Improving mixed-matrix membrane performance via PMMA grafting from functionalized NH_2–UiO-66. *J. Mater. Chem. A* **2018**, *6*, 2775–2791. [CrossRef]
87. Hou, L.; Wang, L.; Zhang, N.; Xie, Z.; Dong, D. Polymer brushes on metal–organic frameworks by UV-induced photopolymerization. *Polym. Chem.* **2016**, *7*, 5828–5834. [CrossRef]
88. Rowe, M.D.; Chang, C.-C.; Thamm, D.H.; Kraft, S.L.; Harmon, J.F.; Vogt, A.P.; Sumerlin, B.S.; Boyes, S.G. Tuning the magnetic resonance imaging properties of positive contrast agent nanoparticles by surface modification with RAFT polymers. *Langmuir* **2009**, *25*, 9487–9499. [CrossRef]
89. Rowe, M.D.; Thamm, D.H.; Kraft, S.L.; Boyes, S.G. Polymer-Modified gadolinium Metal–organic framework nanoparticles used as multifunctional nanomedicines for the targeted imaging and treatment of cancer. *Biomacromolecules* **2009**, *10*, 983–993. [CrossRef]
90. Aykaç, A.; Noiray, M.; Malanga, M.; Agostoni, V.; Casas-Solvas, J.M.; Fenyvesi, É.; Gref, R.; Vargas-Berenguel, A. A non-covalent "click chemistry" strategy to efficiently coat highly porous MOF nanoparticles with a stable polymeric shell. *Biochim. Biophys. Acta (BBA) Gen. Subj.* **2017**, *1861*, 1606–1616. [CrossRef]
91. Li, Y.; Zheng, Y.; Lai, X.; Chu, Y.; Chen, Y. Biocompatible surface modification of nano-scale zeolitic imidazolate frameworks for enhanced drug delivery. *RSC Adv.* **2018**, *8*, 23623–23628. [CrossRef]
92. Hidalgo, T.; Giménez-Marqués, M.; Bellido, E.; Avila, J.; Asensio, M.C.; Salles, F.; Lozano, M.V.; Guillevic, M.; Simón-Vázquez, R.; González-Fernández, A.; et al. Chitosan-coated mesoporous MIL-100(Fe) nanoparticles as improved bio-compatible oral nanocarriers. *Sci. Rep.* **2017**, *7*, 43099. [CrossRef]
93. Filippousi, M.; Turner, S.; Leus, K.; Siafaka, P.I.; Tseligka, E.D.; Vandichel, M.; Nanaki, S.G.; Vizirianakis, I.S.; Bikiaris, D.N.; van Der Voort, P.; et al. Biocompatible Zr-based nanoscale MOFs coated with modified poly(epsilon-caprolactone) as anticancer drug carriers. *Int. J. Pharm.* **2016**, *509*, 208–218. [CrossRef]
94. He, S.; Wang, H.; Zhang, C.; Zhang, S.; Yu, Y.; Lee, Y.; Li, T. A generalizable method for the construction of MOF@polymer functional composites through surface-initiated atom transfer radical polymerization. *Chem. Sci.* **2019**, *10*, 1816–1822. [CrossRef] [PubMed]
95. Márquez, A.G.; Hidalgo, T.; Lana, H.; Cunha, D.; Blanco-Prieto, M.J.; Álvarez-Lorenzo, C.; Boissière, C.; Sánchez, C.; Serre, C.; Horcajada, P. Biocompatible polymer–metal–organic framework composite patches for cutaneous administration of cosmetic molecules. *J. Mater. Chem. B* **2016**, *4*, 7031–7040. [CrossRef]
96. Liu, J.; Yang, Y.; Zhu, W.; Yi, X.; Dong, Z.; Xu, X.; Chen, M.; Yang, K.; Lu, G.; Jiang, L.; et al. Nanoscale metal–organic frameworks for combined photodynamic & radiation therapy in cancer treatment. *Biomaterials* **2016**, *97*, 1–9. [CrossRef] [PubMed]
97. Cai, W.; Gao, H.; Chu, C.; Wang, X.; Wang, J.; Zhang, P.; Lin, G.; Li, W.; Liu, G.; Chen, X. Engineering phototheranostic nanoscale metal–organic frameworks for multimodal imaging-guided cancer therapy. *ACS Appl. Mater. Interfaces* **2017**, *9*, 2040–2051. [CrossRef] [PubMed]
98. Wang, W.; Wang, L.; Li, Y.; Liu, S.; Xie, Z.; Jing, X. Nanoscale polymer metal–organic framework hybrids for effective photothermal therapy of colon cancers. *Adv. Mater.* **2016**, *28*, 9320–9325. [CrossRef] [PubMed]
99. Wu, X.; Yang, C.; Ge, J.; Liu, Z. Polydopamine tethered enzyme/metal–organic framework composites with high stability and reusability. *Nanoscale* **2015**, *7*, 18883–18886. [CrossRef]

100. Feng, J.; Xu, Z.; Dong, P.; Yu, W.; Liu, F.; Jiang, Q.; Wang, F.; Liu, X. Stimuli-responsive multifunctional metal–organic framework nanoparticles for enhanced chemo-photothermal therapy. *J. Mater. Chem. B* **2019**, *7*, 994–1004. [CrossRef]
101. Yang, S.; Peng, L.; Sun, D.T.; Asgari, M.; Oveisi, E.; Trukhina, O.; Bulut, S.; Jamali, A.; Queen, W.L. A new post-synthetic polymerization strategy makes metal–organic frameworks more stable. *Chem. Sci.* **2019**, *10*, 4542–4549. [CrossRef]
102. Castells-Gil, J.; Novio, F.; Padial, N.M.; Tatay, S.; Ruíz-Molina, D.; Martí-Gastaldo, C. Surface functionalization of metal–organic framework crystals with catechol coatings for enhanced moisture tolerance. *ACS Appl. Mater. Interfaces* **2017**, *9*, 44641–44648. [CrossRef]
103. Chun, J.; Kang, S.; Park, N.; Park, E.J.; Jin, X.; Kim, K.-D.; Seo, H.O.; Lee, S.M.; Kim, H.J.; Kwon, W.H.; et al. Metal–Organic framework@microporous organic network: Hydrophobic adsorbents with a crystalline inner porosity. *J. Am. Chem. Soc.* **2014**, *136*, 6786–6789. [CrossRef]
104. Wang, L.; Wang, W.; Zheng, X.; Li, Z.; Xie, Z. Nanoscale fluorescent metal–organic framework@microporous organic polymer composites for enhanced intracellular uptake and bioimaging. *Chem. Eur. J.* **2017**, *23*, 1379–1385. [CrossRef] [PubMed]
105. Wang, W.; Wang, L.; Liu, S.; Xie, Z. Metal–organic frameworks@polymer composites containing cyanines for near-infrared fluorescence imaging and photothermal tumor therapy. *Bioconj. Chem.* **2017**, *28*, 2784–2793. [CrossRef] [PubMed]
106. Ostermann, R.; Cravillon, J.; Weidmann, C.; Wiebcke, M.; Smarsly, B.M. Metal–organic framework nanofibers via electrospinning. *Chem. Commun.* **2011**, *47*, 442–444. [CrossRef] [PubMed]
107. Gamage, N.-D.H.; McDonald, K.A.; Matzger, A.J. MOF-5-Polystyrene: Direct production from monomer, improved hydrolytic stability, and unique guest adsorption. *Angew. Chem.* **2016**, *128*, 12278–12282. [CrossRef]
108. Zhang, W.; Hu, Y.; Ge, J.; Jiang, H.L.; Yu, S.H. A facile and general coating approach to moisture/water-resistant Metal–organic frameworks with intact porosity. *J. Am. Chem. Soc.* **2014**, *136*, 16978–16981. [CrossRef]
109. Pastore, V.J.; Cook, T.R.; Rzayev, J. Polymer–MOF hybrid composites with high porosity and stability through surface-selective ligand exchange. *Chem. Mater.* **2018**, *30*, 8639–8649. [CrossRef]

© 2020 by the authors. Licensee MDPI, Basel, Switzerland. This article is an open access article distributed under the terms and conditions of the Creative Commons Attribution (CC BY) license (http://creativecommons.org/licenses/by/4.0/).

Review

Extraction of Metal Ions with Metal–Organic Frameworks

Natalia Manousi [1,*], Dimitrios A. Giannakoudakis [2], Erwin Rosenberg [3] and George A. Zachariadis [1,*]

1. Laboratory of Analytical Chemistry, Department of Chemistry, Aristotle University of Thessaloniki, 54124 Thessaloniki, Greece
2. Institute of Physical Chemistry, Polish Academy of Sciences, Kasprzaka 44/52, 01-224 Warsaw, Poland; DAGchem@gmail.com
3. Institute of Chemical Technology and Analytics, Vienna University of Technology, 1060 Vienna, Austria; egon.rosenberg@tuwien.ac.at
* Correspondence: nmanousi@chem.auth.gr (N.M.); zacharia@chem.auth.gr (G.A.Z.); Tel.: +30-2310-997707 (G.A.Z.)

Academic Editors: Victoria Samanidou and Eleni Deliyanni
Received: 15 November 2019; Accepted: 13 December 2019; Published: 16 December 2019

Abstract: Metal–organic frameworks (MOFs) are crystalline porous materials composed of metal ions or clusters coordinated with organic linkers. Due to their extraordinary properties such as high porosity with homogeneous and tunable in size pores/cages, as well as high thermal and chemical stability, MOFs have gained attention in diverse analytical applications. MOFs have been coupled with a wide variety of extraction techniques including solid-phase extraction (SPE), dispersive solid-phase extraction (d-SPE), and magnetic solid-phase extraction (MSPE) for the extraction and preconcentration of metal ions from complex matrices. The low concentration levels of metal ions in real samples including food samples, environmental samples, and biological samples, as well as the increased number of potentially interfering ions, make the determination of trace levels of metal ions still challenging. A wide variety of MOF materials have been employed for the extraction of metals from sample matrices prior to their determination with spectrometric techniques.

Keywords: MOFs; metals; extraction; sample preparation; microextraction; spectrometry; environmental samples; food samples; biological samples

1. Introduction

The terminology of metal–organic frameworks (MOFs) was initially introduced in 1995, when Yaghi and Li reported the synthesis of a new "zeolite-like" crystalline structure upon the polymeric coordination of Cu ions with 4,4′-bipyridine and nitrate ions, resulting to large rectangular channels [1]. MOFs are known to have superior characteristics, such as high surface area (theoretically up to 14.600 $m^2 g^{-1}$) [2], porosity of uniform in structure and topology nanoscaled cavities, and satisfactory thermal and mechanical stability. Therefore, metal–organic frameworks were established as successful candidates for various applications like environmental remediation, detoxification media of toxic vapors, heterogeneous catalysis, gas storage, imaging and drug delivery, fuel cells, supercapacitors, and sensors [2–13].

In the field of analytical chemistry, MOFs have been employed in various analytical sample preparation methods including solid-phase extraction (SPE), dispersive solid-phase extraction (d-SPE), magnetic solid-phase extraction (MSPE), stir bar sorptive extraction (SBSE), and pipette tip solid-phase extraction (PT-SPE) [14–18]. Metal–organic frameworks have been also tested as stationary phases for high-performance liquid chromatography (HPLC), capillary electrochromatography (CEC), and gas chromatography (GC) with many advantages. Moreover, with the use of chiral MOFs, separation of chiral compounds has been also reported [19–22].

Metal–organic frameworks have been synthesized and successfully applied for the preconcentration of heavy metals from environmental samples prior to their detection/analysis with a spectroscopic technique. The most common metal ions used in MOFs are Zn(II), Cu(II), Fe(III), and Zr(IV), while terephthalic acid, trimesic acid, or 2-methylimidazole have been excessively used as organic linkers [23]. Many efforts have been made in order to overcome the low water stability of MOFs toward the preparation of suitable sorbents for the extraction of metal ions [24]. Examples of MOFs are presented in Figure 1 [25]. Compared with other sorbent materials, MOFs have a significant advantage of stable and homogeneous pores of specific sizes [26].

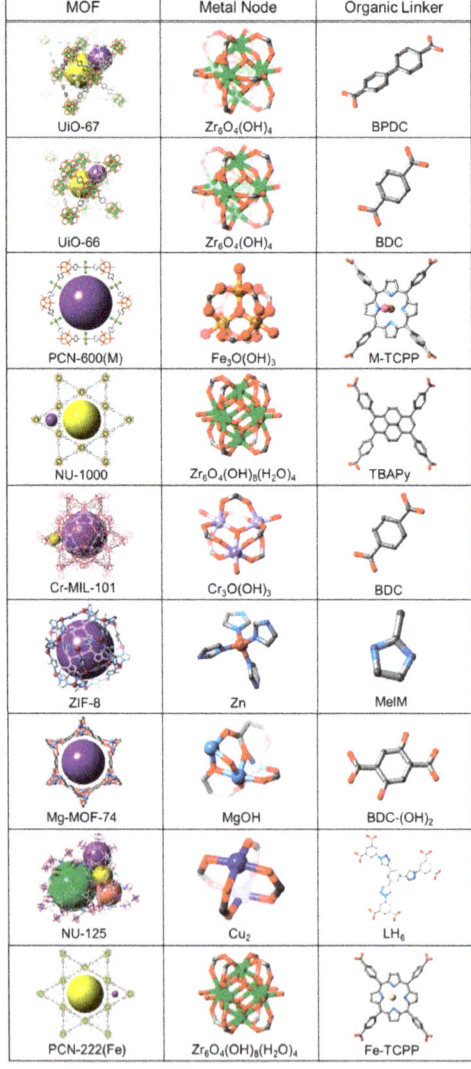

Figure 1. Examples of Metal–Organic Frameworks. Adapted with permission from Reference [25]. Copyright (2016) American Chemical Society.

The effect of trace heavy metals on human health has attracted worldwide attention. Their increasing industrial, domestic, agricultural, and technological utilization has resulted in wide distribution in the environment. Metals such as cadmium, lead, mercury, chromium, and arsenic are considered as systemic toxicants and it, therefore, is essential to determine their levels in environmental samples [27]. Among the different analytical techniques that are widely used for the determination of metal ions are flame atomic absorption spectroscopy (FAAS), electrothermal atomic absorption spectroscopy (ETAAS), inductively coupled plasma optical emission spectrometry (ICP-OES), and inductively coupled plasma mass spectrometry (ICP-MS) [28–30].

Due to the low concentrations of metals and the presence of various interfering ions in complex matrices, the direct determination of such ions at trace levels is still challenging. Various novel materials including graphene oxide, activated carbon, carbon nanotubes, porous oxides, and metal–organic frameworks have been successfully employed for this purpose [31–34].

Until now, a plethora of articles discuss the perspective of the use of MOFs in analytical chemistry [19,20,24,26,34–38]. Most of the reported review articles are focused on the extraction of organic compounds from food, biological, and environmental matrices. Herein, we aim to discuss the applications of MOFs as potential sorbents for the extraction of metal ions prior to their determination from environmental, biological, and food samples. Application of subfamilies of MOFs, such as zeolitic imidazole frameworks (ZIFs) or covalent organic frameworks (COFs), will also be discussed.

2. Stability of MOFs in Aquatic Environment

The stability of the framework in aqueous solutions depends on the strength of the metal–ligand coordination bonds [39]. The collapse of MOFs in the presence of water is linked to the competitive coordination of water and the organic linkers with the metal ions/nodes. The stability of the structure is also associated with other factors like the geometry of the coordination between metal-ligand, the surface hydrophobicity, the crystallinity, and the presence of defective sites [40]. The use of additives like graphite oxide, graphitic carbon nitride, nanoparticles, or the deposition on substrates such as carbon, fibers, or textiles, can have a positive effect on the framework stability [41–47]. In order to evaluate the stability and as a result the properness of utilizing a MOF for adsorption application, the pH and the temperature under which the preconcentration of the metal will take place, must be considered.

The strength of the coordination between the organic moieties and the metal ions can be described in general according to the HSAB (hard/soft acid/base) principles [9,47]. Zr^{4+}, Fe^{3+}, Cr^{3+}, and Al^{3+} are regarded as hard acidic metal ions, while Cu^{2+}, Zn^{2+}, Ni^{2+}, Mn^{2+}, and Ag^+ as soft ones [39]. On the other hand, carboxylate-based linkers act as hard bases, while azolate ligands (such as pyrazolates, triazolates, or imidazolates) as soft bases. For that reason, most of the Zr-based UiO (University of Oslo) and MIL-53(Fe) (Material Institut Lavoisier) series possess remarkable water stability, while for instance one of the most known and studied MOF, HKUST-1 (Hong Kong University of Science and Technology) does not. On representative paradigm of Zn-based water-stable structure is the zeolitic imidazolate framework (ZIF), formed from imidazolate ligands and Zn^{2+}.

When used in analytical chemistry, MOFs must be stable both under adsorption and under desorption conditions. Usually, adsorption of metal ions takes place under weakly acidic conditions (pH = 5–6), while desorption is performed predominately with the addition of a strong acid. However, even though many MOFs are stable under adsorption conditions, they are decomposed with the addition of strong acids like nitric, hydrochloric, and sulfuric acid [24,29]. Other reagents that have been employed for the elution of metal ions without decomposing the MOF material are ethylenediaminetetraacetic acid (EDTA), sodium chloride (NaCl), or sodium hydroxide (NaOH) solution in EDTA or in thiourea.

3. Mechanisms of Metal Ions Extraction with Metal–Organic Frameworks

MOFs, as well as their composites, have been successfully applied as adsorbents for various heavy metal/metalloid species. The adsorption of the latter from aquatic environments is still among the ultimate research targets, and there are plenty of reports in which adsorption/removal of heavy metals was a success story [48–50]. Although, not all MOFs are water-stable as discussed above. The most widely reported interactions/mechanisms are collected in Figure 2 [51]. In many cases, more than one mechanism is responsible for the high adsorptive capability of MOFs. The binding/interaction sites can be either the metal or the clusters as well as the linkers. In order to enhance the adsorptive capability and/or selectivity, the functionalization of the linkers, with groups as hydroxyl, thiol, or amide, is a well-explored and successive strategy.

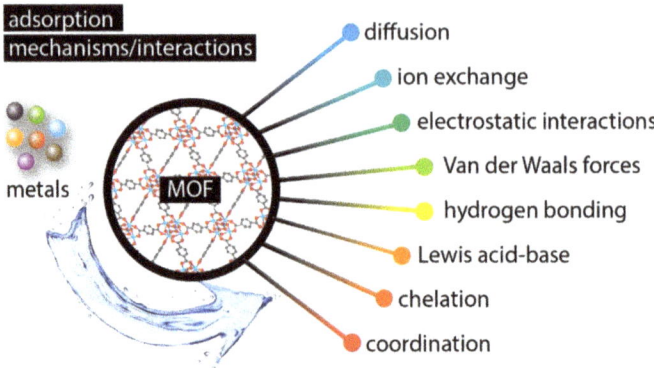

Figure 2. A schematic illustration of the interactions/mechanisms involved in the adsorption of metals by metal–organic frameworks (MOFs).

Lewis acid–base interactions are the most common adsorption mechanism of metal ions by metal–organic frameworks [52]. The presence of O-, S-, and N-containing groups that act as Lewis bases is very important for the preconcentration of the various ionic species from aqueous solution since metal ions act as Lewis acids. The donor atoms of the MOFs are present in the molecules of the organic linkers. Pre- or post-synthesis functionalization of the frameworks can increase the number of O-, S-, or N-containing groups in order to enhance the adsorption selectivity and efficiency of the target metal ions. Since Lewis acid–base interactions are critical for metal adsorption onto the donor atoms of the MOFs, it is obvious that the pH of the solution plays the most critical role, influencing the adsorption process and kinetics. In low pH value, those atoms are protonated, and adsorption cannot take place due to the repulsive forces of the cationic form of metal with the positively charged adsorption sites [53]. However, by increasing the pH of the aqueous samples that contain the metal ions, the donor atoms of the adsorbent are deprotonated and they become favorable for complex formation and sorption of the target analytes. In basic solutions, the addition of hydroxide may lead to complex formation and precipitation of many metals, therefore, after a certain pH value, any further increase can lead to a decrease of the sorption efficiency [54,55].

Adsorption by coordination is another adsorption mechanism in which the functionalization plays a key role. For instance, Liu et al. showed that the post-synthetic modification of Cr-MIL-101 with incorporation of -SH functionalities led to an improvement of Hg(II) removal, even at ultra-low concentrations [56]. This improvement was linked to the coordination between Hg(II) with the -SH groups. The incorporation of thiol-containing benzene-1,4-dicarboxylic acid (BDC) linkers in the case of UiO-66 MOF resulted in a material capable of simultaneously adsorbing As(III) and As(V) oxyanions. The adsorption of the former occurred via coordination to the -SH groups, while of the latter by the binding of the oxyanions to the $Zr_6O_4(OH)_4$ cluster via hydroxyl exchange [57]. The hydroxyl exchange mechanism was also proposed as the predominant capturing pathway in the study of Howard and co-workers [58], in which they studied the adsorption of Se(IV) and Se(VI) in water by seven Zr-based MOFs (UiO-66, UiO-66-NH_2, UiO-66-$(NH_2)_2$, UiO-66-$(OH)_2$, UiO-67, NU-1000, and NU-1000BA).

Additionally, the adsorption mechanism with metal–organic frameworks can be enhanced via the chelation mechanism, after functionalization of MOFs with compounds that can form chelating complexes with the metal ions [59]. For example, functionalization of metal–organic frameworks with dithizone can enhance Pb extraction by forming penta-heterocycle chelating complex compounds. In this case, the binding sites of the chelating molecules are also protonated in low pH values and adsorption cannot take place. Adsorption capacity increases with increasing pH until a certain point, normally at a pH value of 5 to 6. Further increase in pH value can lead to precipitation of the target analytes, due to hydrolysis [60].

In the case of the physical-based adsorption, various interactions can be responsible for the elevated adsorptive capability of MOFs as mentioned above. The net charge of the framework and the presence of specific functional groups have a positive impact on the extent of the physical interactions [61]. The manipulation of the above can be achieved by grafting of particular species/groups into the framework or by tuning the net charge as a result of the solution pH in which the adsorption takes place.

The electrostatic interactions between the negatively charged adsorption sites of MOFs with the oppositely charged adsorbates are the most widely reported pathway [62]. The diffusion of the metal ions toward the active sites prior to the blockage of the outer entrances of the channels is also an important aspect and so, the volume, geometry, and size of the pores are of paramount importance [63].

4. Sample Preparation Techniques for the Extraction of Metal Ions

Solid-phase extraction (SPE) is a well-established analytical technique that has been widely used for the extraction, preconcentration, clean-up, and class fractionation of various pollutants from environmental, biological, and food samples. Different sorbents have been evaluated for the SPE procedure usually placed into cartridges [64]. MOFs have been employed as sorbents for the solid-phase extraction. In a typical SPE application, the sorbent is conditioned to increase the effective surface area and to minimize potential interferences, prior to the loading of the sample solution onto a solid-phase [65–67]. The analytes are retained onto the active sites of the sorbent and the undesired components are washed out. Finally, elution of the analytes with the desired solvent is carried out [54].

SPE and other conventional sample preparation techniques like protein precipitation and liquid–liquid extraction (LLE) have fundamental drawbacks such as time-consuming complex steps, difficulty in automation, and need for large amounts of sample and organic solvents. Novel extraction techniques, including MSPE, d-SPE, SBSE, and PT-SPE, have been developed in order to overcome these problems. Figure 3 shows the typical steps of MSPE and d-SPE. Recently, MOFs have been used as sorbents for these extraction techniques [68].

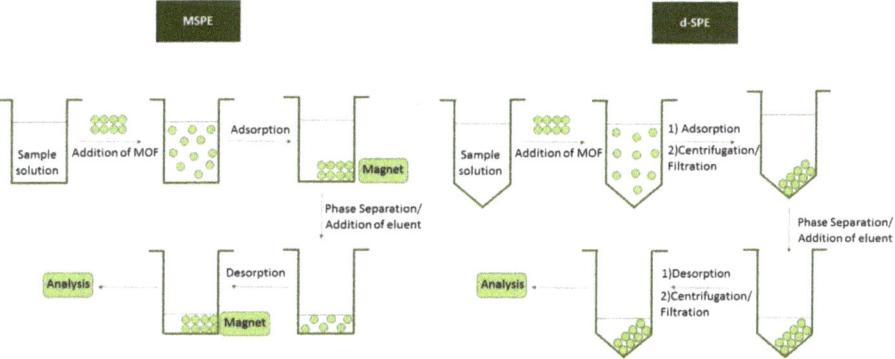

Figure 3. Typical magnetic solid-phase extraction (MSPE) and dispersive solid-phase extraction (d-SPE) procedures for the enrichment and analysis of trace metal ions.

Dispersive solid-phase extraction is performed by direct addition of the sorbent into the solution that contains the target analytes. Various MOF materials have been employed for the d-SPE of metal ions from complex sample matrices. After a certain time, the sorbent is retrieved from the solution with centrifugation or filtration and the solution is discarded. Elution with an appropriate solvent is performed and the liquid phase is isolated for instrumental analysis. The dispersion is often enhanced by stirring, vortex mixing, or ultrasound irradiation, in order to enable an efficient transfer of the target analytes to the active sites of the sorbent. Therefore, several devices including shakers, vortex mixers, and ultrasonic probes and baths have been implemented for sorbent dispersion. Until today, the ultrasound-assisted dispersive solid-phase microextraction is the most common d-SPE approach [24,69].

MSPE is based on the use of sorbents with magnetic properties. There are several different procedures to fabricate magnetic MOFs that have been employed to prepare sorbents for MSPE. The most common approaches are the direct post-synthesis of magnetic MOF materials with magnetic nanoparticles and the second one, in situ growth of magnetic nanoparticles during the synthesis of the framework. In the first case, the desired MOF and the magnetic nanoparticles (Fe_3O_4) are synthesized separately and mixed under sonication. For the in situ approach, the MOF is added to a solution containing the reagents for the synthesis of Fe_3O_4 in order to give a magnetic material. Moreover, single-step MOF coating can take place by adding the Fe_3O_4 nanoparticles into a mixture of inorganic and organic precursors for MOF synthesis. Carbonization of some MOFs can shape magnetic nanoparticles due to aggregation of the metallic component of the MOF. At the same time, the organic linker is converted to a porous carbon. Finally, the layer-by-layer approach is based on the sequential immobilization of the different components of the MOFs into a functionalized support.

For the typical MSPE procedure, a magnetic sorbent is added to the sample for sufficient time in order to ensure a quantitative extraction. After this period of time, an external magnet is employed to retrieve the sorbent and the sample is discarded. The sorbent is washed and an appropriate solvent is added in order to desorb the analytes. After magnetic separation, the eluent can be directly analyzed or it can be evaporated and reconstitute in an appropriate solvent prior to the analysis [70,71].

Other extraction techniques that can be coupled with MOFs in order to extract different analytes from complex matrices are stir bar sorptive extraction (SBSE) and pipette tip solid-phase extraction (PT-SPE). SBSE is an equilibrium technique, initially introduced by Baltussen et al. In this technique, extraction of the analytes takes place onto the surface of a coated stir bar [72–74]. PT-SPE is a miniaturized form of SPE in which ordinary pipette tips act as the extracting column and small amount of sorbent is packed inside the tip [75,76]. Only a small range of SBSE and PT-SPE sorbents are commercially available, which limits the possible applications of those techniques. MOF materials have been successfully used as coatings for stir bars and as packed sorbents in pipette tips [72–76].

Although MOFs pose several benefits as extraction sorbents for SPE, MSPE d-SPE, SBSE, and PT-SPE, their water stability and selectivity have to be enhanced with appropriate functional groups or pore functionalization. Therefore, the type of metal–organic framework and the possible functionalization should be carefully chosen. Other parameters that should be thoroughly investigated are the pH value of the sample solution, the extraction and desorption time, the desorption solvent, etc.

As mentioned before, the pH of the sample solution is one of the most critical parameters for the extraction of heavy metals from aqueous samples. Therefore, the pH value has to be optimized carefully in order to allow the Lewis acid–base interactions between the sorbent and the target analytes and to prevent precipitation due to hydrolysis.

The mass of the MOF material, as well as the extraction time, are other parameters that can influence the extraction step and require optimization. First of all, an optimum adsorbent amount is necessary in order to maximize the extraction efficiency. Certain extraction time is also required to facilitate the interaction between the analytes and adsorption sites of the MOF material. Finally, the sample volume and the volume of the eluent has to be optimized in order to provide a higher enrichment factor that is possible.

Regarding the desorption step, among the parameters that should be thoroughly investigated are the type, the volume, and the concentration of the eluent. In most cases, elution can be achieved with acidic solutions of nitric or hydrochloric acid. The presence of H^+ ions weakens the interaction between the analyte and the MOF, as it competes for binding with the active sites of the adsorbent. However, decomposition of most MOFs has been observed in acidic conditions. Other reagents that have been used for the elution of metal ions without decomposing the MOF material are EDTA, NaCl, NaOH in EDTA, NaOH in thiourea, etc. Furthermore, enough desorption time should be provided in order to enable the quantitative elution of the adsorbed analytes.

Other parameters that can be investigated are the stirring speed, salt addition, the use of ultrasonic radiation, etc., depending on the extraction procedure [74–78]. The optimization of the experimental parameters can be performed by evaluating one-factor-at-a-time or by performing Design of Experiments (DoE), such as Box–Behnken experimental design [79].

Finally, the effect of potentially interfering ions that naturally occur in the various sample matrices, the adsorption capacity of the MOF material, as well as the reusability of the sorbent should be also evaluated [74–78].

5. Applications of Metal–Organic Frameworks for the Extraction of Metal Ions

The applications of MOFs for the extraction of metal ions from environmental, biological, and food samples, as well as the obtained recoveries and limits of detection (LODs), are summarized in Table 1.

Table 1. Applications of metal–organic frameworks for the extraction of metal ions.

Analyte	Organic Linker of MOF	Metal of MOF	Modification	Matrix	Sample Preparation Technique	Detection Technique	Recovery (%)	LOD (ng mL^{-1})	Reusability	Ref.
Pd(II)	Trimesic acid	Cu	Fe$_3$O$_4$@Py	Fish, sediment, soil, water,	MSPE	FAAS	96.8–102.5	0.37	-	[80]
	Malonic acid	Ag	-	Water	SPE	FAAS	>95	0.5	Up to 5 times	[65]
	Trimesic Acid	Cu	DHz, Fe$_3$O$_4$	Water	MSPE	ETAAS	97–102	0.0046	At least 80 times	[81]
Pb(II)	Trimesic Acid	Cu	Fe$_3$O$_4$@SH	Rice, pig liver, tea, water	MSPE	FAAS	>95	0.29–0.97	-	[77]
	meso-tetra(4- carboxyphenyl) porphyrin	Zr	-	Cereal, beverage, water	d-SPE	FAAS	90–107	1.78	Up to 42 times	[82]
	Trimesic acid	Cu	Fe$_3$O$_4$@4-(5)-imidazole-dithiocarboxylic acid	Fish, canned tuna	MSPE	CVAAS	95–102	10	At least 12 times	[83]
	Trimesic acid	Cu	Thiol-modified silica	Fish, sediment, water	d-SPE	CV-AAS	91–102	0.02	-	[78]
Hg(II)	3′,5,5′-azobenzenetetracarboxylic acid	Cu	-	Tea, mushrooms	d-SPE	AFS	Average 93.3	>0.58 mg kg^{-1}	Up to 3 times	[84]
	Benzoic acid and meso-tetrakis(4-Carboxyphenyl)porphyrin	Zr	-	Fish	PT-SPE	CVAAS	74.3–98.7	20 × 10^{-3}	At least 15 times	[76]
Cu (II)	Aminoterephthalic acid	Zn	Fe$_3$O$_4$	Water	MSPE	ETAAS	98–102	0.073	-	[29]
Cd(II)	Terephthalic acid	Fe	Fe$_3$O$_4$@MAA, AMSA	Water	MSPE	FAAS	>96	0.04	Up to 10 times	[85]
	2-hydroxyterephthalic acid	Zr	-	Water	d-SPE	Spectrophotometry	>90	0.35	At least 25 times	[86]
Th(IV)	[1,1′-biphenyl]-4- carboxylic acid	Eu	-	Water	Probe	UV	N.A.	24.2	N.A.	[87]

Table 1. Cont.

Analyte	Organic Linker of MOF	Metal of MOF	Modification	Matrix	Sample Preparation Technique	Detection Technique	Recovery (%)	LOD (ng mL^{-1})	Reusability	Ref.
U(VI)	4,4′,4″-(1,3,5- triazine-2,4,6-triyltriimino)tris- benzoic acid	Te	-	Water	d-SPE	ICP-MS	94.2–98.0	0.9	At least 3 times	[88]
Se(IV), Se(VI)	Terephthalic acid	Cr	Fe$_3$O$_4$@dithiocarbamate	Water, agricultural samples	MSPE	ETAAS	>92	0.01	Up to 12 times	[89]
Cd(II), Pb(II)	Trimesic acid	Cu	Fe$_3$O$_4$@Py	Fish, sediment water	MSPE	FAAS	92.0–103.3	0.2–1.1	-	[90]
Cd(II) Pb(II) Ni(II)	Trimesic acid	Cu	Fe$_3$O$_4$@TAR	Sea food, agricultural samples	MSPE	FAAS	83–112	0.15–0.8	-	[91]
Cd(II), Pb(II), Zn(II) Cr(III)	Trimesic acid	Cu	Fe$_3$O$_4$-benzoyl isothiocyanate	Vegetables	MSPE	FAAS	80–114	0.12–0.7	-	[54]
	Terephthalic acid	Fe	Fe$_3$O$_4$-ethylenediamine	Agricultural samples	MSPE	FAAS	87.3–110	0.15–0.8	-	[92]
Cd(II), Pb(II), Ni(II), Zn(II)	Trimesic Acid	Cu	Fe$_3$O$_4$@DHz	Fish, sediment, soil, water	MSPE	FAAS	88–104	0.12–1.2	-	[60]
Pb(II), Cu(II)	Trimesic acid	Dy	-	Water	d-SPE	FAAS	95–105	0.26–0.40	At least 5 times	[55]
Cd(II), Co(II), Cr(III), Cu(II), Pb(II)	4-bpmb	Zn	-	Water	d-SPE	ICP-OES	90–110	0.01–1	-	[24]
Co(II), Cu(II), Pb(II), Cd(II), Ni(II), Cr(III), Mn(II)	4,4′-oxybisbenzoic acid	Cd	Fe$_3$O$_4$	Water	MSPE	ICP-OES	>90	0.3–1	-	[93]
Hg(II), Cr(VI) Pb(II) Cd(II)	Terephthalic acid	Cu	Dithioglycol	Tea	d-SPE	AFS, AAS	95–99	Not mentioned	Up to 3 times	[94]

131

5.1. Extraction of Palladium

In 2012, Bagheri et al. [80] synthesized a MOF material using trimesic acid and copper nitrate trihydrate. The metal–organic framework was modified with pyridine functionalized Fe_3O_4 (Fe_3O_4@Py) nanoparticles and used for the preconcentration of Pd (II) from aqueous samples prior to its determination by FAAS. Modification with pyridine was performed to increase selectivity toward palladium. Optimization of extraction and elution steps was performed with the Box–Behnken experimental design through response surface methodology [79]. The developed method was used for the analysis of fish, sediment, soil, tap water, river water, distilled water, and mineral water. Acid digestion with nitric acid (for fish samples) and nitric acid with hydrochloric acid (for soil and sediment) was carried out prior to the MSPE procedure. The researchers observed that hydrochloric acid and nitric acid decomposed the structure of the magnetic MOF sorbent; however, 0.01 mol L^{-1} NaOH in potassium sulfate provided quantitative recovery without any decomposition. The developed method showed high sample clean-up as well as satisfactory recovery values and enhancement factors [79].

5.2. Extraction of Lead

Lead(II) has been extracted from water samples with the implementation of a metal–organic framework sustained by a nanosized Ag12 cuboctahedral node [65]. The MOF material was prepared from silver nitrate, melamine, and malonic acid. The sorbent was packed in a glass column and secured with polypropylene frits. For the extraction, the sample was loaded onto the column and lead was desorbed with EDTA prior to its determination by FAAS. Due to the cage-like structure of the MOF material and the presence of melamine and malonic acid, rapid and selective adsorption of lead was achieved resulting in a SPE method with low LODs, high extraction recoveries, and good enhancement factors. No significant decrease in binding affinity was observed for the repeated use of the sorbent (up to five times).

A dithizone-functionalized magnetic metal–organic framework was synthesized by Wang et al. and applied for the magnetic solid-phase extraction of lead from environmental water samples prior to its determination by ETAAS [81]. For the synthesis of the material, a Fe_3O_4 functionalized copper benzene-1,3,5-tricarboxylate was further functionalized with dithizone (DHz). The dithizone functionalized MOF exhibited good adsorption efficiency and selectivity toward lead via chelation mechanism. Elution was performed with 2.0 mol L^{-1} HNO_3 and even though nitric acid is known to decompose many MOF materials, the prepared sorbent was found to be reusable for at least 80 times under acidic condition. Furthermore, with the use of the developed MSPE sorbent, a rapid, reliable and highly selective method for lead quantification was developed.

Lead has been also extracted from food samples with metal–organic framework adsorbent modified with mercapto groups prior to determination by FAAS [77]. The MOF material was prepared from copper nitrate trihydrate and trimesic acid and was subsequently modified with Fe_3O_4 nanoparticles functionalized with mercapto groups (Fe_3O_4@SH). Elution was performed with 1 mol L^{-1} of HNO_3, however, no sorbent reusability or data about sorbent decomposition was reported. The developed MSPE method was successfully applied for the analysis of rice, pig liver, tea, and water samples. The presence of thiol groups in combination with the high surface area of the sorbent enhanced significantly the sensitivity of the determination.

Finally, lead has been determined in cereal, beverages, and water samples, using the highly porous zirconium-based MOF-545 [82]. The novel sorbent was implemented for the vortex-assisted d-SPE of lead prior to determination by FAAS. The material was prepared from zirconyl chloride octahydrate and meso-tetra(4-carboxyphenyl) porphyrin in dimethylformamide (DMF). Prior to the extraction procedure, cereals, legumes, and juices (chickpeas, beans, wheat, lentils, and cherry juice) were dried and digested with nitric acid and hydrogen peroxide while mineral water was used without digestion. High adsorption capacity achieved as well as low LOD values. The sorbent demonstrated good stability after the elution with 1 mol L^{-1} HCl and was found to be reusable for up to 42 times.

5.3. Extraction of Mercury

Mercury has been extracted from fish samples with HKUST-1 prior to its determination of Hg(II) using cold vapor atomic absorption spectroscopy (CVAAS) [83]. The MOF material was prepared from trimesic acid and copper acetate and was subsequently modified with Fe_3O_4 nanoparticles functionalized with 4-(5)-imidazole-dithiocarboxylic acid. After elution of mercury with 0.01 mol L^{-1} thiourea solution, the sorbent was found to be reusable for up to 12 times. This novel method was used for the extraction of mercury from fish and canned tuna samples providing low LODs as well as satisfactory recovery values.

A porous metal–organic framework was prepared from thiol-modified silica nanoparticles and copper complex of trimesic acid and used for the extraction of Hg(II) from water and fish samples [78]. For this purpose, SH@SiO_2 nanoparticles were prepared from SiO_2 and (3-mercaptopropyl)-trimethoxysilane. The thiol-modified nanoparticles were mixed with trimesic acid and copper acetate monohydrate in a DMF/ethanol solution to give the desired MOF sorbent. The optimum elution solvent was found to be 0.01 mol L^{-1} NaOH since it provided satisfactory recoveries without structure decomposition. The copper benzene-1,3,5-tricarboxylate sorbent was used for the d-SPE of Hg(II) from tap, river, sea and wastewater, fish, and sediment samples prior to cold vapor atomic absorption spectrometry. The developed method was simple, selective, rapid, low-cost, environment- friendly and provided high enrichment factor. Although the preparation of MOF material was complicated, large quantity of sorbent can be prepared at once.

Mercury was also extracted from tea and mushroom samples with a JUC-62, prepared from 3,3′5,5′-azobenzenetetracarboxylic acid and copper nitrate trihydrate. [84] Tea samples were dried and digested with nitric acid prior to the d-SPE procedure. The novel sorbent was studied in both static and kinetic adsorption mode, and the static mode showed excellent adsorption capacity. Acetate buffer (0.02 M, pH 4.6) was chosen for elution and the sorbent was found to be reusable for up to 3 times. Mercury was finally measured by atomic fluorescence spectrometry (AFS).

A mesoporous porphyrinic zirconium metal–organic framework (PCN-222/MOF-545) was synthesized and used for the pipette-tip solid-phase extraction of Hg ions from fish samples prior to their determination by cold vapor atomic absorption spectrometry [76]. For the preparation of the MOF, 200 mg of zirconyl chloride octahydrate, benzoic acid, and meso-tetrakis(4-carboxyphenyl)porphyrin were used. For the extraction procedure, two milligrams of the sorbent were placed into a pipette-tip and 1.8 mL of the sample were aspirated and dispensed into a tube for 10 repeated cycles, while elution was performed with 15 µL of hydrochloric acids (10% *v/v*) at 15 cycles. The total analysis time was less than 7 min, the novel MOF material could be used for at least 15 extractions–desorption cycles without any change in its extraction efficiency and the preconcentration method provided 120-fold enhancement for mercury.

5.4. Extraction of Copper

In 2014, Wang et al. [29] synthesized a superparamagnetic Fe_3O_4-functionalized metal–organic framework from Fe_3O_4 nanoparticles zinc nitrate hexahydrate and 2-aminoterephthalic acid in DMF with the hydrothermal approach. The reaction mixture was heated to 110 °C for 24 h in a Teflon liner. The obtained IRMOF-3 material was used to determine Cu(II) ions by electrothermal atomic absorption spectrometry. Sulfuric, nitric, and hydrochloric acids were found to decompose the sorbent, therefore, 0.1 mol L^{-1} NaCl solution (pH = 2) was used for the elution of the adsorbed analytes. After optimization of the extraction procedure, the novel sorbent was successfully applied for the analysis of tap and lake water. The novel sorbent was found to be reusable for at least 10 times without any significant decrease in recovery. Due to the presence of abundant amine groups in the MOF material, high adsorption capacity and extraction efficiency toward the target analyte was achieved.

5.5. Extraction of Cadmium

Cadmium (II) ions have been preconcentrated from environmental water samples with a sulfonated MOF loaded onto iron oxide nanoparticles (Fe_3O_4@MOF235(Fe)-OSO_3H) [85]. For the synthesis of the sorbent, mercaptoacetic acid functionalized Fe_3O_4 (Fe_3O_4@MAA) nanoparticles were mixed with terephthalic acid and iron chloride hexahydrate in DMF. The reaction mixture was placed into an autoclave and heated at 85 °C for 24 h. Finally, the sulphonated MOF loaded onto the magnetic nanoparticles was prepared by the suspension of Fe_3O_4@MOF-235(Fe) in aminomethanesulfonic acid (AMSA). A solution of 0.5 mol L^{-1} EDTA was used to elute the adsorbed analyte. The obtained functionalized MOF material exhibited good stability, reusability (up to 10 times), as well as low toxicity. The novel sorbent was used for MSPE of cadmium prior to FAAS determination and enhancement factor of 195 was achieved. The Langmuir isotherm indicated that cadmium was adsorbed as the monolayer on the homogenous adsorbent surface.

5.6. Extraction of Thorium

UiO-66-OH metal–organic framework has been successfully applied for the selective d-SPE and trace determination of thorium from water samples prior to its determination by spectrophotometry [86]. The MOF material was prepared from zinc chloride and 2-hydroxyterephthalic acid in DMF after heating at 80 °C for 12 h. The developed method showed high extraction efficiency and capacity toward Th after its chelating with morin. The developed metal–organic framework exhibited low toxicity, reusability for more than 25 times (after elution with 0.2 mol L^{-1} HNO_3) as well as high stability. The d-SPE method showed high accuracy, low LODs, and high tolerance to co-existing ions.

Thorium has been also monitored in natural water with a dual-emission luminescent europium organic framework. The MOF material was synthesized from europium(III) acetate hexahydrate and [1,1'-biphenyl]-4-carboxylic acid in DMF with a solvothermal approach. After thorium uptake, the emission spectrum of the metal–organic framework was excited by UV irradiation. The LOD of the reported procedure was found to be 24.2 μg L^{-1} [87].

5.7. Extraction of Uranium

Uranium has been extracted from natural water samples with a hydrolytically stable mesoporous terbium(III)-based luminescent mesoporous MOF equipped with abundant Lewis basic sites [88]. High sensitivity and selectivity were achieved in real lake samples, where there is a huge excess of potentially interfering ions. The MOF material was prepared from terbium nitrate hexahydrate and 4,4',4''-(1,3,5-triazine-2,4,6-triyltriimino)tris-benzoic acid in DMF. The reaction mixture was at 100 °C for 72 h into a Teflon-lined reactor. Desorption of uranium was performed with nitric acid and the sorbent was found to be stable under acidic conditions since it was found to be reusable for at least 3 times. The novel sorbent was successfully used for the d-SPE of uranyl ions prior to its determination by ICP-MS. Uranium uptake by MOFs has been also studied by Zheng et al. [95]. Uptake of strontium and technetium with metal–organic frameworks has been also reported [96,97].

5.8. Extraction of Selenium

Selenium(IV) and selenium(VI) have been extracted from agricultural samples prior to their determination by electrothermal AAS with a nanocomposite consisting of MIL-101(Cr) and magnetite nanoparticles modified with dithiocarbamate [89]. The sorbent was found to be stable at acidic conditions since a solution of 0.064 mol L^{-1} HCl was chosen for elution and reusability for up to 12 times was reported. The herein developed method was successfully applied to water and agricultural samples for the determination of total selenium.

5.9. Multielement Extraction

HKUST-1 (MOF-199) material was used for the preconcentration of Cd(II) and Pb(II) ions from fish, sediment, and water samples prior to their determination by FAAS [90]. Trimesic acid and copper nitrate trihydrate were used for the synthesis of the material, which was further functionalized with Fe$_3$O$_4$@Py nanoparticles. Modification with pyridine was performed to increase selectivity toward the examined metal ions. The researchers came to the same conclusion regarding the material decomposition with hydrochloric acid and nitric acid. Therefore, elution was performed with 0.01 mol L^{-1} NaOH in EDTA solution. High adsorption capacity, low limit of detection, and high enrichment factor were achieved with the proposed MSPE sample preparation method.

HKUST-1 have been also employed for the extraction of Cd(II), Pb(II), and Ni(II) ions from seafood (fish and shrimps) and agricultural samples after modification with magnetic nanoparticles carrying covalently immobilized 4-(thiazolylazo) resorcinol (Fe$_3$O$_4$@TAR) [91]. TAR was utilized in this work as a chelator to show more selectivity toward the target analytes. Nitric acid was used for the acidic digestion of the samples and FAAS was used for the determination of the analytes. The adsorption and desorption steps were optimized with Box–Behnken experimental design [79]. Since HKUST-1 is not stable at acidic solutions, elution with EDTA was performed. The developed MSPE method was simple, selective, rapid, reproducible, and able to provide low LOD values and good extraction recoveries.

A magnetic copper benzene-1,3,5-tricarboxylate metal–organic framework functionalized with Fe$_3$O$_4$-benzoyl isothiocyanate nanoparticles was employed for the MSPE of Cd(II), Pb(II), Zn(II), and Cr(III) from vegetable samples prior to their determination by FAAS [54]. Modification with benzoyl isothiocyanate was performed to increase the selectivity toward the examined metals. Box–Behnken experimental design in combination with response surface methodology was used for the optimization of the adsorption and desorption steps [79]. The MSPE method was successfully used for the analysis of leek, parsley, fenugreek, beetroot leaves, garden cress, coriander, and basil. For the elution step, decomposition of the MOF was observed with hydrochloric acid, nitric acid, and sodium hydroxide, while EDTA and thiourea provided satisfactory recoveries without structure decomposition. Compared with Fe$_3$O$_4$-benzoyl isothiocyanate sorbent, the developed MOF material exhibited higher extraction efficiency. The novel method was simple and rapid while it provided good extraction efficiency and high enhancement factors.

The same elements have been extracted from agricultural samples with MIL-101(Fe) functionalized with Fe$_3$O$_4$-ethylenediamine prior to their determination by FAAS [92]. The presence of ethylenediamine in the sorbent enhances the selectivity of the sorbent toward the reported metals. Adsorption and desorption steps were optimized with Box–Behnken experimental design and response surface methodology [79]. MIL-101(Fe) was synthesized by iron chloride hexahydrate and terephthalic acid. Elution with EDTA was performed to avoid decomposition of the material. Leek, fenugreek, parsley, radish, radish leaves, beetroot eaves, garden cress, basil, and coriander were successfully analyzed with the developed MSPE method. Trace amounts of metal ions can be determined in a relatively high volume of samples due to the high preconcentration factor of the MSPE procedure.

A copper-(benzene-1,3,5-tricarboxylate) MOF material functionalized with dithizone-modified Fe$_3$O$_4$ nanoparticles (Fe$_3$O$_4$@DHz) and used for the preconcentration of Cd(II), Pb(II), Ni(II), and Zn(II) ions [60]. The modification with dithizone enhances the selectivity toward the examined metals. For the synthesis of the MOF material, trimeric acid in DMF/ethanol (1:1 v/v) was mixed with an ethanol solution of Fe$_3$O$_4$@DHz and copper acetate monohydrate and the mixture was heated at 70 °C under stirring for 4 h. Box–Behnken design through response surface methodology was used for the extraction optimization and FAAS was implemented for the detection of the analytes [79]. Elution of the adsorbed analytes was performed with 0.01 mol L^{-1} NaOH in thiourea to avoid any structure decomposition. The developed method provided low LODs, good recovery values, and high enhancement factors for the examined heavy metal ions.

Lanthanide Metal–Organic Frameworks have been also evaluated for their suitability as sorbents for the adsorption of heavy metal ions. In 2016, Jamali et al. [55] synthesized MOF materials using terbium hexahydrate, dysprosium nitrate hexahydrate, erbium nitrate hexahydrate, and ytterbium nitrate hexahydrate with trimesic acid in DMF. The reaction mixture was heated into a Teflon-lined reactor at 105 °C for 24 h. The novel dysprosium MOF exhibited high surface area as well as high dispersibility in aqueous solutions and it was found to be the most selective among the four examined material and it was further employed for the d-SPE of Pb(II) and Cu(II) ions from environmental water samples prior to FAAS analysis. Elution of the adsorbed analytes was performed with 0.1 mol L^{-1} HNO$_3$. The MOF material was found to be stable after the desorption process and it could be used for at least 5 times without loss of functionality. The developed method showed low LODs, good linearity, selectivity, and satisfactory recovery values.

Mechanosynthesized azine decorated zinc(II) organic frameworks have been evaluated for the extraction of Cd(II), Co(II), Cr(III), Cu(II), and Pb(II) from water samples prior to their determination by flow injection ICP-OES [24]. The TMU-4, TMU-5, and TMU-6 examined metal–organic frameworks were prepared by the mechanochemical of zinc acetate dihydrate, 4,4′-oxybisbenzoic acid and an N-donor ligand. The ligands were 1,4-bis(4-pyridyl)-2,3-diaza1,3-butadiene, 2,5-bis(4-pyridyl)-3,4-diaza-2,4-hexadiene(4-bpmb), and N1,N4-bis((pyridin-4-yl)methylene)-benzene-1,4-diamine for TMU-4, TMU-5, and TMU-6, respectively. The novel sorbents were stable in water and a wide range of pH values while they provided high adsorption capacity. A solution of 0.4 mol L^{-1} EDTA was used for the elution of the analytes. It was indicated that for trace amounts of heavy metals, the basicity of the N-donor ligands in the groups of the MOF material is critical for the adsorption efficiency, while for high concentrations of metal ions the main factor that influences the adsorption process is the void space of the MOFs.

Safari et al. [93] prepared metal–organic frameworks with and without modification with azine groups and used them for the MSPE of Co(II), Cu(II), Pb(II), Cd(II), Ni(II), Cr(III), and Mn(II). For this purpose, TMU-8 and TMU-9 metal–organic frameworks were prepared from cadmium nitrate tetrahydrate, 4,4′-oxybisbenzoic acid, and a ligand. TMU-8 contained 1,4-bis(4-pyridyl)-2,3-diaza-1,3-butadiene as a ligand, while TMU-9 contained 4,4′-bipyridine. It was found that the azine-containing TMU-8 showed better adsorption capability compared to TMU-9 that did not have azine groups. Finally, magnetic TMU-8 was prepared by the in-situ synthesis of a magnetic core-shell nanocomposite. Adsorption and desorption steps were optimized with central composite design (CCD) in combination with a Bayesian regularized artificial neural network technique. The novel sorbent was prepared from a cadmium complex compound, 4,4′-oxybisbenzoic acid, and a ligand and it was successfully applied for the analysis of environmental water samples prior to ICP-OES detection. Elution of the analytes was performed with 0.5 mol L^{-1} HNO$_3$, however, no data about material stability or sorbent reusability were provided.

Wu et al. [94] used a crystalline highly porous copper terephthalate MOF for the sample preparation of samples containing heavy metal ions after its post-synthetic modification. The MOF material was prepared from copper nitrate trihydrate and terephthalic acid in DMF with at 100 °C for 24 h. The material was dispersed dehydrated alcohol and the thiol-functionalized copper terephthalate nanoparticles were obtained with the addition of dithioglycol after stirring at room temperature for 24 h. The novel sorbent was used for the extraction of four heavy metals Hg(II), Cr(VI), Pb(II), and Cd(II) showing remarkable extraction efficiency, especially for mercury. EDTA was used to desorb the analytes. However, the addition of EDTA caused a structural collapse to the sorbent, which limited its reusability to up to three times. The d-SPE method was successfully applied for the preconcentration of the metal ions from tea samples prior to their determination with AFS (for mercury) and AAS (for chromium, lead and cadmium).

5.10. Application of ZIFs for the Extraction of Metal Ions

Zeolitic imidazolate frameworks are a subclass of metal–organic frameworks structured with Zn(II) or Co(II) ions and imidazolate and its derivatives, combining the benefits of zeolites and MOFs [98,99]. In 2016, Zou et al. [100] used a magnetic ZIF-8 material for the ultrasensitive determination of inorganic arsenic by hydride generation-atomic fluorescence spectrometry. ZIF-8 was synthesized from zinc nitrate hexahydrate and 2-methylimidazole. The obtained nanoparticles were functionalized with Fe_3O_4. The adsorption of arsenic took place in 6 h, following by dissolution of the sorbent in hydrochloric acid to assist the desorption procedure. The novel MSPE method was successfully employed for the extraction of inorganic arsenic from water and urine samples. It has been reported that unlike other metal–organic frameworks, ZIF-8 has exceptional thermal and chemical stability in water and aqueous alkaline solutions, which makes it an appropriate sorbent for sample preparation [101]. However, in the hydrochloric acid solution, the sorbent was completely dissolved, indicating low stability in acidic solution and no potential sorbent reusability [100].

5.11. Application of COFs for the Extraction of Metal Ions

Covalent organic frameworks (COFs) are structurally related materials with MOFs that consist of light elements (H, O, C, N, B, Si) connected with organic monomers through strong covalent bonds [102,103]. COFs are a novel type of ordered crystalline porous polymers that exhibit superior properties such as low crystal density, high specific surface area, tunable pore size, and very good thermal stability [104]. In 2018, Liu et al. [105] fabricated porous covalent organic frameworks and used them as a selective advanced adsorbent for the on-line preconcentration of trace elements against from complex sample matrices. For this purpose, two different COF materials were synthesized. The first COF was prepared from 1,3,5-triformylphloroglucinol and benzidine. For the preparation of the second COF, 1,3,5-triformylphloroglucinol was functionalized with diglycolic anhydride to decorate the carboxylic groups. Accordingly, the functionalized 1,3,5-triformylphloroglucinol was mixed with benzidine 1,4-dioxane and mesitylene. The two COF materials were packed into cartridges and were employed for the on-line solid-phase extraction of Cr (III), Mn (II), Co (II), Ni (II), Cd (II), V (V), Cu (II), As (III), Se (IV), and Mo (VI) prior to their determination by ICP-MS. Due to the presence of the carboxylic groups, the second COF showed effective adsorption behavior for more than 10 metal ions, while the non-functionalized COF showed effective adsorption behavior for only five metal ions. The porous COFs exhibited superior chemical and thermal stability as well as a large surface area. The developed method was successfully applied for the analysis of milk and wastewater samples.

6. Conclusions

For the use of metal–organic frameworks in the field of analytical chemistry, we conclude that they offer a further interesting possibility by enriching the analytical toolbox for trace metal analysis. One advantage of MOFs is their high surface area, which leads to high extraction efficiency and enrichment factors. Compared with other sorbent materials (including activated carbon and graphite-based materials), MOFs have also the advantage of tunable and homogeneous pores of specific sizes. However, until now there are only a few research articles regarding the extraction of metal ions with MOFs prior to their determination by a spectrometric technique.

On the other hand, a significant disadvantage of various MOFs is their instability in aqueous solution. In contrast with the environmental remediation applications of MOFs in which the researchers focus on the stability of the material only under adsorption conditions, in analytical chemistry quantitative desorption of the metal ion is essential in order to provide satisfactory recovery, no carry-over effect, and satisfactory sorbent reusability. Even though most of the studied MOFs were found to be stable under adsorption conditions at intermediate pH values, acidic desorption was found to cause their structure decomposition. In order to overcome this problem, milder eluents including EDTA, NaCl, and NaOH in EDTA or thiourea were evaluated. However, only a few sorbents were found to be reusable after the desorption step, which is considered a significant drawback for MOF sorbents.

Recent advances in the preparation of MOFs include chemical pre- or post-synthetic modification and functionalization in order to overcome their well-known limitation of water instability, which reduce their possible application to real sample analysis.

Moreover, the selectivity of MOFs toward specific metal ions is considered relatively low. This limitation can be overcome with functionalization of metal–organic frameworks with compounds like dithizone that can form chelating complex compounds with the target analytes and extract metal ions through chelation.

Until today, MOFs have been used for the extraction of metal ions by a limited number of extraction techniques. Future applications of MOFs as sorbents in other extraction formats such as SBSE or on-line techniques should be investigated. Since only a limited amount of metal ions have been extracted with MOFs, in-depth study using extraction formats such as SPE, MSPE, d-SPE, and PT-SPE also need to be performed. Finally, metal–organic frameworks have to be evaluated for the sample preparation of more sample matrices including agricultural, biological, environmental, and food samples.

Funding: The research work was supported by the Hellenic Foundation for Research and Innovation (HFRI) under the HFRI PhD Fellowship grant (Fellowship Number: 138).

Conflicts of Interest: The authors declare no conflict of interest.

References

1. Yaghi, O.; Li, H. Hydrothermal synthesis of a Metal-Organic Framework containing large rectangular channels. *J. Am. Chem. Soc.* **1995**, *117*, 10401–10402. [CrossRef]
2. Farha, O.K.; Eryazici, I.; Jeong, N.C.; Hauser, B.G.; Wilmer, C.E.; Sarjeant, A.A.; Snurr, R.Q.; Nguyen, S.T.; Yazaydin, A.Ö.; Hupp, J.T. Metal-Organic Framework materials with ultrahigh surface areas: Is the sky the limit? *J. Am. Chem. Soc.* **2012**, *134*, 15016–15021. [CrossRef] [PubMed]
3. Furukawa, H.; Cordova, K.E.; O'Keeffe, M.; Yaghi, O.M. The chemistry and applications of Metal-Organic Frameworks. *Science* **2013**, *341*, 1230444. [CrossRef] [PubMed]
4. Mueller, U.; Schubert, M.; Teich, F.; Puetter, H.; Schierle-Arndta, K.; Pastréa, J. Metal-Organic Frameworks—Prospective industrial applications. *J. Mater. Chem.* **2006**, *16*, 626–636. [CrossRef]
5. Taylor-Pashow, K.; Della Rocca, J.; Xie, Z.; Tran, S.; Lin, W. Postsynthetic modifications of iron-carboxylate nanoscale Metal–Organic Frameworks for imaging and drug delivery. *J. Am. Chem. Soc.* **2009**, *131*, 14261–14263. [CrossRef] [PubMed]
6. Corma, A.; Garcia, H.; Llabres i Xamena, F.X.L.I. Engineering Metal Organic Frameworks for heterogeneous catalysis. *Chem. Rev.* **2010**, *110*, 4606–4655. [CrossRef]
7. Getman, R.; Bae, Y.; Wilmer, C.; Snurr, R. Review and analysis of molecular simulations of methane, hydrogen, and acetylene storage in Metal–Organic Frameworks. *Chem. Rev.* **2011**, *112*, 703–723. [CrossRef]
8. Lu, K.; Aung, T.; Guo, N.; Weichselbaum, R.; Lin, W. Nanoscale Metal-Organic Frameworks for Therapeutic, Imaging, and Sensing Applications. *Adv. Mater.* **2018**, *30*, 1707634. [CrossRef]
9. Dhakshinamoorthy, A.; Asiri, A.; Garcia, H. 2D Metal–Organic Frameworks as multifunctional materials in heterogeneous catalysis and electro/photocatalysis. *Adv. Mater.* **2019**, *31*, 1900617. [CrossRef]
10. Giannakoudakis, D.A.; Bandosz, T.J. *Detoxification of Chemical Warfare Agents*, 1st ed.; Springer International Publishing: Cham, Switzerland, 2018.
11. Hashemi, B.; Zohrabi, P.; Raza, N.; Kim, K. Metal-Organic Frameworks as advanced sorbents for the extraction and determination of pollutants from environmental, biological, and food media. *TrAC Trends Anal. Chem.* **2017**, *97*, 65–82. [CrossRef]
12. Raza, W.; Kukkar, D.; Saulat, H.; Raza, N.; Azam, M.; Mehmood, A.; Kim, K. Metal-Organic Frameworks as an emerging tool for sensing various targets in aqueous and biological media. *TrAC Trends Anal. Chem.* **2019**, *120*, 115654. [CrossRef]
13. DeCoste, J.B.; Peterson, G.W. Metal-organic frameworks for air purification of toxic chemicals. *Chem. Rev.* **2014**, *114*, 5695–5727. [CrossRef] [PubMed]
14. Wang, G.; Lei, Y.; Song, H. Evaluation of $Fe_3O_4@SiO_2$–MOF-177 as an advantageous adsorbent for magnetic solid-phase extraction of phenols in environmental water samples. *Anal. Methods* **2014**, *6*, 7842–7847. [CrossRef]
15. Dai, X.; Jia, X.; Zhao, P.; Wang, T.; Wang, J.; Huang, P.; He, L.; Hou, X. A combined experimental/computational study on metal-organic framework MIL-101(Cr) as a SPE sorbent for the determination of sulphonamides in environmental water samples coupling with UPLC-MS/MS. *Talanta* **2016**, *15*, 581–588. [CrossRef] [PubMed]

16. Hu, C.; He, M.; Chen, B.; Zhong, C.; Hu, B. Polydimethylsiloxane/metal-organic frameworks coated stir bar sorptive extraction coupled to high performance liquid chromatography-ultraviolet detector for the determination of estrogens in environmental water samples. *J. Chromatogr. A* **2013**, *1310*, 21–30. [CrossRef] [PubMed]
17. Lv, Z.; Sun, Z.; Song, C.; Lu, S.; Chen, G.; You, J. Sensitive and background-free determination of thiols from wastewater samples by MOF-5 extraction coupled with high-performance liquid chromatography with fluorescence detection using a novel fluorescence probe of carbazole-9-ethyl-2-maleimide. *Talanta* **2016**, *161*, 228–237. [CrossRef]
18. Yan, Z.; Wu, M.; Hu, B.; Yao, M.; Zhang, L.; Lu, Q.; Pang, J. Electrospun UiO-66/polyacrylonitrile nanofibers as efficient sorbent for pipette tip solid-phase extraction of phytohormones in vegetable samples. *J. Chromatogr. A* **2018**, *1542*, 19–27. [CrossRef]
19. Rocío-Bautista, P.; Taima-Mancera, T.; Pasán, J.; Pino, V. Metal-Organic Frameworks in green analytical chemistry. *Separations* **2019**, *6*, 33. [CrossRef]
20. Yusuf, K.; Aqel, A.; Alothman, Z. Metal-Organic Frameworks in chromatography. *J. Chromatogr. A* **2014**, *1348*, 1–16. [CrossRef]
21. Fei, Z.; Zhang, M.; Zhang, J.; Yuan, L. Chiral metal–organic framework used as stationary phases for capillary electrochromatography. *Anal. Chim. Acta* **2014**, *830*, 49–55. [CrossRef]
22. González-Rodríguez, G.; Taima-Mancera, I.; Lago, A.; Ayala, J.; Pasán, J.; Pino, V. Mixed functionalization of organic ligands in UiO-66: A tool to design Metal–Organic Frameworks for tailored microextraction. *Molecules* **2019**, *24*, 3656. [CrossRef] [PubMed]
23. Rocío-Bautista, P.; González-Hernández, P.; Pino, V.; Pasán, J.; Afonso, A. Metal-Organic Frameworks as novel sorbents in dispersive-based microextraction approaches. *TrAC Trends Anal. Chem.* **2017**, *90*, 114–134. [CrossRef]
24. Tahmasebi, E.; Masoomi, M.; Yamini, Y.; Morsali, A. Application of mechanosynthesized azine-decorated Zinc(II) Metal–Organic Frameworks for highly efficient removal and extraction of some heavy-metal ions from aqueous samples: A comparative study. *Inorg. Chem.* **2014**, *54*, 425–433. [CrossRef] [PubMed]
25. Howarth, A.; Peters, A.; Vermeulen, N.; Wang, T.; Hupp, J.; Farha, O. Best Practices for the Synthesis, Activation, and Characterization of Metal-Organic Frameworks. *Chem. Mater.* **2016**, *29*, 26–39. [CrossRef]
26. Rocío-Bautista, P.; Pacheco-Fernández, I.; Pasán, J.; Pino, V. Are Metal-Organic Frameworks able to provide a new generation of solid-phase microextraction coatings?—A review. *Anal. Chim. Acta* **2016**, *939*, 26–41.
27. Tchounwou, P.B.; Yedjou, C.G.; Patlolla, A.K.; Sutton, D.J. Heavy metals toxicity and the environment. *EXS* **2012**, *101*, 133–164.
28. Anthemidis, A.; Kazantzi, V.; Samanidou, V.; Kabir, A.; Furton, K. An automated flow injection system for metal determination by flame atomic absorption spectrometry involving on-line fabric disk sorptive extraction technique. *Talanta* **2016**, *156*, 64–70. [CrossRef]
29. Wang, Y.; Xie, J.; Wu, Y.; Hu, X. A magnetic Metal-Organic Framework as a new sorbent for solid-phase extraction of copper(II), and its determination by electrothermal AAS. *Microchim. Acta* **2014**, *181*, 949–956. [CrossRef]
30. Samanidou, V.; Sarakatsianos, I.; Manousi, N.; Georgantelis, D.; Goula, A.; Adamopoulos, K. Detection of mechanically deboned meat in cold cuts by inductively coupled plasma-mass spectrometry. *Pak. J. Anal. Environ. Chem.* **2018**, *19*, 115–121. [CrossRef]
31. Zhang, Y.; Zhong, C.; Zhang, Q.; Chen, B.; He, M.; Hu, B. Graphene oxide–TiO_2 composite as a novel adsorbent for the preconcentration of heavy metals and rare earth elements in environmental samples followed by on-line inductively coupled plasma optical emission spectrometry detection. *RSC Adv.* **2015**, *5*, 5996–6005. [CrossRef]
32. Narin, I.; Soylak, M.; Elçi, L.; Doğan, M. Determination of trace metal ions by AAS in natural water samples after preconcentration of pyrocatechol violet complexes on an activated carbon column. *Talanta* **2000**, *52*, 1041–1046. [CrossRef]
33. Sitko, R.; Zawisza, B.; Malicka, E. Modification of carbon nanotubes for preconcentration, separation and determination of trace-metal ions. *TrAC Trends Anal. Chem.* **2012**, *37*, 22–31. [CrossRef]
34. Manousi, N.; Zachariadis, G.; Deliyanni, E.; Samanidou, V. Applications of Metal-Organic Frameworks in food sample preparation. *Molecules* **2018**, *23*, 2896. [CrossRef] [PubMed]

35. Gu, Z.; Yang, C.; Chang, N.; Yan, X. Metal–Organic Frameworks for analytical chemistry: From sample collection to chromatographic separation. *Acc. Chem. Res.* **2012**, *45*, 734–745. [CrossRef]
36. Yu, Y.; Ren, Y.; Shen, W.; Deng, H.; Gao, Z. Applications of Metal-Organic Frameworks as stationary phases in chromatography. *TrAC Trends Anal. Chem.* **2013**, *50*, 33–41. [CrossRef]
37. Gutiérrez-Serpa, A.; Pacheco-Fernández, I.; Pasán, J.; Pino, V. Metal–Organic Frameworks as key materials for solid-phase microextraction devices—A Review. *Separations* **2019**, *6*, 47. [CrossRef]
38. Wang, X.; Ye, N. Recent advances in metal-organic frameworks and covalent organic frameworks for sample preparation and chromatographic analysis. *Electrophoresis* **2017**, *38*, 3059–3078. [CrossRef]
39. Feng, M.; Zhang, P.; Zhou, H.; Sharma, V.K. Water-Stable Metal-Organic Frameworks for aqueous removal of heavy metals and radionuclides: A review. *Chemosphere* **2018**, *209*, 783–800. [CrossRef]
40. Zuluaga, S.; Fuentes-Fernandez, E.M.A.; Tan, K.; Xu, F.; Li, J.; Chabal, Y.J.; Thonhauser, T. Understanding and controlling water stability of MOF-74. *J. Mater. Chem. A.* **2016**, *4*, 5176–5183. [CrossRef]
41. Giannakoudakis, D.A.; Bandosz, T.J. Defective UiO-66 MOF Nanocomposites as Reactive Media of Superior Protection against Toxic Vapors. *ACS Appl. Mater. Interfaces* **2019**, in press. [CrossRef]
42. Giannakoudakis, D.A.; Hu, Y.; Florent, M.; Bandosz, T.J. Smart textiles of MOF/g-C_3N_4 nanospheres for the rapid detection/detoxification of chemical warfare agents. *Nanoscale Horiz.* **2017**, *2*, 356–364. [CrossRef]
43. Giannakoudakis, D.A.; Travlou, N.A.; Secor, J.; Bandosz, T.J. Oxidized g-C_3N_4 Nanospheres as Catalytically Photoactive Linkers in MOF/g-C_3N_4 Composite of Hierarchical Pore Structure. *Small* **2017**, *13*, 1601758. [CrossRef] [PubMed]
44. Xiang, W.; Zhang, Y.; Lin, H.; Liu, C.J. Nanoparticle/Metal–Organic Framework Composites for Catalytic Applications: Current Status and Perspective. *Molecules* **2017**, *2*, 2103. [CrossRef] [PubMed]
45. Ahmed, I.; Jhung, S.H. Composites of metal-organic frameworks: Preparation and application in adsorption. *Mater. Today* **2014**, *17*, 136–146. [CrossRef]
46. Petit, C.; Bandosz, T.J. Engineering the surface of a new class of adsorbents: Metal-organic framework/graphite oxide composites. *J. Coll. Interface Sci.* **2015**, *447*, 139–151. [CrossRef]
47. Alfarra, A.; Frackowiak, E.; Beguin, F. The HSAB concept as a means to interpret the adsorption of metal ions onto activated carbons. *Appl. Surf. Sci.* **2004**, *228*, 84–92. [CrossRef]
48. Howarth, A.J.; Liu, Y.; Hupp, J.; Farha, O.K. Metal–Organic Frameworks for applications in remediation of oxyanion/cation-contaminated water. *CrystEngComm* **2015**, *17*, 7245–7253. [CrossRef]
49. Li, S.; Chen, Y.; Pei, X.; Zhang, S.; Feng, X.; Zhou, J.; Wang, B. Water purification: Adsorption over Metal-Organic Frameworks. *Chin. J. Chem.* **2016**, *34*, 175–185. [CrossRef]
50. Hasan, Z.; Jhung, S.H. Removal of hazardous organics from water using Metal-Organic Frameworks (MOFs): Plausible mechanisms for selective adsorptions. *J. Hazard. Mater.* **2015**, *283*, 329–339. [CrossRef]
51. Khan, N.A.; Hasan, Z.; Jhung, S.H. Adsorptive removal of hazardous materials using Metal-Organic Frameworks (MOFs): A review. *J. Hazard. Mater.* **2013**, *244*, 444–456. [CrossRef]
52. Vu, T.A.; Le, G.H.; Dao, C.D.; Dang, L.Q.; Nguyen, K.T.; Nguyen, Q.K.; Dang, P.T.; Tran, H.T.K.; Duong, Q.T.; Nguyen, T.V.; et al. Arsenic removal from aqueous solutions by adsorption using novel MIL-53(Fe) as a highly efficient adsorbent. *RSC Adv.* **2015**, *5*, 5261–5268. [CrossRef]
53. Zhang, J.; Xiong, Z.; Li, C.; Wu, C. Exploring a thiol-functionalized MOF for elimination of lead and cadmium from aqueous solution. *J. Mol. Liq.* **2016**, *221*, 43–50. [CrossRef]
54. Hassanpour, A.; Hosseinzadeh-Khanmiri, R.; Babazadeh, M.; Abolhasani, J.; Ghorbani-Kalhor, E. Determination of heavy metal ions in vegetable samples using a magnetic Metal–Organic Framework nanocomposite sorbent. *Food Addit. Contam. Part. A* **2015**, *32*, 725–736. [CrossRef] [PubMed]
55. Jamali, A.; Tehrani, A.; Shemirani, F.; Morsali, A. Lanthanide Metal–Organic Frameworks as selective microporous materials for adsorption of heavy metal ions. *Dalton Trans.* **2016**, *45*, 9193–9200. [CrossRef]
56. Ke, F.; Qiu, L.G.; Yuan, Y.P.; Peng, F.M.; Jiang, X.; Xie, A.J.; Shen, Y.H.; Zhu, J.F. Thiol-Functionalization of Metal-Organic Framework by a facile coordination-based postsynthetic strategy and enhanced removal of Hg^{2+} from water. *J. Hazard. Mater.* **2011**, *196*, 36–43. [CrossRef]
57. Audu, C.O.; Nguyen, H.G.T.; Chang, C.; Katz, M.J.; Mao, L.; Farha, O.K.; Hupp, J.T.; Nguyen, S.T. The dual capture of AsV and AsIII by UiO-66 and analogues. *Chem. Sci.* **2016**, *7*, 6492–6498. [CrossRef]
58. Howarth, A.J.; Katz, M.J.; Wang, T.C.; Platero-Prats, A.E.; Chapman, K.W.; Hupp, J.T.; Farha, O.K. High efficiency adsorption and removal of selenate and selenite from water using Metal-Organic Frameworks. *J. Am. Chem. Soc.* **2015**, *137*, 7488–7494. [CrossRef]

59. Fang, Q.-R.; Yuan, D.-Q.; Sculley, J.; Li, J.-R.; Han, Z.-B.; Zhou, H.-C. Functional mesoporous metal-organic frameworks for the capture of heavy metal ions and size-selective catalysis. *Inorg. Chem.* **2010**, *49*, 11637–11642. [CrossRef]
60. Taghizadeh, M.; Asgharinezhad, A.; Pooladi, M.; Barzin, M.; Abbaszadeh, A.; Tadjarodi, A. A novel magnetic Metal-Organic Framework nanocomposite for extraction and preconcentration of heavy metal ions, and its optimization via experimental design methodology. *Microchim. Acta* **2013**, *180*, 1073–1084. [CrossRef]
61. Liu, B.; Jian, M.P.; Liu, R.P.; Yao, J.F.; Zhang, X.W. Highly efficient removal of arsenic(III) from aqueous solution by zeolitic imidazolate frameworks with different morphology. *Coll. Surf. A Physicochem. Eng. Asp.* **2015**, *481*, 358–366. [CrossRef]
62. Rahimi, E.; Mohaghegh, N. Removal of toxic metal ions from sungun acid rock drainage using mordenite zeolite, graphene nanosheets, and a Novel Metal–Organic Framework. *Mine Water Environ.* **2015**, *35*, 18–28. [CrossRef]
63. Jian, M.P.; Liu, B.; Zhang, G.S.; Liu, R.P.; Zhang, X.W. Adsorptive removal of arsenic from aqueous solution by zeolitic imidazolate framework-8 (ZIF-8) nanoparticles. *Coll. Surf. A Physicochem. Eng. Asp.* **2014**, *465*, 67–76. [CrossRef]
64. Andrade-Eiroa, A.; Canle, M.; Leroy-Cancellieri, V.; Cerdà, V. Solid-Phase extraction of organic compounds: A critical review (Part I). *TrAC Trends Anal. Chem.* **2016**, *80*, 641–654. [CrossRef]
65. Salarian, M.; Ghanbarpour, A.; Behbahani, M.; Bagheri, S.; Bagheri, A. A Metal-Organic Framework sustained by a nanosized Ag12 cuboctahedral node for solid-phase extraction of ultra traces of lead(II) ions. *Microchim. Acta* **2014**, *181*, 999–1007. [CrossRef]
66. Li, X.; Xing, J.; Chang, C.; Wang, X.; Bai, Y.; Yan, X.; Liu, H. Solid-Phase extraction with the Metal-Organic Framework MIL-101(Cr) combined with direct analysis in real time mass spectrometry for the fast analysis of triazine herbicides. *J. Sep. Sci* **2014**, *37*, 1489–1495. [CrossRef] [PubMed]
67. Chatzimichalakis, P.F.; Samanidou, V.F.; Verpoorte, R.; Papadoyannis, I.N. Development of a validated HPLC method for the determination of B-complex vitamins in pharmaceuticals and biological fluids after solid phase extraction. *J. Sep. Sci.* **2004**, *27*, 1181–1188. [CrossRef] [PubMed]
68. Manousi, N.; Raber, G.; Papadoyannis, I. Recent advances in microextraction techniques of antipsychotics in biological fluids prior to liquid chromatography analysis. *Separations* **2017**, *4*, 18. [CrossRef]
69. Ghorbani, M.; Aghamohammadhassan, M.; Chamsaz, M.; Akhlaghi, H.; Pedramrad, T. Dispersive solid-phase microextraction. *TrAC Trends Anal. Chem.* **2019**, *118*, 793–809. [CrossRef]
70. Giakisikli, G.; Anthemidis, A.N. Magnetic materials as sorbents for metal/metalloid preconcentration and/or separation. A review. *Anal. Chim. Acta* **2013**, *789*, 1–16. [CrossRef]
71. Maya, F.; Cabello, C.P.; Frizzarin, R.M.; Estela, J.M.; Palomino, G.T.; Cerdà, V.; Turnes, G. Magnetic solid-phase extraction using Metal-Organic Frameworks (MOFs) and their derived carbons. *TrAC Trends Anal. Chem.* **2017**, *90*, 142–152. [CrossRef]
72. Baltussen, E.; Sandra, P.; David, F.; Cramers, C. Stir bar sorptive extraction (SBSE), a novel extraction technique for aqueous samples: Theory and principles. *J. Microcolumn Sep.* **1999**, *11*, 737–747. [CrossRef]
73. Hu, C.; He, M.; Chen, B.; Zhong, C.; Hu, B. Sorptive extraction using polydimethylsiloxane/metal–organic framework coated stir bars coupled with high performance liquid chromatography-fluorescence detection for the determination of polycyclic aromatic hydrocarbons in environmental water samples. *J. Chromatogr. A* **2014**, *1356*, 45–53. [CrossRef] [PubMed]
74. Xiao, Z.; He, M.; Chen, B.; Hu, B. Polydimethylsiloxane/metal-organic frameworks coated stir bar sorptive extraction coupled to gas chromatography-flame photometric detection for the determination of organophosphorus pesticides in environmental water samples. *Talanta* **2016**, *156*, 126–133. [CrossRef] [PubMed]
75. Hashemi, S.H.; Kaykhaii, M.; Keikha, A.J.; Mirmoradzehi, E.; Sargazi, G. Application of response surface methodology for optimization of metal-organic framework based pipette-tip solid phase extraction of organic dyes from seawater and their determination with HPLC. *BMC Chem.* **2019**, *13*, 59. [CrossRef] [PubMed]
76. Rezaei Kahkha, M.; Daliran, S.; Oveisi, A.; Kaykhaii, M.; Sepehri, Z. The mesoporous porphyrinic zirconium Metal-Organic Framework for pipette-tip solid-phase extraction of mercury from fish samples followed by cold vapor atomic absorption spectrometric determination. *Food Anal. Methods* **2017**, *10*, 2175–2184. [CrossRef]

77. Wang, Y.; Chen, H.; Tang, J.; Ye, G.; Ge, H.; Hu, X. Preparation of magnetic Metal-Organic Frameworks adsorbent modified with mercapto groups for the extraction and analysis of lead in food samples by flame atomic absorption spectrometry. *Food Chem.* **2015**, *181*, 191–197. [CrossRef] [PubMed]
78. Sohrabi, M. Preconcentration of mercury(II) using a thiol-functionalized Metal-Organic Framework nanocomposite as a sorbent. *Microchim. Acta* **2013**, *181*, 435–444. [CrossRef]
79. Box, G.E.P.; Behnken, D.W. Some new three level designs for the study of quantitative variables. *Technometrics* **1960**, *2*, 455–475. [CrossRef]
80. Bagheri, A.; Taghizadeh, M.; Behbahani, M.; Akbar Asgharinezhad, A.; Salarian, M.; Dehghani, A.; Ebrahimzadeh, H.; Amini, M. Synthesis and characterization of magnetic Metal-Organic Framework (MOF) as a novel sorbent, and its optimization by experimental design methodology for determination of palladium in environmental samples. *Talanta* **2012**, *99*, 132–139. [CrossRef]
81. Wang, Y.; Xie, J.; Wu, Y.; Ge, H.; Hu, X. Preparation of a functionalized magnetic Metal–Organic Framework sorbent for the extraction of lead prior to electrothermal atomic absorption spectrometer analysis. *J. Mater. Chem. A* **2013**, *1*, 8782–8789. [CrossRef]
82. Tokalıoglu, S.; Yavuz, E.; Demir, S.; Patat, S. Zirconium-Based highly porous Metal-Organic Framework (MOF-545) as an efficient adsorbent for vortex assisted-solid-phase extraction of lead from cereal, beverage and water samples. *Food Chem.* **2017**, *237*, 707–715. [CrossRef] [PubMed]
83. Tadjarodi, A.; Abbaszadeh, A. A magnetic nanocomposite prepared from chelator-modified magnetite (Fe_3O_4) and HKUST-1 (MOF-199) for separation and preconcentration of mercury(II). *Microchim. Acta* **2016**, *183*, 1391–1399. [CrossRef]
84. Wu, Y.; Xu, G.; Wei, F.; Song, Q.; Tang, T.; Wang, X.; Hu, Q. Determination of Hg (II) in tea and mushroom samples based on Metal-Organic Frameworks as solid-phase extraction sorbents. *Microporous Mesoporous Mater.* **2016**, *235*, 204–210. [CrossRef]
85. Moradi, S.E.; Dadfarnia, S.; Emami, S.; Shabani, A.M.H. Sulfonated metal organic framework loaded on iron oxide nanoparticles as a new sorbent for the magnetic solid phase extraction of cadmium from environmental water samples. *Anal. Methods* **2016**, *8*, 6337–6346. [CrossRef]
86. Moghaddam, Z.; Kaykhaii, M.; Khajeh, M.; Oveisi, A. Synthesis of UiO-66-OH zirconium Metal-Organic Framework and its application for selective extraction and trace determination of thorium in water samples by spectrophotometry. *Spectrochim. Acta A* **2018**, *194*, 76–82. [CrossRef]
87. Liu, W.; Dai, X.; Wang, Y.; Song, L.; Zhang, L.-J.; Zhang, D.; Xie, J.; Chen, L.; Diwu, J.; Wang, J.; et al. Ratiometric monitoring of thorium contamination in natural water using a dual-emission luminescent europium organic framework. *Environ. Sci. Technol.* **2018**, *53*, 332–341. [CrossRef]
88. Liu, W.; Dai, X.; Bai, Z.; Wang, Y.; Yang, Z.; Zhang, L.; Xu, L.; Chen, L.; Li, Y.; Gui, D.; et al. Highly sensitive and selective uranium detection in natural water systems using a luminescent mesoporous Metal–Organic Framework equipped with abundant lewis basic sites: A combined batch, X-ray absorption spectroscopy, and first principles simulation investigation. *Environ. Sci. Technol* **2017**, *51*, 3911–3921.
89. Kalantari, H.; Manoochehri, M. A nanocomposite consisting of MIL-101(Cr) and functionalized magnetite nanoparticles for extraction and determination of selenium(IV) and selenium(VI). *Microchim. Acta* **2018**, *185*, 196. [CrossRef]
90. Sohrabi, M.; Matbouie, Z.; Asgharinezhad, A.; Dehghani, A. Solid-Phase extraction of Cd(II) and Pb(II) using a magnetic metal-organic framework, and their determination by FAAS. *Microchim. Acta* **2013**, *180*, 589–597. [CrossRef]
91. Ghorbani-Kalhor, E. A metal-organic framework nanocomposite made from functionalized magnetite nanoparticles and HKUST-1 (MOF-199) for preconcentration of Cd(II), Pb(II), and Ni(II). *Microchim. Acta* **2016**, *183*, 2639–2647. [CrossRef]
92. Babazadeh, M.; Hosseinzadeh-Khanmiri, R.; Abolhasani, J.; Ghorbani-Kalhor, E.; Hassanpour, A. Solid-phase extraction of heavy metal ions from agricultural samples with the aid of a novel functionalized magnetic metal–organic framework. *RSC Adv.* **2015**, *5*, 19884–19892. [CrossRef]
93. Safari, M.; Yamini, Y.; Masoomi, M.; Morsali, A.; Mani-Varnosfaderani, A. Magnetic Metal-Organic Frameworks for the extraction of trace amounts of heavy metal ions prior to their determination by ICP-AES. *Microchim. Acta* **2017**, *184*, 1555–1564. [CrossRef]

94. Wu, Y.; Xu, G.; Liu, W.; Yang, J.; Wei, F.; Li, L.; Zhang, W.; Hu, Q. Postsynthetic modification of copper terephthalate metal-organic frameworks and their new application in preparation of samples containing heavy metal ions. *Microporous Mesoporous Mater.* **2015**, *210*, 110–115. [CrossRef]
95. Zheng, T.; Yang, Z.; Gui, D.; Liu, Z.; Wang, X.; Dai, X. Overcoming the crystallization and designability issues in the ultrastable zirconium phosphonate framework system. *Nat. Commun.* **2017**, *8*, 15369. [CrossRef] [PubMed]
96. Zhang, J.; Chen, L.; Dai, X.; Zhu, L.; Xiao, C.; Xu, L.; Zhang, Z.; Alekseev, E.V.; Wang, Y.; Zhang, C.; et al. Distinctive two-step intercalation of Sr^{2+} into a coordination polymer with record high 90Sr uptake capabilities. *Chem* **2019**, *5*, 977–994. [CrossRef]
97. Sheng, D.; Zhu, L.; Xu, C.; Xiao, C.; Wang, Y.; Wang, Y.; Chen, L.; Diwu, J.; Chen, J.; Chai, Z.; et al. Efficient and selective uptake of TcO_4 by a cationic Metal-Organic Framework material with open Ag^+ sites. *Environ. Sci. Technol.* **2017**, *51*, 3471–3479. [CrossRef]
98. Chen, B.; Yang, Z.; Zhu, Y.; Xia, Y. Zeolitic imidazolate framework materials: Recent progress in synthesis and applications. *J. Mater. Chem. A* **2014**, *2*, 16811–16831. [CrossRef]
99. Tan, J.-C.; Bennett, T.D.; Cheetham, A.K. Chemical structure, network topology, and porosity effects on the mechanical properties of Zeolitic Imidazolate Frameworks. *Proc. Natl. Acad. Sci. USA* **2010**, *107*, 9938–9943. [CrossRef]
100. Zou, Z.; Wang, S.; Jia, J.; Xu, F.; Long, Z.; Hou, X. Ultrasensitive determination of inorganic arsenic by hydride generation-atomic fluorescence spectrometry using Fe_3O_4@ZIF-8 nanoparticles for preconcentration. *Microchem. J.* **2016**, *124*, 578–583. [CrossRef]
101. Ge, D.; Lee, H.K. Water stability of zeolite imidazolate framework 8 and application to porous membrane-protected micro-solid-phase extraction of polycyclic aromatic hydrocarbons from environmental water samples. *J. Chromatogr. A* **2011**, *1218*, 8490–8495. [CrossRef]
102. Ding, S.; Wang, W. Covalent Organic Frameworks (COFs): From design to applications. *Chem. Soc. Rev.* **2013**, *42*, 548–568. [CrossRef] [PubMed]
103. Feng, X.; Ding, X.; Jiang, D. Covalent organic frameworks. *Chem. Soc. Rev.* **2012**, *18*, 6010–6022. [CrossRef] [PubMed]
104. Li, N.; Du, J.; Wu, D.; Liu, J.; Li, N.; Sun, Z.; Li, G.; Wu, Y. Recent advances in facile synthesis and applications of covalent organic framework materials as superior adsorbents in sample pretreatment. *TrAC Trends Anal. Chem.* **2018**, *108*, 154–166. [CrossRef]
105. Liu, J.; Wang, X.; Zhao, C.; Hao, J.; Fang, G.; Wang, S. Fabrication of porous covalent organic frameworks as selective and advanced adsorbents for the on-line preconcentration of trace elements against the complex sample matrix. *J. Hazard. Mater.* **2018**, *344*, 220–229. [CrossRef]

© 2019 by the authors. Licensee MDPI, Basel, Switzerland. This article is an open access article distributed under the terms and conditions of the Creative Commons Attribution (CC BY) license (http://creativecommons.org/licenses/by/4.0/).

Review

Metal Organic Frameworks as Desulfurization Adsorbents of DBT and 4,6-DMDBT from Fuels

Zoi-Christina Kampouraki [1], Dimitrios A. Giannakoudakis [2,*], Vaishakh Nair [3], Ahmad Hosseini-Bandegharaei [4,5], Juan Carlos Colmenares [2] and Eleni A. Deliyanni [1,*]

1. Laboratory of Chemical and Environmental Technology, Chemistry Department, Aristotle University of Thessaloniki, GR–541 24 Thessaloniki, Greece; zoiikamp@gmail.com
2. Institute of Physical Chemistry, Polish Academy of Sciences, Kasprzaka 44/52, 01-224 Warsaw, Poland; jcarloscolmenares@ichf.edu.pl
3. Department of Chemical Engineering, National Institute of Technology Karnataka (NITK), Surathkal, Srinivasanagar P.O. Mangalore 575025, India; vaishakhchem@gmail.com
4. Department of Environmental Health Engineering, Faculty of Health, Sabzevar University of Medical Sciences, Sabzevar POB 319, Iran; ahoseinib@yahoo.com
5. Department of Engineering, Kashmar Branch, Islamic Azad University, PO Box 161, Kashmar, Iran
* Correspondence: dagchem@gmail.com (D.A.G.); lenadj@chem.auth.gr (E.A.D.)

Academic Editor: Nigel T. Lucas
Received: 8 November 2019; Accepted: 26 November 2019; Published: 10 December 2019

Abstract: Ultradeep desulfurization of fuels is a method of enormous demand due to the generation of harmful compounds during the burning of sulfur-containing fuels, which are a major source of environmental pollution. Among the various desulfurization methods in application, adsorptive desulfurization (ADS) has low energy demand and is feasible to be employed at ambient conditions without the addition of chemicals. The most crucial factor for ADS application is the selection of the adsorbent, and, currently, a new family of porous materials, metal organic frameworks (MOFs), has proved to be very effective towards this direction. In the current review, applications of MOFs and their functionalized composites for ADS are presented and discussed, as well as the main desulfurization mechanisms reported for the removal of thiophenic compounds by various frameworks. Prospective methods regarding the further improvement of MOF's desulfurization capability are also suggested.

Keywords: metal organic framework (MOF); adsorptive desulfurization of fuels; thiophenic compounds; dibenzothiophene (DBT); 4,6-dimethyldibenzothiophene (4,6-DMDBT)

---

1. Introduction

Fossil fuels are the most commonly used source of energy all around the world; however, the emission of hazardous and dangerous chemical substances during their use is an important threat to the human society as well as the environment [1]. Crude oil, gasoline, diesel, jet fuel, and furnace oil are some of the fossil-derived fuels which contain nitrogen and sulfur compounds (NCs and SCs, respectively), which, during combustion, produce hazardous oxides such as SO_x, NO_x, and CO_2. The major SCs found in these fuels, collected in Figure 1, are thiophene (TP) and its derivatives like benzothiophene (BT), 2-methylbenzothiophene (2-MBT), 5-methylbenzothiophene (5-MBT), dibenzothiophene (DBT), 4,6-dimethyldibenzothiophene (4,6-DMDBT), 3,7-dimethyldibenzothiophene (3,7-DMDBT), and 2,8-dimethyldibenzothiophene (2,8-DMDBT) [2]. In addition, some gaseous sulfur-containing moieties can be found, mainly H_2S, SO_2, and SO_3, produced after burning or degradation of thiophenic compounds [2].

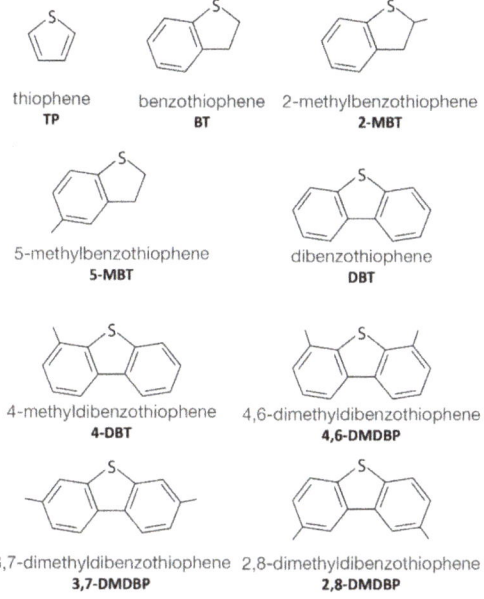

Figure 1. The most important thiophene derivatives.

Sulfur oxides, especially SO_2, which is the dominant oxide, are emitted in the environment upon combustion of S-containing fuels and can cause dangerous effects on health and the environment. The emitted SO_2 can react with rainwater or air moisture and cause acid rain that can be transferred to soils, destroy foliage, cause corrosion of historical buildings, and decrease the pH of water bodies [3]. Besides, SO_2 is known to have poisoning effects on the cars' catalysts (TWC) due to the sulfates produced by sulfur-containing fuel, which lowers the catalyst efficiency. Sulfate aerosol particles formation, at a diameter of around 2.5 µm, can also be responsible for respiratory illnesses since they are able to penetrate into the lungs [4].

In order to control and prevent SO_2 emissions, international agreements have been established from 1979 [5]. USA, Canada, and the EU have developed regulations primarily for transport fuels since they are the prime source for most of the SO_2 emission. In 1993, the Clean Air Act (CAA), the comprehensive federal law of USA that regulates air emissions from stationary and mobile sources, stated a limit of 0.5 g kg^{-1} for sulfur concentration in diesel oil, while in EU, the limits were set in 1998 at the levels of 0.35 and 0.05 g kg^{-1} for the years 2000 and 2005, respectively [6]. From 2006, new regulations in USA targeted to reduce the sulfur content of on-road diesel fuel and gasoline from 0.5 g kg^{-1} and 0.35 g kg^{-1} to 0.015 g kg^{-1} and 0.03 g kg^{-1}, respectively, targeting a maximum sulfur content limit in diesel of 0.01 g kg^{-1} by 2010. In spite of these regulations, the SO_2 emissions will continue to increase, especially due to countries such as China that still depend on coal to fulfill their high energy demands, thereby contributing to air pollution [7,8]. Hence, in order to prevent the generation of these hazardous contaminants (SCs), exploring and developing various highly efficient, economical, and environmentally friendly methods is required.

2. Desulfurization Methods

There are generally two different approaches for eliminating SO_x emissions: precombustion and postcombustion treatment methods. The precombustion treatment method is applicable in the case of flue gas treatment and reduction of the SO_x emissions and of the sulfur present in the fossil fuel [9]. However, it is not a viable method due to the use of hot and corrosive effluents, the generation of carbon dioxide (CO$_2$), and the produced refractory organic sulfur that is difficult to remove. For all these reasons, it is essential for additional methods to be developed, that can decrease the operation cost, minimize CO$_2$ emission, and be feasible for the removal of the refractory part even under extremely invasive conditions. The main methods that have been developed for the desulfurization of fuels besides hydrodesulfurization (HDS) [10–13] include oxidative desulfurization (ODS) [12–14], biodesulfurization (BDS) [15], extractive desulfurization (EDS) [16], and adsorptive desulfurization (ADS) [17]. Among them, ODS [13,14] and BDS [15], present advantages that ingrain on the fact that by their application, fuel sulfur is removed under ambient conditions based on the property of organic sulfur compounds to form oxidized products that can be extracted.

The HDS process is the most widely used industrial desulfurization method [16], in which sulfur containing compounds (SCCs) are hydrogenated to H$_2$S for the ease of separation. On the contrary, this method is not efficient for the elimination of aromatic SCCs, such as thiophenes and their derivatives, and only a minimum limit of 50 ppm of sulfur content can be removed [17–19]. During the operation, high temperature, pressure, and hydrogen are also required [20].

With the ODS process, the removable amount of SCCs can reach an ultralow level [13,14,21–23]. Since SCCs and their oxidized counterparts (i.e., sulfones and sulfoxides) are polar, they can be selectively removed after oxidation. During the ODS process, initially, SCCs with the aid of oxidizing agent are transformed to sulfones and sulfoxides and then, these oxidation products are extracted by a solvent. The drawbacks of this method are: (a) the fact that it is a multistep process, (b) the extraction part consumes energy, and (c) the use of oxidizing agents that may be corrosive or hazardous [13,21,23].

In the EDS process, the removal of SCCs is due to the higher solubility of the compounds in some solvents compared to hydrocarbons [16]. In addition, the selective removal by solvent extraction can be performed multiple times until the desired level of desulfurization is achieved. EDS can be performed at ambient conditions, resulting in lower consumption of energy. However, the use of expensive and nongreen solvents and the need for regeneration stages are the major drawbacks of this method [16].

Adsorptive Desulfurization (ADS) is an important method based on liquid-phase adsorption applied for ultralow-level desulfurization with important advantages, such as ambient operating conditions (near to ambient temperatures and atmospheric pressure) without the use of oxygen or hydrogen. Since ADS mainly depends on the adsorptive capacity of the material, the selection of the adsorbent is crucial. The main qualities for an effective adsorbent include a simple synthesis route, adsorption at ambient conditions, high porosity, regeneration capability, and low environmental footprint. The removal of SCCs from fuels using adsorbents has been successfully tried, and some of the best performing and promising materials include materials such as activated carbons (ACs) [24–33], zeolites [34–39], mesoporous silica, alumina and related materials [40–45], and ion exchange resins [46,47].

Recently, metal-organic frameworks (MOFs) have been stated to be a new category of prosperous materials that can be utilized as adsorbents for removal of SCCs. In this review, the main focus is to showcase the up-to-date research that has been carried out on the utilization of MOFs as sorbent materials in adsorptive desulfurization (ADS). Besides, the functionalization of these materials with their linkers, as well as the adsorption mechanisms that are proposed, are also discussed in order to illustrate the chemistry involved using MOFs during ADS.

3. Metal-Organic Frameworks (MOFs) as Efficient Adsorbents for Desulfurization

There has been a significant progress in the development of novel porous materials during the last few decades due to the rise in the importance of research in the field of materials science [42,43]. Among various new materials designed and synthesized during the past few years, metal-organic frameworks (MOFs) have been found to be promising candidates for a wide range of applications [48–51], due to high porosity, high surface area, and availability of active sites. In general, a MOF can be regarded as a coordination network of organic ligands and metal ion or metal clysters, containing potential voids, with one-, two-, or three-dimensional extended structures [52,53]. They consist of an inorganic center that can be either metal ions, a cluster of metal ions, or, in more advanced cases, a multinuclear complex. These inorganic centers, referred to as metal clusters/subunits or secondary building units (SBUs), are coordinated/linked each other via di- or poly-dentate chelating organic bridges/molecules, called linkers. Some typical linkers are benzenetricarboxylic acid (BTC), benzenedicar-boxylic acid (BDC) or imidazole. During MOF synthesis, the main template is the solvent, which has weak interactions with the framework, an important factor for obtaining products with neutral frameworks and accessible pores [54,55].

These hybrid inorganic-organic framework materials are known for their very high adsorption capacity from gaseous or liquid phases, with a characteristic paradigm of hydrogen adsorption/storage at moderate operation conditions. Even though the possibility to synthesize solid highly-porous materials based on coordination between metal ions and organic linkers was a well-explored research topic, in 1995, Yaghi and Li reported a hydrothermal protocol to obtain, via polymeric coordination between copper with 4,4'-bipyridine and nitrate ions, a "zeolite-like" crystalline structure [56]. Since it was a new class of hybrid materials, different names were proposed that are still in use [54,55], such as porous coordination networks [57], porous coordination polymers [58], microporous coordination polymers [59], zeolite-like MOFs [60], and isoreticular MOFs [61].

Due to the wide availability of potential metals and linkers, the number of possible structures of MOFs is virtually infinite. Some of the characteristic MOFs are collected in Figure 2. Interestingly, different frameworks can be developed since many metals of the periodic table can be involved, which can be in their singlet form or in the form of clusters, and various organic compounds can be utilized as linkers. For these reasons, MOFs can present a variety of physical and chemical properties, making them important materials. Clearly these features establish MOFs as promising materials for gas storage [62–64], separation of chemicals [54,65,66], catalysis [67], drug delivery [68], polymerization [69], magnetism [70], luminescence [71], reactive detoxification of toxic compounds [72], and especially adsorption [55,73], including sulfur containing compounds (SCCs) [74], due to their properties, i.e., large pore volume and high surface area [75,76].

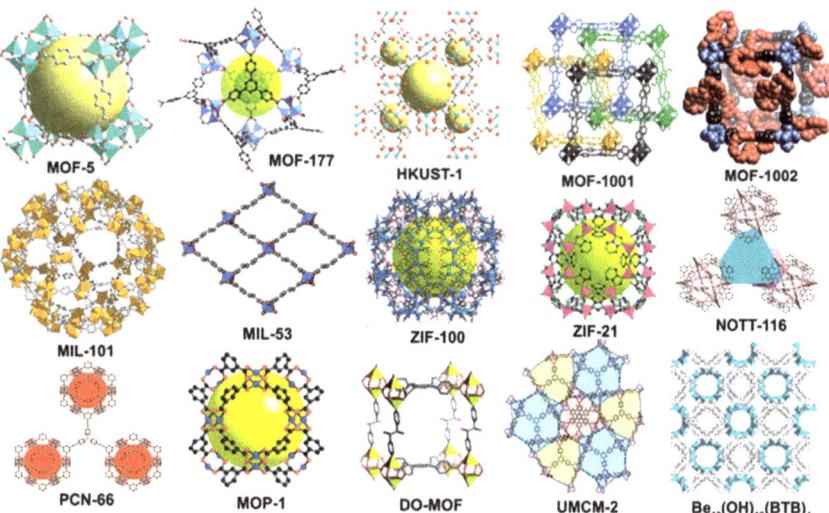

Figure 2. Metal-organic framework (MOF) structures (reproduced from [49] with permission from the Royal Society of Chemistry).

4. Desulfurization with MOFs

Fluid catalytic cracking (FCC)-obtained naphtha [77] contains sulfur content on the order of 200–7000 ppmw, while the large majority of the sulfur content of the gasoline originates from FCC naphtha. FCC naphtha contains hydrogen sulfide, thiols, disulfide, thiophene, and its alkyl derivatives, with the two latest representing 60–70 wt% of the total sulfur compounds [78,79]. The adsorption process to remove thiophene and alkylthiophenes in FCC naphtha [80,81] has to be highly selective for adsorption of thiophenic molecules versus the major components of FCC naphtha, that is, paraffins (20%–40%), naphthenes (5%–15%), olefins (20%–40%), and aromatics (20%–40%). Thiophene (TP), 3-methylthiophene (3-MT) and 2,5-dimethylthiophene (2,5-DMT), in the order of 2,5-dimethylthiophene < 3-methylthiophene < thiophene < benzothiophene, were found to be adsorbed on Cu^+-13X zeolites [82]. The potential of MOFs for being successful sulfur-selective adsorbents for thiophenic molecules from model feed was reported for HKUST-1, CPO-27-Ni, RHO-ZMOF, ZIF-8, and ZIF-76 by Perlada et al. [77]. Besides, four different MOFs consisting of two different metals (Cu^{2+} and Cr^{3+}) proved to be promising adsorbents for 3-methylthiophene (3-MT) from model oil [83]. A double adsorption mechanism by physisorption and chemisorption was proposed as the main mechanism [83]. HKUST-1 (or Cu-BTC) was also examined for thiophene and tetrahydrothiophene (THT) adsorption, achieving 78 wt% sulfur content removal from thiophene-containing model oils [84]. An even higher removal of up to 86 wt% was obtained for THT-containing model oils [84]. Three conjugated polycarbazole porous organic frameworks, named o-Cz-POF, m-Cz-POF, and p-Cz-POF, that possessed ortho, meta, and para steric configuration, were also examined for adsorption of 3-methylthiophene [85]. The highest uptake amount of 3-methylthiophene was observed in m-Cz-POF, which could reach 7.762 mmol/g (248.4 mg of S/g) at 298 K. This value is far beyond those of the porous absorbents previously reported [85]. Various MOFs have been used to selectively adsorb organo-sulfur compounds [59,86]. The first reported work highlighting the use of MOFs as adsorbents for adsorptive desulfurization (ADS) was carried out by the research group of Matzger in 2008, where the removal of BT, DBT, and DMDBT was successfully carried out using various MOFs such as HKUST-1 (also known as Cu-BTC), UMCM-150, MOF-5, MOF-505, and MOF-177 [41]. During the adsorption, MOFs interact with S-compounds present in the fuel predominantly by π–π interactions [40,41]. Moreover, they also develop metal-S coordination bonds through unsaturated coordination sites of selected metal ions such as Cu^{2+}, Zn^{2+}, Co^{2+}, Ni^{2+}, and Cu^+ [41]. Some of the other MOFs that have been reported to be highly efficient for ADS are: HKUST-1 [22,47,87], UMCM-152 [88], CuCl/MIL-47(V) [89], MIL-101(Cr) [89] MIL-100(Fe)], MOF-505 [84], PWA/HKUST-1 [90], and Cu_2O/MIL-100(Fe) [74].

In the work carried out by Matzger and coworkers [20,59,91], the maximum ADS capacities were reported to be 0.38 mmol/g (51 mg/g) for BT using MOF-5 while using UMCM-150 the adsorption capacity was 0.45 mmol/g (83 mg/g) and 0.19 mmol/g (35 mg/g) for DBT and DMDBT, respectively. In addition, the maximum adsorption capacities were reported to be higher than those presented by Na-Y zeolite [37,88]. Commonly, the high surface area and pore volume of the adsorbent is stated to be the main reason for a good adsorption. However, Matzger was able to show that the adsorption studies using MOFs showed opposite trend, indicating that the porosity of MOFs is not the key governing factor for SCCs' adsorption. MOF-177, which had the highest porosity among the materials tested in his study, showed the lowest maximum adsorption capacity, thereby implying the fact that the chemical properties of the active sites are much more important.

Similarly, adsorption studies of aromatic sulfur compounds have been investigated by various research groups with different types of MOFs for obtaining low-sulfur liquid fuels [58,77,87,92–98]. A list of ADS results in the liquid phase using different MOFs is presented for the adsorption of benzothiophene (BT) in Table 1 and the adsorption of dibenzothiophene (DBT) in Table 2. In Table 3, adsorption of BT, DBT, and 4,6-dimethyldibenzothiophene (4,6-DMDBT) using different MOFs for varying experimental conditions are shown.

Table 1. MOFs as adsorbents for benzothiophene (BT).

Adsorbent	Conditions or Remarks	Adsorption Capacity (mmol/g)	Ref.
MIL-53(Cr)		0.60	[89]
MIL-53(Al)	n-Octane solvent, 298 K	0.26	[89]
MIL-47(V)		1.6	[89]
NENU-511	i-Octane solvent, 298 K	2.2	[99]
NENU-512		1.4	[99]
NENU-513	n-octane	1.1	[99]
NENU-514		1.0	[99]
Zr(BTC)	liquid fuel	290 mg/g	[100]
ZIF-8		45	[101]
MIL-100(Fe)		114	[101]
MIL-101(Cr)		35.77%	[92]
MIL-100(Fe)	n-octane	20.76%	[92]
MOF-74(Ni)		76.97	[102]
MIL-101		36.4	[103]
UiO-66		19.83	[104]
HKUST-1		18.2	[92]

Table 2. MOFs as adsorbents for dibenzothiophene (DBT).

Adsorbent	Conditions or Remarks Solvent, Temperature (K)	Adsorption Capacity	Ref.
NENU-511		2.6 mmol/g	[99]
NENU-512		2.2 mmol/g	[99]
NENU-513	i-Octane	2.0 mmol/g	[99]
NENU-514		1.9 mmol/g	[99]
HKUST-1		7.7 mgS/g	[92]
MIL-101(Cr)		32.5 mgS/g	[92]
ZIF-8		45 mgS/g	[101]
MIL-100(Fe)		114 mgS/g	[101]
MOF-101	n-octane	52.4 mg/g	[101]
MIL-100(Fe)		35.77%	[101]
MIL-101(Cr)		20.76%	[101]
MOF-74(Ni)		85.05%	[102]
MOF-505		39.2%	[91]
MOF-199	dodecane	90%	[105]

Table 3. MOFs as adsorbents for BT, DBT, and 4,6-dimethyldibenzothiophene (4,6-DMDBT).

Adsorbent	Adsorbate (SCC)	Conditions or Remarks Solvent, Temperature (K)	Adsorption Capacity (mmol/g)	Ref.
UMCM-152		i-Octane, 298 K	1.8, 2.6	[38]
UMCM-153			2.8, 1.2	[38]
MIL-101(Cr)	DBT/DMDBT		0.20/0.17	[65]
MIL-100(Fe)		Octane, 298 K	0.20/0.25	[65]
HKUST-1			0.57/0.28	[65]
MOF-505			0.38/0.21/0.13	[91]
UMCM-150	BT/DBT/DMDBT	i-Octane, 298 K	0.30/0.45/0.19	[65]
HKUST-1			0.19/0.24/0.08	[65]

HKUST-1, which is one of the most influential frameworks presented in 1999 by Chui et al., is assumed as a benchmark MOF, especially for gaseous adsorption-oriented applications. Its secondary building unit (SBU) consists of a paddle wheel shaped metal cluster of $Cu_2(CO_2)_4$ that is formed by a dimer of copper ions, with each Cu^{2+} ion being coordinated with four benzene-1,3,5-tricarboxylic acid (BTC) groups, thereby acting as a tritopic linker. The adsorption isotherms and capacities of HKUST-1 for BT, DBT, and 4,6-DMDBT from iso-octane were studied using batch experiments at room temperature [66]. For an initial sulfur content of 1500 ppmw in the model fuel, the adsorption capacity was found to be 25 g S/kg sorbent for BT, while for DBT, the capacity reached 45 g S/kg sorbent. In the case of adsorption of 4,6-DMDBT from a sulfur content of 600 ppmw S in the model fuel, the adsorption capacity was found to be 16 g S/kg of sorbent. Similar studies for adsorption of BT, DBT, and 4,6- DMDBT using C300 Basolite MOF (HKUST-1 commercially available and produced by BASF) in iso-octane have been carried out [76] for an equilibrium time of 72 h at 304 K. For initial concentration of 1724 ppmw, the adsorption capacity was found to be 40 g S/kg for BT, 45 g S/kg for DBT and 13 g S/kg for 4,6-DMDBT. In another study using the same MOF, C300 Basolite, the adsorption capacity of BT, and DBT, in iso-octane after 24 h, for an initial concentration of 370 ppmw S, was found to be 81 g S/kg for BT and 32 g S/kg for DBT [76]. MOF-199 was also examined as an adsorbent for the removal of DBT in dodecane, as the model fuel. For an initial DBT concentration of 50 ppmw and a dosage of 5 wt% of MOF-199, the final DBT concentration in the outlet of the two-stage hydrocyclones was 8.79 ppmw with a separation efficiency as high as 99.75% within 30 s [99]. Similarly, MOF-14 was used for the removal of BT, DBT, and 4,6-DMDBT. The experimental results indicated that MOF-14 possesses high selectivity for the organosulphur compounds, a characteristic feature that was not found for other adsorbents [106].

5. Functionalization of MOFs

MOFs can be upgraded using various modification techniques such as grafting, impregnation, addition of functional groups at the linkers, or making composites materials [107–110]. Important advances were made in obtaining more complex structures having higher order of structures using nanocrystals of MOFs as building units [111]. Functional materials can also be grafted to the Lewis acid CUSs of the MOFs. As discussed above, the importance of the chemical features was proposed by the research group of Matzger, who reported that adsorption of SCCs was highly correlated with the functional groups rather than the porosity of MOFs. Metal salts, $CuCl_2$, Cu_2O, γ-Al_2O_3, heteropolyacids, different MOFs, pyrazine, NH_2 and SO_3H groups, graphite or graphite oxide, or are some of the functionalities reported in the literature, and some of them are collected and presented in Table 4.

Metal salts presenting Lewis acidity have been proven to enhance the adsorption of basic contaminants after being impregnated on a MOF surface. Optimization of the impregnation is always needed in order to overcome the decrease in porosity upon modifications, which can have impact on their adsorption capacity. A $CuCl_2$-loaded vanadium terephthalate framework (MIL-47) presented an increase of the adsorption capacity for the adsorption of benzothiophene (BT) from n-octane by 122% compared to the pristine MIL-47. The increase could be due to the π-complexation mechanism between the thiophene ring of the BT molecule and Cu(I) [74].

However, the MIL-53s (Al and Cr) loaded with $CuCl_2$ did not present improvement on the adsorption of BT [90]. Unlike V(III) of MIL-47, Al(III) and Cr(III) were not capable of the reduction of Cu(II) to Cu(I). Cu(I) species in the Cu_2O-loaded MIL-100(Fe) and MIL-101(Cr) [101] were used for the adsorption of BT from n-octane. The presence of Cu(I) species in the porous network of MIL100(Fe) decreased the porosity by 9% but showed a 16% increase in the adsorption capacity compared to initial MIL-100(Fe). The formation of π-complexes during the adsorption of SCCs contributed to the higher adsorption capacity of the metal loaded MOF than the virgin ones [101].

A bimetallic MOF (Zn/Cu-1,3,5- benzenetricarboxylate (BTC)) was examined for the adsorption of DBT by Wang et al. [105]. The bimetallic MOF presented an increase in the adsorption capacity for DBT than the virgin Cu-BTC due to the interaction of the Zn(II) π-complex with the π-electrons of DBT [105]. Hasan et al. reported the adsorption of BT and DBT adsorption from liquid fuel using a composite of two different MOFs via π-complexation [101]. In another work, the surface acidity of a MOF was enhanced by using heteropolyacids (HPAs) [101]. These strategies led to an enhance removal of basic SCCs. Huang et al. functionalized MIL-101(Cr) (chromium terephthalate) with –SO_3H groups to form AgO_3S-MIL-101(Cr), and it was further used for BT and DBT adsorption from liquid fuel [97].

Table 4. Functionalized MOFs as adsorbents for adsorptive desulfurization (ADS).

Adsorbent	Functionalizing Group	Adsorbate (SCC)	Solvent	Adsorption Capacity (mgS/g)	Ref.
MIL-53(Al)	Al			8.3	[112,113]
MIL-53(Cr)	Cr			23.6	[112,113]
IL/MIL-101(Cr)	Cr			0.65	[112,113]
MIL-101	-			36.4	[102]
Cu/MIL-101	Cu			52.0	[102]
Ce/MIL-101	Ce			45.6	[102]
Cu-Ce/MIL-101	Cu-Ce			62.1	[102]
Cu-MIL-100-Fe	Cu		n-octane	-	[101]
Cu$_2$O/MIL-100(Fe)	Cu$_2$O	BD		1.1	[101]
CuCl/MIL-47(V)	CuCl			2.3	[74]
MOF-74(Ni)@γ-Al$_2$O$_3$	γ-Al$_2$O$_3$			87.77	[102]
UiO-66-NH$_2$	–NH$_2$			-	[104]
UiO-66-COOH	–COOH			22.6	[104]
HPA/IL@ZIF-8	HPA			68	[101]
HPA/IL@MIL-100(Fe)	HPA			167	[101]
PWA/HKUST-1	PWA			1.1	[110]
HPW(1.5)/Zr(BTC)	HPW		liquid fuel	238	[100]
Al(OH)(1,4-NDC)@γ-AlOOH	γ-AlOOH			-	[112]
MIL-101(Cr)-SO$_3$H	–SO$_3$H			9.95, 2.14	[92]
MIL-101(Cr)- SO$_3$Ag	–SO$_3$Ag	BT, DBT		28.8, 31	[92]
MIL-101(Cr)-NH$_2$	–NH$_2$			2.6, 5.6	[97]
MIL-101(Cr)-NO$_2$	–NO$_2$			1.2, 2.1	[97]
Ag$^+$/MOF-101(L)	Ag$^+$			50.9	[92]
Ag$^+$/MOF-101(M)	Ag$^+$			47.8	[92]
Ag$^+$/MOF-101(H)	Ag$^+$			42.7	[92]
Cu-BTC/Gr			n-octane	46.2	[105]
CuCl/MOF-5	CuCl			3.4	[105]
MOF-74(Ni)@-γAl$_2$O$_3$	γ-Al$_2$O$_3$	DBT		76.97%	[114,115]
MOF-74(Ni)@γ-Al$_2$O$_3$	γ-Al$_2$O$_3$			93.43%	[114,115]
HPA/IL@MIL-100(Fe)	HPA			167	[101]
HPA/IL@ZIF-8	HPA			65	[101]
PTA@MIL-101(Cr)	PTA			136.5	[116,117]
PWA/MIL-101(Cr)	PWA			0.35	[65]

Pristine and functionalized UiO-66 (Zr) was also tried for removal of thiophene (TP) and benzothiophene (BT). The functionalization involved the introduction of amino (–NH$_2$) groups at the linker and introduced carboxylic (–COOH) groups at or as the defectous sites. Even though the functionalized MOFs presented decreased porosity compared to the pristine one, they showed increased adsorption capacity. The authors linked this effect to the hydrogen bond sites in their surface as well as acid–base interactions [104]. MOF-74(Ni) was impregnated on γ-Al$_2$O$_3$ beads for the synthesis of MOF-74(Ni)@γ-Al$_2$O$_3$ composite [114,115], which showed excellent DBT and BT adsorption. This enhanced adsorption was attributed to strong metal-S bonding between the adsorbent and SCCs [102]. Similarly, using an in situ green synthesis method, a composite, Al(OH)(1,4-NDC)@γ-AlOOH, was prepared from 1,4-H$_2$NDC (1,4-naphthalene dicarboxylic acid) and porous γ-Al$_2$O$_3$ beads and was tried for the adsorption of SCCs. The composite presented maximum adsorption capacities with the following order: benzothiophene > dibenzothiophene > 4,6-dimethyldibenzothiophene > thiophene. The main adsorption mechanism was due to the presence of Lewis acid sites on the metal (Al) [112].

Composites of metal organic frameworks (HKUST-1) with graphite oxide (GO) were also reported to be efficient adsorbents. With a minimal content of GO (~1.75%), the composite MOF (GO/HKUST-1) showed sufficient desulfurization results for TP (adsorption capacity 60.67 mg S/g) that were attributed to the improved porosity [118]. With the green solvothermal method, a MOF (Cu-BTC) and a MOF/Graphene (Gr) hybrid nanocomposite were also prepared and used as adsorbents for DBT removal. The experimental results showed that MOF/Gr (9:1 wt ratio) presented a high dibenzothiophene adsorption capacity for DBT, 46.2 mg S/g, compared to the unmodified MOF sample that presented an adsorption capacity of 35 mg S/g [119].

Metal organic frameworks decorated on fabric composites, (MOF)@fabric, such as MIL-53(Al)-NH$_2$ in-situ prepared within fabrics (cotton or/and wool), have been used for thiophene adsorption from n-heptane [120]. The Q_{max} followed the order of MIL-53(Al)-NH$_2$ (739.0 mg/g) > MIL-53(Al)-NH$_2$@fabric (469.4–516.5 mg/g) >>> fabric (83.1–153.8 mg/g) [120].

6. Mechanisms of Desulfurization

Adsorptive desulfurization has been attributed to different mechanisms/interactions. The main adsorption mechanisms are collected in Figure 3, include the acid–base interactions (Lewis acid–base), coordination bond formation, π–π complexation, Van der Waals force, and H-bonding [113,121,122].

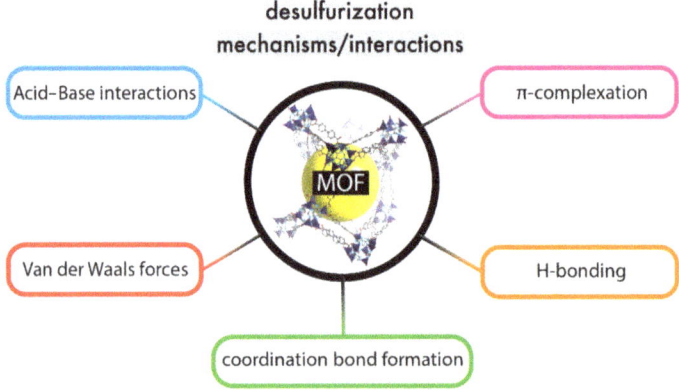

Figure 3. The predominant desulfurization interactions/mechanisms of MOFs.

6.1. Effect of Porosity

The porosity of the adsorbents plays a crucial role in adsorption because it influences their adsorption capability. Moreover, during adsorptive desulfurization, there is the opportunity of selective separation of molecules that have different molecular size. This is true when the molecular size of the thiophenic derivative is smaller than the MOF pores. During adsorption, molecules can diffuse into the porous channels and become anchored at the active adsorption sites [92]. On the contrary, if the molecular size is similar or smaller than the pore sizes of the MOF, steric hindrances forbid the penetration inside the framework and adsorption cannot take place [122].

6.2. Acid–Base Interactions

Acid–base interactions are the most common mechanism involved in ADS. Many MOFs can act as Lewis acids due to coordinatively unsaturated metal sites (CUSs) that are able to accept a pair of electrons by forming coordination bonds with molecules having a lone pair of electrons (Lewis acid sites) [123]. Hence, the adsorption of the majority of SCCs can be attributed to interactions with the Lewis acidic metal ion sites by coordination.

Thiophenic compounds, due to their solitary electrons, can be regarded as bases and thus can be easily adsorbed onto MOFs' CUSs via acid–base interactions. Bases can be classified into polarizable and nonpolarizable, and after the Pearson's hard and soft acid–base characterization, these types are denoted as "soft" and "hard" bases, respectively. Acids can also be classified as hard or soft based on

6.3. Coordination Bond Formation (Lewis Acid–Base Interaction)

Several MOFs, such as HKUST-1, MOF-74 (Ni, Mn, Co, etc.), MIL-100(Cr, Fe), and MIL-101(Cr) [61, 76,91,102,124,125] etc., have proved to be promising adsorbents due to the fact that their CUSs are surrounded by regular pore channels that can be used to induce region-selective interactions. This is not possible with adsorbents like zeolites, activated carbons, or mesoporous silica [126].

Adsorption of thiophenic compounds via hydrogen bonding has also been reported; however, this kind of bonding is not common in adsorption of SCCs. Voorde et al. studied the adsorption of heterocyclic SCCs by MIL-53(Fe) and reported that the adsorbates have the capability to form hydrogen bonds (as acceptor for hydrogen bonds) with MIL-53(Fe) [127].

6.4. π-Complexation

Some metal ions, such as Cu^{2+}, Ag^+, Pd^{2+}, and Pt^{2+}, have shown adsorption ability for SCCs through π-complex formation [89]. The complexes formed via electronic interaction between some metal cations and π-electron clouds of the chemicals, known as π-complexes [89]. Metals with empty s-orbitals can be π-complexed with the sulfur of the thiophenic compound, thereby creating a σ-bond [48]. Adsorbates with high π-electron densities (i.e., polyaromatic hydrocarbons) are greatly favorable for π-complexation. Adsorptive removal of SCCs by π-complexation was first reported in 2003 by Yang et al., in which adsorption studies where carried out by metal modified Y zeolites [19] and were described as "back-donation effects". A weak interaction can also be created between electron-rich and electron-poor aromatic groups, leading to the aromatic compounds being adsorbed via π–π stacking. This is a widely occurring mechanism in aromatic systems but with limited selectivity, especially for ADS.

Based on computational and experimental results, Wu et al. studied the nature (sites, configuration, and energies) of the adsorption for thiophenic compounds over HKUST-1 [92]. The results derived from DFT calculation revealed three possible adsorption sites; via coordinative unsaturated copper sites from the cluster (M-site), via oxygen from the coordinated to copper carboxylic group (O-site), and via the phenyl part of the linker (L-site). The adsorption energy and configuration upon adsorption of DBT at the above-mentioned three sites are demonstrated in Figure 4. Adsorption at the L-sites is not feasible, because of the presence of the neighboring clusters and the bulky in size DBT, which is bigger than the linker and the space in between the clusters (8.0 Å between neighboring M-sites). In order to overcome these steric hindrances, a possible strategy is to increase the size of the linker, resulting in an increment of the distance and space between the metallic cluster, and based on this, the adsorption via the L-sites it will be feasible. More details of this are discussed herein after.

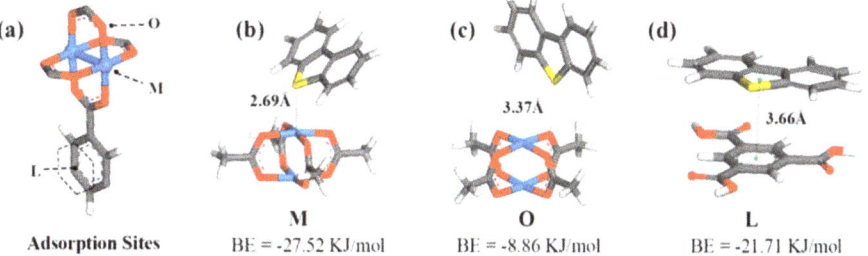

Figure 4. (a) Three adsorption sites (M-, O-, L-) on Cu-BTC. Adsorption configuration and BEs of DBT on (b) M-, (c) O-, and (d) L-sites of Cu-BTC. adapted with permission from [92]. Copyright (2014) American Chemical Society.

The authors concluded that adsorption can take place only via the coordinatively unsaturated metal sites (CUS) or M-sites, as presented in Figure 4b, which is consistent to that reported previously for adsorption of H_2O and CO_2. The CSUs can interact with either the lone electron pair of sulfur atom (σ-M interaction) or the conjugated π systems of the two rings of DBT (π-M interaction). For 4,6-DMDBT, the presence of the alkyl groups increases the adsorbate's electron density, which increases the π–M interaction. On the contrary, the alkyl groups introduce steric hindrance, which has a negative impact on the σ–M interaction. The comparison of the results obtained for the adsorption of DBT and 4,6-DMDBT from the DFT calculations can be seen in Figure 5.

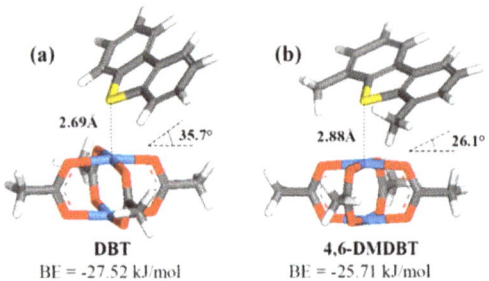

Figure 5. Adsorption configuration and energies of DBT and 4,6-DMDBT adsorption on Cu coordinatively unsaturated metal sites (CUS), adapted with permission from [92]. Copyright (2014) American Chemical Society.

Functionalities also play an important role in the π-complexation mechanism; π-electron-rich compounds with no functionalities have no adsorption ability since the adsorbent–adsorbate interactions (coordination, acid–base, and H-bonding) are difficult to occur. Among functional metals, Cu(I) functional sites, when π-complexed into MOF materials, increased their adsorption capacity. For example, $CuCl_2$-loaded MIL-47 presented a higher adsorptive performance in the adsorption of benzothiophene (BT) from n-octane than the pristine MOF [20], due to a π-complex between Cu(I) sites and porous MIL-47, which resulted in a BT adsorption capacity of 122%. Cu(I) sites, when functionalized on MIL-101(Cr), MIL-100(Fe), and CuBTC, presented higher adsorbed amounts of SCCs than the virgin ones [91]. Cu_2O-loaded MIL-100(Fe) introduced Cu(I) into the network of MIL100(Fe) and caused a 16% increase of the maximum adsorption capacity (Q_0) compared to the initial MIL-100(Fe), although the porosity was decreased by 9% [76]. Dai et al., examined MOF-5-based π-complexing adsorbents with different concentrations of Cu(I) for the adsorption of DBT from n-octane at dynamic mode [89]. With the increase of Cu(I) content, the breakthrough and saturation sulfur capacity of the adsorbents increased from 2.11 and 5.05 wt% for MOF-5, respectively, to 5.89 and 8.59 wt% for 2 mmol of Cu(I) into 1 g of MOF-5 and to 9.42 and 10.94 wt% for 3 mmol of Cu(I) into 1 g of MOF-5.

6.5. Van der Waals Forces

Van der Waals interactions are generally very weak interactions in molecules and play an important role in adsorption, since they are only applicable at low temperatures. Besides, if no special chemical interactions are created between compounds, they are usually adsorbed through van der Waals forces. For MOFs, due to their high porosity, when other mechanisms are unfavored, adsorption can be achieved mainly by van der Waals forces [113]. Other interactions, such as electrostatic interactions, that have been frequently applied to explain contaminant removals during water purification have not been reported in ADS. This might be due to the low possibility for cationic or anionic SCCs.

7. Drawbacks

Apart from all the advantages, MOFs also have some drawbacks, which have to be considered during their application in ADS. MOFs are not very stable at high temperatures and for this reason they are not appropriate for applications at elevated temperatures. Additionally, some frameworks such as HKUST-1, are known to be sensitive to humidity, while others, such as UiO-66, are stable. Detailed analysis of the stability of MOF structure after synthesis and utilization is crucial for maintaining higher adsorption and selectivity of SCCs during adsorptive desulfurization. Synthesized pristine MOF materials are usually in fine powder form with poor mechanical strength, making it difficult for recycling and reuse in actual operation. Another important drawback is particle aggregation, which may take place during the process of adsorption and recycling treatment, leading to decreased accessibility of the reactive sites, and consequently diminishing the adsorbent activity. Finally, the small pore apertures of many MOFs cause high diffusion resistance due to steric and dynamic hindrance, thus restricting movement of S-containing molecules into the pores and thereby resulting in adsorption on the surface rather inside the framework. This may result in low utilization of unsaturated metal sites, surface area, and, ultimately, the adsorption capacity. In catalysis, the problem of fast deactivation of the catalyst due to incomplete desorption of reaction products is well understood. Similarly, regeneration of "spent" sorbent is of importance, since its efficiency determines the usable lifetime of the sorbent, operating costs, and practical aspects of the scale-up of the desulfurization process.

Although there are drawbacks, the utilization of MOFs as adsorbents is of great importance due to the fact that adsorption is performed at mild conditions, and, at these conditions, MOFs can be successfully used for adsorptive removal due to their large porosity and great functionalization properties.

8. Perspectives

Expect the above-mentioned drawbacks that should be explored and be overwhelmed, a great challenge is to enhance the adsorptive desulfurization capability from fuels by MOFs. In this direction, the most crucial aspect is the enhancement of the availability and density of the coordinatively unsaturated metal sites (CUS). Increase of the porosity and, more importantly, the aperture and size of the pores will lead to beneficial effects. A followed strategy in order to enhance the surface area and pores volume is the use of longer organic linkers in order to expand the structure, while the underlying topology (structure/net of the framework) remains the same [48,61,119,128–132] A typical example is the exchange of the bicarboxylate linker BDC^{2-} with $TPDC^{2-}$ (terphenyl-4,4''-dicarboxylate), as can be seen in Figure 6a. The longer linker leads to a bigger unit cell edge and cages/pores. These two MOFs, as well as all of this family, are called isoreticular (IR) and have the same primitive cubic net shape. MOF-5, which is the parent framework of this isoreticular family, is abbreviated as IRMOF-1, while the one with $TPDC^{2-}$ is abbreviated as IRMOF-16. A smaller isoreticular to MOF-5 is $Zn_4O(fumarate)_3$ (furamate: $^-OOCCH=CHCOO^-$), reported in 2009 by Xue et al. [61]. The latter has half the unit cell edge and the volume of the cage is decreased by eight-fold compared to IRMOF-16, while the surface area is almost half of that of MOF-5 (1120 m^2/g BET surface area) [131,132]. The theoretically geometrically calculated (by Monte Carlo integration approach) surface area of IRMOF-16 was reported as 6074 m^2 g^{-1} [130]. The experimentally calculated BET surface area for this MOF was reported to be between 472 and 1912 m^2 g^{-1}, depending on the solvent evacuation/activation process. This can be linked also to the instability of the structure upon exposure to humidity (collapse of interparticle voids), and, for that reason, experimental research regarding this series of isoreticular MOFs is limited [130]. HKUST-1 or MOF-199 is the smallest member of its isoreticular family, while the largest member is $Cu_3(BBC)_2$ or MOF-399, as shown in Figure 6b. The latter has a 17.4-fold larger cell volume than HKUST-1. It is worth mentioning that MOF-399 has the lowest density (0.13 g cm^{-3}) and greatest void fraction (94%) reported of any MOF to date [48,131]. These values are 0.88 g cm^{-3} and 72%, respectively, for HKUST-1.

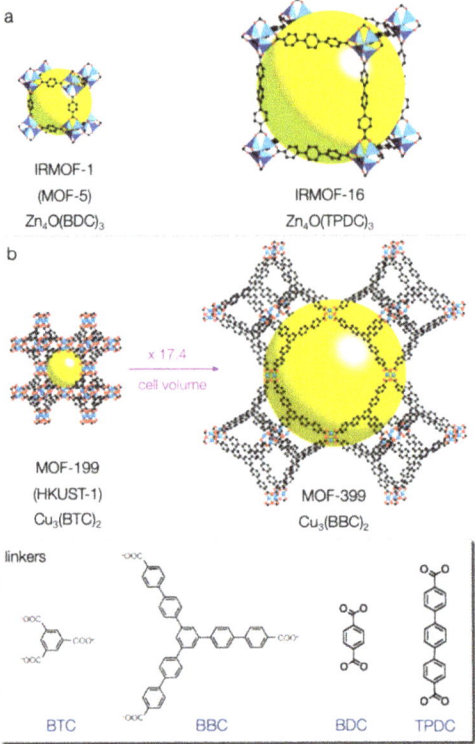

Figure 6. A scaled comparison of the single crystal structure of (**a**) MOF-5 and (**b**) HKUST-1 and the largest representatives of their isoreticular family (the yellow spheres represent the maximum volume of the biggest cavity of each structure (**a**: adapted with permission from [61]. Copyright (2002) The American Association for the Advancement of Science, **b**: adapted with permission from [48]. Copyright (2011) American Chemical Society).

The expansion of the linker with alkyne rather than only phenylene units led to an even higher increase of surface area and total pore volume. The characteristic example is Cu_3(BHEHPI) or NU-110 (NU stands for Northwestern University in Chicago, USA), which has the highest reported surface area and total pore volume up to date [132]. This copper-based MOF with a hexacarboxylate macromolecule as a ligand (BHEHPI^{6-} stands for 5,5′,5″-((((benzene-1,3,5-triyltris(benzene-4,1-diyl)) tris(ethyne-2,1-diyl))-tris(benzene-4,1-diyl)) tris(ethyne-2,1-diyl)) triisophthalate) was reported by O. Farha, J. Hupp and coworkers in 2012 and is shown in Figure 7 [132]. Using the N_2 sorption experiments, the ligand-modified Cu-MOF revealed a BET surface area of 7140 m^2g^{-1} and a total pore volume of 4.4 cm^3g^{-1}. Interestingly, the obtained nitrogen isotherm was closer to type IV rather than type I, revealing multiple sizes of pores, a fact which is consistent with the different types of cages illustrated in Figure 7. The authors showed also that the MOF's theoretical surface area can reach up to 14,600 $m^2 g^{-1}$ [132].

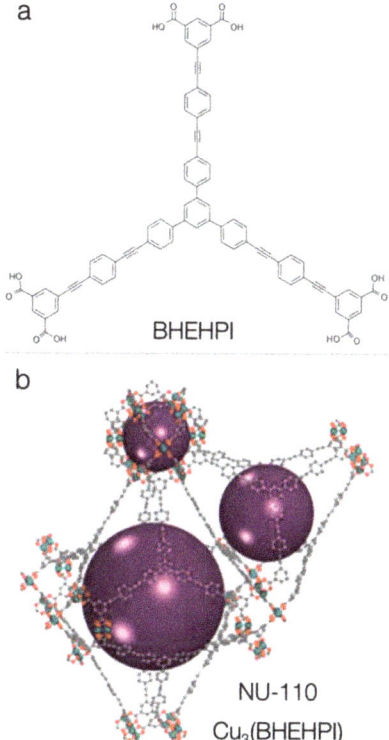

Figure 7. (a) The chemical structure of the ligand and (b) the different cages of the NU-110 framework. Adapted with permission from [132]. Copyright (2012) American Chemical Society.

Another strategy for the upgrading of MOF desulfurization performance is the formation of nanocomposites with metal-free fillers such as graphite (GR), graphite oxide (GO), graphitic carbon nitride (g-C_3N_4), or oxidized graphitic carbon nitride nanospheres (g-CNOx) [133–139]. Petit, Bandosz, and coworkers showed that the addition of a limited amount of GO (5 wt% of the final composite's mass) during the synthesis of HKUST-1 led to an enhancement of the hydrogen sulfide adsorption capacity by more than two-fold compared to pure HKUST-1, reaching a very high capacity of almost 200 mg/g [134–137]. The composite formation led to a wide range of positive aspects, such as increment of the porosity and surface chemistry heterogeneity, improved dispersion, density, and availability of the active adsorption sites, redox reactivity, and more [133,134]. The interactions were linked to the interactions of sulfur with the copper of the cluster, as was also reported in the case of copper hydroxide/oxide [137], while the addition of GO increases their availability through the formed defects effect. Ahmned et al. synthesized a highly porous MOF composite, consisting of Cr-benzenedicaboxyate and graphite oxide [115]. The addition of the GO resulted in an increase of the porosity, which had a positive effect on the removal of nitrogen- and sulfur-containing compounds from model fuel. Chen et al. reported that HKUST-1 composite with GO (1.75 wt%) has 61% increased adsorption capacity against thiophene (0.72 mmol/g) compared to virgin HKUST-1 [139].

Various reports have shown that the incorporation of g-C_3N_4 inside the MOF matrix leads to elevation of the removal and reactivity capabilities of the composites compared to the pristine MOF [140–142]. Going a step further, Bandosz and co-workers synthesized a HKUST-1-based nanocomposite with nanospheres of oxidized graphitic carbon nitride (gCNox) [143–145]. The latter were incorporated inside the matrix of the framework and were dispersed on the outer surface of each particle. Due to the enhanced chemistry heterogeneity, the gCNox acted as linkers, predominately via the carboxylic groups. The formed nanocomposites revealed dramatic alterations of the optical, structural, textural, and chemical features, while formation of mesoporosity was also determined. The proposed illustration of the composite structure can be seen in Figure 8. This MOF-based composite showed significantly higher adsorptive and catalytic reactivity compared to pristine MOF, which was linked to the enhancement of the uncoordinated active copper sites' availability and the formation of defectous sites in the framework. UiO-66 based composite with gCNox was also reported [145]. The growth of the framework in this case was around these nanospheres, with the final composite materials possessing higher catalytic activity compared to pristine UiO-66.

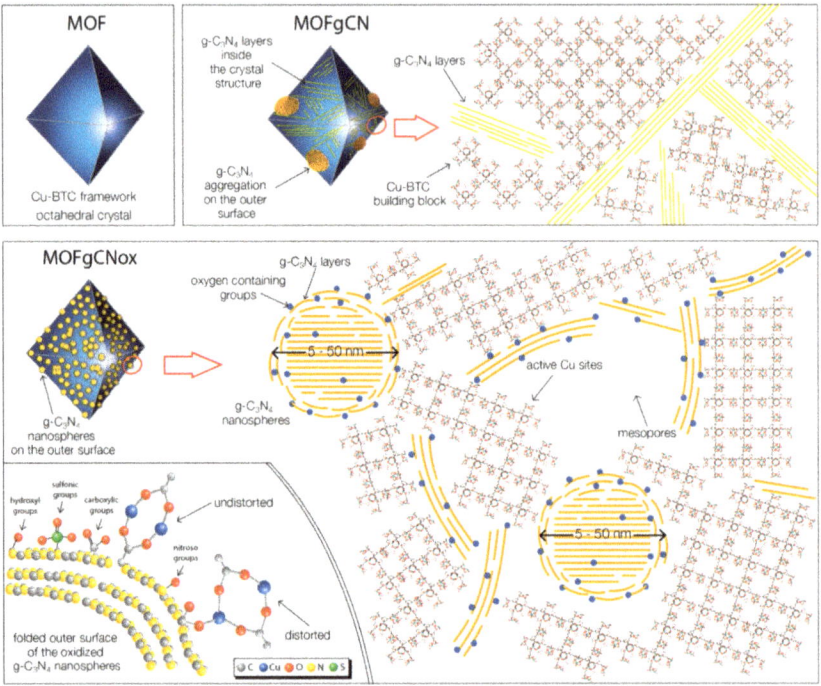

Figure 8. A schematic illustration of the HKUST-1-based nanocomposite with nanospheres of oxidized graphitic carbon nitride as filler. Reproduced from [143] with permission from the Wiley.

9. Conclusions

Adsorptive desulfurization (ADS) presents many advantages for application in fuel purification/desulfurization. The main advantages are that the ADS can be performed at ambient conditions of pressure and temperature, as well as the non-requirement of hazardous additives/chemicals, such as hydrogen. Since thiophene derivatives are difficult to remove by hydrodesulfurization (HDS), ADS present a promising alternative procedure for ultradeep desulfurization. Among the different adsorbents, MOFs have been demonstrated as a promising new class of materials for deep desulfurization applications due to high ADS capabilities. The current results reveal that the selectivity and adsorption capacity are further enhanced after functionalization of the MOF surface. Sulfur compounds can diffuse into the MOF's channels and can be adsorbed into the MOF's pore system via π- complexation, acid–base interactions, etc. The promising application of MOFs for adsorption-based desulfurization is expected to increase their use in industry. For the elimination of the drawbacks and for elevating MOF performance, further research of isoreticular MOFs with larger linkers, and of composite formation should be performed.

Author Contributions: Conceptualization and supervision was by E.A.D.; writing—original draft preparation and writing—review and editing, all the authors equally; graphical abstract by D.A.G.

Funding: This research received no external funding. APC was sponsored by MDPI.

Acknowledgments: D.A.Giannakoudakis and J.C.Colmenares are very grateful for the support from the National Science Centre in Poland within OPUS-13 project nr 2017/25/B/ST8/01592 (http://photo-catalysis.org).

Conflicts of Interest: The authors declare no conflict of interest.

References

1. Colvile, R.N.; Hutchinson, E.J.; Mindell, J.S.; Warren, R.F. The transport sector as a source of air pollution. *Atmos. Env.* **2001**, *35*, 1537–1565. [CrossRef]
2. Brandy, R.F.; Benjamin, L.B.; Jonathan, L.M.; Robert, L.H.; Mohammad, N.S.; Tawfik, A.S.T. Enhanced oxidative desulfurization in a film-shear reactor. *Fuel* **2015**, *156*, 142–147.
3. Air quality guidelines: Global update 2005: Particulate matter, ozone, nitrogen dioxide, and sulfur dioxide. Available online: https://apps.who.int/iris/handle/10665/69477 (accessed on 25 November 2019).
4. Subramani, V. Hydrogen and syngas production and purification technologies. John Wiley & Sons Inc: Hoboken, NJ, USA, 2010.
5. Samokhvalov, A.; Tatarchuk, B.J. Review of experimental characterization of active sites and determination of molecular mechanisms of adsorption, desorption and regeneration of the deep and ultra deep desulfurization sorbents for liquid fuels. *Catal. Rev. Sci. Eng.* **2010**, *52*, 381–410. [CrossRef]
6. Stanislaus, A.; Marafi, A.; Rana, M.S. Recent advances in the science and technology of ultra low sulfur diesel (ULSD) production. *Catal. Today* **2010**, *153*, 1–68. [CrossRef]
7. Pearson, R.G. Hard and soft acids and bases. *J. Am. Chem. Soc.* **1963**, *85*, 3533–3539. [CrossRef]
8. Global, BP Statistical Review of World Energy. Available online: http://www.usaee.org/usaee2013/submissions/presentations/SR%202013%20US%20events.pdf (accessed on 25 November 2019).
9. Brieva, G.B.; Campos-Martin, J.M.; Al-zahrani, S.; Fierro, J.L.G. Removal of refractory organic sulfur compounds in fossil fuels using mof sorbents. *Glob. Nest J.* **2010**, *12*, 296–304.
10. González-García, O.; Cedeño-Caero, L. V-Mo based catalysts for oxidative desulfurization of diesel fuel. *Catal. Today* **2009**, *148*, 42–48. [CrossRef]
11. Kulkarni, P.S.; Afonso, C.A.M. Deep desulfurization of diesel fuel using ionic liquids: Current status and future challenges. *Green Chem.* **2010**, *12*, 1139–1149. [CrossRef]
12. Capel-Sanchez, M.C.; Perez-Presas, P.; Campos-Martin, J.M.; Fierro, J.L.G. Highly efficient deep desulfurization of fuels by chemical oxidation. *Catal. Today* **2010**, *157*, 390–396. [CrossRef]
13. Campos-Martin, J.M.; Capel-Sanchez, M.C.; Perez-Presas, P.; Fierro, J.L.G. Oxidative Processes of Desulfurization of Liquid Fuels. *J. Chem. Tech. Biotech.* **2010**, *85*, 879–890. [CrossRef]
14. Capel-Sanchez, M.C.; Campos-Martin, J.M.; Fierro, J.L.G. Removal of refractory organosulfur compounds via oxidation with hydrogen peroxide on amorphous Ti/SiO_2 catalysts. *Energy Env. Sci.* **2010**, *3*, 328–333. [CrossRef]
15. Monticello, D.J. Riding the fossil fuel bio-desulfurization wave. *Chem. Tech.* **1998**, *28*, 38–45.
16. Babich, I.V.; Moulijn, J.A. Science and technology of novel processes for deep desulfurization of oil refinery streams: A review. *Fuel* **2003**, *82*, 607–631. [CrossRef]
17. Khan, N.A.; Hasan, Z.; Jhung, S.H. Adsorption and removal of sulfur or nitrogen-containing compounds with metal-organic frameworks (MOFs). *Adv. Porous Mater.* **2013**, *1*, 91–102. [CrossRef]
18. Bruneta, S.; Meya, D.; Pérot, G.; Bouchy, C.; Diehl, F. On the hydrodesulfurization of FCC gasoline: A review. *Appl. Catal. A: Gen.* **2005**, *278*, 143–172. [CrossRef]
19. Yang, R.T.; Hernandez-Maldonado, A.J.; Yang, F.H. Desulfurization of transportation fuels with zeolites under ambient conditions. *Science* **2003**, *301*, 79–81. [CrossRef]
20. Park, T.-H.; Cychosz, K.A.; Wong-Foy, A.G.; Dailly, A.; Matzger, A.J. Gas and liquid phase adsorption in isostructural Cu_3[biaryltricarboxylate]$_2$ microporous coordination polymers. *Chem. Commun.* **2011**, *47*, 1452–1454. [CrossRef]
21. Jia, Y.; Li, G.; Ning, G. Efficient oxidative desulfurization (ODS) of model fuel with H_2O_2 catalyzed by $MoO_3/-Al_2O_3$ under mild and solvent free conditions. *Fuel Process. Technol.* **2011**, *92*, 106–111. [CrossRef]

22. Dhir, S.; Uppaluri, R.; Purkait, M.K. Oxidative desulfurization: Kinetic modeling. *J. Hazard. Mater.* **2009**, *16*, 1360–1368. [CrossRef]
23. Hasan, Z.; Jeon, J.; Jhung, S.H. Oxidative desulfurization of benzothiophene and thiophene with WO x/ZrO$_2$ catalysts: Effect of calcination temperature of catalysts. *J. Hazard. Mater.* **2012**, *205–206*, 216–221. [CrossRef]
24. Almarri, M.; Ma, X.; Song, C. Role of surface oxygen-containing functional groups in liquid-phase adsorption of nitrogen compounds on carbon-basedadsorbents. *Energy Fuels* **2009**, *23*, 3940–3947. [CrossRef]
25. Deliyanni, E.; Seredych, M.; Bandosz, T.J. Interactions of 4,6-dimethyldibenzothiophene with the surface of activated carbons. *Langmuir* **2009**, *25*, 9302–9312. [CrossRef] [PubMed]
26. Xiong, J.; Zhu, W.; Li, H.; Yang, L.; Chao, Y.; Wu, P.; Xun, S.; Jiang, W.; Zhang, M.; Li, H. Carbon-doped porous boron nitride: Metal-free adsorbents for sulfur removal from fuels. *J. Mater. Chem. A* **2015**, *3*, 12738–12747. [CrossRef]
27. Zhou, A.; Ma, X.; Song, C. Liquid-phase adsorption of multi-ring thiophenic sulfur compounds on carbon materials with different surface properties. *J. Phys. Chem. B* **2006**, *110*, 4699–4707. [CrossRef] [PubMed]
28. Jeon, H.J.; Ko, C.H.; Kim, S.H.; Kim, J.N. Removal of refractory sulfur compounds in diesel using activated carbon with controlled porosity. *Energy Fuels* **2009**, *23*, 2537–2543. [CrossRef]
29. Seredych, M.; Bandosz, T.J. Adsorption of dibenzothiophenes on nanoporous carbons: Identification of specific adsorption sites governing capacity and selectivity. *Energy and Fuels* **2010**, *24*, 3352–3360. [CrossRef]
30. Seredych, M.; Bandosz, T.J. Investigation of the enhancing effects of sulfur and/or oxygen functional groups of nanoporous carbons on adsorption of dibenzothiophenes. *Carbon* **2011**, *49*, 1216–1224. [CrossRef]
31. Seredych, M.; Bandosz, T.J. Removal of dibenzothiophenes from model diesel fuel on sulfur rich activated carbons. *Appl. Catal. B Environ.* **2011**, *106*, 133–141. [CrossRef]
32. Seredych, M.; Messali, L.; Bandosz, T.J. Analysis of factors affecting visible and UV enhanced oxidation of dibenzothiophenes on sulfur-doped activated carbons. *Carbon* **2013**, *62*, 356–364. [CrossRef]
33. Kampouraki, Z.C.; Giannakoudakis, D.A.; Triantafyllidis, K.S.; Deliyanni, E.A. Catalytic oxidative desulfurization of a 4,6-DMDBT containing model fuel by metal-free activated carbons: the key role of surface chemistry. *Green Chem.* **2019**. [CrossRef]
34. Liu, D.; Gui, J.; Sun, Z. Adsorption structures of heterocyclic nitrogen compounds over Cu(I)Y zeolite: A first principle study on mechanism of the denitrogenation and the effect of nitrogen compounds on adsorptive desulfurization. *J. Mol. Catal. A Chem.* **2008**, *291*, 17–21. [CrossRef]
35. Xiao, J.; Li, Z.; Liu, B.; Xia, Q.; Yu, M. Adsorption of benzothiophene and dibenzothiophene on ion-impregnated activated carbons and ion-exchanged Y zeolites. *Energy Fuels* **2008**, *22*, 3858–3863. [CrossRef]
36. Hernández-Maldonado, A.J.; Qi, G.; Yang, R.T. Desulfurization of commercial fuels by π-complexation: Monolayer CuCl/γ-Al$_2$O$_3$. *Appl. Catal. B Env.* **2005**, *61*, 212–218. [CrossRef]
37. Zhang, Z.Y.; Shi, T.B.; Jia, C.Z.; Ji, W.J.; Chen, Y.; He, M.Y. Adsorptive removal of aromatic organosulfur compounds over the modified Na-Y zeolites. *Appl. Catal. B Env.* **2008**, *82*, 1–10. [CrossRef]
38. Yang, K.; Yan, Y.; Chen, W.; Kang, H.; Han, Y.; Zhang, W.; Fan, Y.; Li, Z. The high performance and mechanism of metal–organic frameworks and their composites in adsorptive desulfurization. *Polyhedron* **2018**, *152*, 202–215. [CrossRef]
39. Hernández-Maldonado, A.J.H.; Yang, R.T. Desulfurization of Transportation Fuels by Adsorption Desulfurization of Transportation Fuels. *Catal. Rev.* **2004**, *46*, 111–150. [CrossRef]
40. Kwon, J.M.; Moon, J.H.; Bae, Y.S.; Lee, D.G.; Sohn, H.C.; Lee, C.H. Adsorptive desulfurization and denitrogenation of refinery fuels using mesoporous silica adsorbents. *ChemSusChem* **2008**, *1*, 307–309. [CrossRef]
41. Li, Z.; Barnes, J.C.; Bosoy, A.; Stoddart, J.F.; Zink, J.I. Mesoporous silica nanoparticles in biomedical applications. *Chem. Soc. Rev.* **2012**, *41*, 2590–2605. [CrossRef]
42. Vinu, A.; Ariga, K. New ideas for mesoporous materials. *Adv. Porous Mater.* **2013**, *1*, 63–71. [CrossRef]
43. Ariga, K.; Ji, Q.; Mcshane, M.J.; Lvov, Y.M.; Vinu, A.; Hill, J.P. Inorganic nanoarchitectonics for biological applications. *Chem. Mater.* **2012**, *24*, 728–737.
44. Bae, Y.S.; Kim, M.B.; Lee, H.J.; Lee, C.H.; Ryu, J.W. Adsorptive denitrogenation of light gas oil by silica-zirconia cogel. *Aiche J.* **2006**, *52*, 510–521. [CrossRef]
45. Ko, C.H.; Park, J.G.; Park, J.C.; Song, H.; Han, S.S.; Kim, J.N. Surface status and size influences of nickel nanoparticles on sulfur compound adsorption. *Appl. Surf. Sci.* **2007**, *253*, 5864–5867. [CrossRef]

46. Xie, L.L.; Favre-Reguillon, A.; Wang, X.X.; Fu, X.; Lemaire, M. Selective adsorption of neutral nitrogen compounds from fuel using ion-exchange resins. *J. Chem. Eng. Data* **2010**, *55*, 4849–4853. [CrossRef]
47. Cronauer, D.C.; Young, D.C.; Solash, J.; Seshadrl, K.S.; Dannerx, D.A. Shale Oil Denitrogenation with Ion Exchange. 3. Characterization of Hydrotreated And Ion-Exchange Isolated Products. *Ind. Eng. Chem. Process Des. Dev.* **1986**, *25*, 756–762. [CrossRef]
48. Furukawa, H.; Cordova, K.E.; O'Keeffe, M.; Yaghi, O.M. The chemistry and applications of metal-organic frameworks. *Science* **2013**, *341*, 1230444. [CrossRef] [PubMed]
49. Qiu, S.; Xue, M.; Zhu, G. Metal-organic framework membranes: From synthesis to separation application. *Chem. Soc. Rev.* **2014**, *43*, 6116–6140. [CrossRef]
50. Liu, J.; Chen, L.; Cui, H.; Zhang, J.; Zhang, L.; Su, C.Y. Applications of metal-organic frameworks in heterogeneous supramolecular catalysis. *Chem. Soc. Rev.* **2014**, *43*, 6011–6061. [CrossRef]
51. Xiang, Z.; Cao, D.; Lan, J.; Wang, W.; Broom, D.P. Multiscale simulation and modelling of adsorptive processes for energy gas storage and carbon dioxide capture in porous coordination frameworks. *Energy Env. Sci.* **2010**, *3*, 1469–1487. [CrossRef]
52. Meek, S.T.; Greathouse, J.A.; Allendorf, M.D. Metal-organic frameworks: A rapidly growing class of versatile nanoporous materials. *Adv. Mater.* **2011**, *23*, 249–267. [CrossRef]
53. Zhou, H.C.; Long, J.R.; Yaghi, O.M. Introduction to metal-organic frameworks. *Chem. Rev.* **2012**, *112*, 673–674. [CrossRef]
54. Li, J.-R.; Sculley, J.; Zhou, H.-C. Metal–Organic Frameworks for Separations. *Chem. Rev.* **2012**, *112*, 869–932. [CrossRef] [PubMed]
55. Férey, G. Hybrid porous solids: Past, present, future. *Chem. Soc. Rev.* **2008**, *37*, 191–214. [CrossRef] [PubMed]
56. Yaghi, O.M.; Li, G.; Li, H. Selective binding and removal of guests in a microporous metal-organic framework. *Nature* **1995**, *378*, 703–706. [CrossRef]
57. Kitagawa, S.; Kitaura, R.; Noro, S.I. Functional porous coordination polymers. *Angew. Chem. Int. Ed.* **2004**, *43*, 2334–2375. [CrossRef]
58. Ma, S.; Zhou, H.C. A metal-organic framework with entatic metal centers exhibiting high gas adsorption affinity. *J. Am. Chem. Soc.* **2006**, *128*, 11734–11735. [CrossRef]
59. Cychosz, K.A.; Wong-Foy, A.G.; Matzger, A.J. Liquid phase adsorption by microporous coordination polymers: Removal of organosulfur compounds. *J. Am. Chem. Soc.* **2008**, *130*, 6938–6939. [CrossRef]
60. Liu, Y.; Kravtsov, V.C.; Larsen, R.; Eddaoudi, M. Molecular building blocks approach to the assembly of zeolite-like metal-organic frameworks (ZMOFs) with extra-large cavities. *Chem. Commun.* **2006**, *14*, 1488–1490. [CrossRef]
61. Eddaoudi, M.; Kim, J.; Rosi, N.; Vodak, D.; Wachter, J.; O'Keeffe, M.; Yaghi, O.M. Systematic design of pore size and functionality in isoreticular MOFs and their application in methane storage. *Science* **2002**, *295*, 469–472. [CrossRef]
62. Sumida, K.; Rogow, D.L.; Mason, J.A.; Mcdonald, T.M.; Bloch, E.D.; Herm, Z.R.; Bae, T.; Long, J.R. Carbon dioxide capture in metal-organic frameworks. *Chem. Rev.* **2012**, 724–778. [CrossRef]
63. Li, J.R.; Ma, Y.; McCarthy, M.C.; Sculley, J.; Yu, J.; Jeong, H.K.; Balbuena, P.B.; Zhou, H.C. Carbon dioxide capture-related gas adsorption and separation in metal-organic frameworks. *Coord. Chem. Rev.* **2011**, *255*, 1791–1823. [CrossRef]
64. Suh, M.P.; Park, H.J.; Prasad, T.K.; Lim, D.-W. Hydrogen storage in metal-organic frameworks. *Chem. Rev.* **2012**, *112*, 782–835. [CrossRef] [PubMed]
65. Van De Voorde, B.; Bueken, B.; Denayer, J.; De Vos, D. Adsorptive separation on metal-organic frameworks in the liquid phase. *Chem. Soc. Rev.* **2014**, *43*, 5766–5788. [CrossRef] [PubMed]
66. Yang, Q.; Liu, D.; Zhong, C.; Li, J.R. Development of computational methodologies for metal-organic frameworks and their application in gas separations. *Chem. Rev.* **2013**, *113*, 8261–8323. [CrossRef] [PubMed]
67. Dhakshinamoorthy, A.; Garcia, H. Catalysis by metal nanoparticles embedded on metal-organic frameworks. *Chem. Soc. Rev.* **2012**, *41*, 5262–5284. [CrossRef]
68. Horcajada, P.; Gref, R.; Baati, T.; Allan, P.K.; Maurin, G.; Couvreur, P.; Férey, G.; Morris, R.E.; Serre, C. Metal-organic frameworks in biomedicine. *Chem. Rev.* **2012**, *112*, 1232–1268. [CrossRef]
69. Uemura, T.; Yanai, N.; Kitagawa, S. Polymerization reactions in porous coordination polymers. *Chem. Soc. Rev.* **2009**, *38*, 1228–1236. [CrossRef]
70. Kurmoo, M. Magnetic metal-organic frameworks. *Chem. Soc. Rev.* **2009**, *38*, 1353–1379. [CrossRef]

71. Hu, Z.; Deibert, B.J.; Li, J. Luminescent metal-organic frameworks for chemical sensing and explosive detection. *Chem. Soc. Rev.* **2014**, *43*, 5815–5840. [CrossRef]
72. Giannakoudakis, D.A.; Bandosz, T.J. *Detoxification of Chemical Warfare Agents*, 1st ed.; Springer International Publishing: Cham, Switzerland, 2018; ISBN 978-3-319-70759-4.
73. Li, J.R.; Kuppler, R.J.; Zhou, H.C. Selective gas adsorption and separation in metal-organic frameworks. *Chem. Soc. Rev.* **2009**, *38*, 1477–1504. [CrossRef]
74. Khan, N.A.; Jhung, S.H. Effect of central metal ions of analogous metal-organic frameworks on the adsorptive removal of benzothiophene from a model fuel. *J. Hazard. Mater.* **2013**, *260*, 1050–1056. [CrossRef]
75. Wu, H.; Gong, Q.; Olson, D.H.; Li, J. Commensurate adsorption of hydrocarbons and alcohols in microporous metal organic frameworks. *Chem. Rev.* **2012**, *112*, 836–868. [CrossRef] [PubMed]
76. Decoste, J.B.; Peterson, G.W. Metal−Organic Frameworks for Air Purification of Toxic Chemicals. *Chem. Rev.* **2014**, *114*, 5695–5727. [CrossRef] [PubMed]
77. Peralta, D.; Chaplais, G.; Simon-Masseron, A.; Barthelet, K.; Pirngruber, G.D. Metal−Organic Framework Materials for Desulfurization by Adsorption. *Energy Fuels* **2012**, *26*, 4953–4960. [CrossRef]
78. Yin, C.; Zhu, G.; Xia, D. A study of the distribution of sulfur compounds in gasoline fraction produced in China: Part 2. The distribution of sulfur compounds in full-range FCC and RFCC naphthas. *Fuel Process Technol.* **2002**, *79*, 135–140. [CrossRef]
79. Wen, Y.; Wang, G.; Xu, C.; Gao, J. Study on in Situ Sulfur Removal from Gasoline in Fluid Catalytic Cracking Process. *J. Energy Fuels* **2012**, *26*, 3201–3211. [CrossRef]
80. Song, C. An overview of new approaches to deep desulfurization for ultra-clean gasoline, diesel fuel and jet fuel. *Catal. Today* **2003**, *86*, 211–263. [CrossRef]
81. Gong, Y.; Dou, T.; Kang, S.; Li, Q.; Hu, Y. Deep desulfurization of gasoline using ion-exchange zeolites: Cu(I)- and Ag(I)-beta. *Fuel Process. Technol.* **2009**, *90*, 122–129. [CrossRef]
82. Tong, M.; Shengui, J.U.; Feng, X.U.E. Selectivity adsorption of thiophene alkylated derivatives oveρ modified Cu^+-13X zeolite. *J. Rare Earths* **2012**, *30*, 807–813.
83. Zhang, H.X.; Huang, H.L.; Li, C.X.; Meng, H.; Zhou Lu, Y.Z.; Zhong, C.L.; Liu, D.H.; Yang, Q.Y. Adsorption Behavior of Metal−Organic Frameworks for Thiophenic Sulfur from Diesel Oil. *Ind. Eng. Chem. Res.* **2012**, *51*, 12449–12455. [CrossRef]
84. Achmann, S.; Hagen, G.; Hämmerle, M.; Malkowsky, I.; Kiener, C.; Moos, R. Sulfur Removal from Low-Sulfur Gasoline and Diesel Fuel by Metal-Organic Frameworks. *Chem. Eng. Technol.* **2010**, *33*, 275–280. [CrossRef]
85. Lin, C.; Cheng, Z.; Li, B.; Chen, T.; Zhang, W.; Chen, S.; Yang, Q.; Chang, L.; Che, G.; Ma, H. High-Efficiency Separation of Aromatic Sulfide from Liquid Hydrocarbon Fuel in Conjugated Porous Organic Framework with Polycarbazole Unit. *Acs Appl. Mater. Interfaces* **2019**, *43*, 40970–40979.
86. Mueller, U.; Schubert, M.; Teich, F.; Puetter, H.; Schierle-Arndt, K.; Pastré, J. Metal-organic frameworks—Prospective industrial applications. *J. Mater. Chem.* **2006**, *16*, 626–636. [CrossRef]
87. Sano, Y.; Choi, K.H.; Korai, Y.; Mochida, I. Adsorptive removal of sulfur and nitrogen species from a straight run gas oil over activated carbons for its deep hydrodesulfurization. *Appl. Catal. B Env.* **2004**, *49*, 219–225. [CrossRef]
88. Schnobrich, J.K.; Lebel, O.; Cychosz, K.A.; Dailly, A.; Wong-Foy, A.G.; Matzger, A.J. Linker-directed vertex desymmetrization for the production of coordination polymers with high porosity. *J. Am. Chem. Soc.* **2010**, *132*, 13941–13948. [CrossRef]
89. Dai, W.; Hu, J.; Zhou, L.; Li, S.; Hu, X.; Huang, H. Removal of dibenzothiophene with composite adsorbent MOF-5/Cu(I). *Energy Fuels* **2013**, *27*, 816–821. [CrossRef]
90. Khan, N.A.; Jhung, S.H. Low-temperature loading of Cu+ species over porous metal-organic frameworks (MOFs) and adsorptive desulfurization with Cu^+-loaded MOFs. *J. Hazard. Mater.* **2012**, 180–185. [CrossRef]
91. Cychosz, K.A.; Wong-Foy, A.G.; Matzger, A.J. Enabling cleaner fuels: Desulfurization by adsorption to microporous coordination polymers. *J. Am. Chem. Soc.* **2009**, *131*, 14538–14543. [CrossRef]
92. Wu, L.; Xiao, J.; Wu, Y.; Xian, S.; Miao, G.; Wang, H.; Li, Z. A combined experimental/computational study on the adsorption of organosulfur compounds over metal-organic frameworks from fuels. *Langmuir* **2014**, *30*, 1080–1088. [CrossRef]
93. Li, S.L.; Lan, Y.Q.; Sakurai, H.; Xu, Q. Unusual regenerable porous metal-organic framework based on a new triple helical molecular necklace for separating organosulfur compounds. *Chem. Eur. J.* **2012**, *18*, 16303–16309. [CrossRef]

94. Rui, J.; Liu, F.; Wang, R.; Lu, Y.; Yang, X. Adsorptive Desulfurization of Model Gasoline by Using Different Zn Sources Exchanged NaY Zeolites. *Molecules* **2017**, *22*, 305. [CrossRef]
95. Wang, X.L.; Fan, H.L.; Tian, Z.; He, E.Y.; Li, Y.; Shangguan, J. Adsorptive removal of sulfur compounds using IRMOF-3 at ambient temperature. *Appl. Surf. Sci.* **2014**, *289*, 107–113. [CrossRef]
96. Jin, T.; Yang, Q.; Meng, C.; Xu, J.; Liu, H.; Hu, J.; Ling, H. Promoting desulfurization capacity and separation efficiency simultaneously by the novel magnetic Fe_3O_4@PAA@MOF-199. *RSC Adv.* **2014**, *4*, 41902–41909. [CrossRef]
97. Huang, M.; Chang, G.; Su, Y.; Xing, H.; Zhang, Z.; Yang, Y.; Ren, Q.; Bao, Z.; Chen, B. A metal-organic framework with immobilized Ag(i) for highly efficient desulfurization of liquid fuels. *Chem. Commun.* **2015**, *51*, 12205–12207. [CrossRef] [PubMed]
98. Song, H.S.; Ko, C.H.; Ahn, W.; Kim, B.J.; Croiset, E.; Chen, Z.; Nam, S.C. Selective dibenzothiophene adsorption on graphene prepared using different methods. *Ind. Eng. Chem. Res.* **2012**, *51*, 10259–10264. [CrossRef]
99. He, W.W.; Yang, G.S.; Tang, Y.J.; Li, S.L.; Zhang, S.R.; Su, Z.M.; Lan, Y.Q. Phenyl Groups Result in the Highest Benzene Storage and Most Efficient Desulfurization in a Series of Isostructural Metal-Organic Frameworks. *Chem. Eur. J.* **2015**, *21*, 9784–9789. [CrossRef] [PubMed]
100. Ullah, L.; Zhao, G.; Hedin, N.; Ding, X.; Zhang, S.; Yao, X.; Nie, Y.; Zhang, Y. Highly efficient adsorption of benzothiophene from model fuel on a metal-organic framework modified with dodeca-tungstophosphoric acid. *Chem. Eng. J.* **2019**, 30–40. [CrossRef]
101. Khan, N.A.; Bhadra, B.N.; Jhung, S.H. Heteropoly acid-loaded ionic liquid@metal-organic frameworks: Effective and reusable adsorbents for the desulfurization of a liquid model fuel. *Chem. Eng. J.* **2018**, *334*, 2215–2221. [CrossRef]
102. Zhao, Z.; Zuhra, Z.; Qin, L.; Zhou, Y.; Zhang, L.; Tang, F.; Mu, C. Confinement of microporous MOF-74(Ni) within mesoporous γ-Al_2O_3 beads for excellent ultra-deep and selective adsorptive desulfurization performance. *Fuel Process. Technol.* **2018**, *176*, 276–282. [CrossRef]
103. Khan, N.A.; Kim, C.M.; Jhung, S.H. Adsorptive desulfurization using Cu–Ce/metal–organic framework: Improved performance based on synergy between Cu and Ce. *Chem. Eng. J.* **2017**, *311*, 20–27. [CrossRef]
104. Zhang, X.F.; Wang, Z.; Feng, Y.; Zhong, Y.; Liao, J.; Wang, Y.; Yao, J. Adsorptive desulfurization from the model fuels by functionalized UiO-66(Zr). *Fuel* **2018**, *234*, 256–262. [CrossRef]
105. Wang, T.; Li, X.; Dai, W.; Fang, Y.; Huang, H. Enhanced adsorption of dibenzothiophene with zinc/copper-based metal-organic frameworks. *J. Mater. Chem. A* **2015**, *3*, 21044–21050. [CrossRef]
106. Hasan, Z.; Jhung, S.H. Facile Method to Disperse Nonporous Metal Organic Frameworks: Composite Formation with a Porous Metal Organic Framework and Application in Adsorptive Desulfurization. *Acs Appl. Mater. Interfaces* **2015**, *7*, 10429–10435. [CrossRef] [PubMed]
107. Zhao, X.; Xin, C.; Yin, Y.; Tian, X.; Li, Y.; Bian, W.; Lian, P. Metal organic framework as an adsorbent for desulphurization. *Adsorpt. Sci. Technol.* **2012**, *30*, 483–490. [CrossRef]
108. Jhung, S.H.; Khan, N.A.; Hasan, Z. Analogous porous metal-organic frameworks: Synthesis, stability and application in adsorption. *Cryst. Eng. Comm.* **2012**, *14*, 7099–7109. [CrossRef]
109. Petit, C.; Bandosz, T.J. Exploring the coordination chemistry of MOF-graphite oxide composites and their applications as adsorbents. *Dalt. Trans.* **2012**, *41*, 4027–4035. [CrossRef]
110. Ahmed, I.; Jhung, S.H. Adsorptive desulfurization and denitrogenation using metal-organic frameworks. *J. Hazard. Mater.* **2019**, *301*, 259–276. [CrossRef]
111. Furukawa, S.; Reboul, J.; Diring, S.; Sumida, K.; Kitagawa, S. Structuring of metal-organic frameworks at the mesoscopic/macroscopic scale. *Chem. Soc. Rev.* **2014**, *43*, 5700–5734. [CrossRef]
112. Zuhra, Z.; Zhao, Z.; Qin, L.; Zhou, Y.; Zhang, L.; Ali, S.; Tang, F.; Ping, E. In situ formation of a multiporous MOF(Al)@γ-AlOOH composite material: A versatile adsorbent for both N- and S-heterocyclic fuel contaminants with high selectivity. *Chem. Eng. J.* **2019**, *360*, 1623–1632. [CrossRef]
113. Liu, B.; Peng, Y.; Chen, Q. Adsorption of N/S-Heteroaromatic Compounds from Fuels by Functionalized MIL-101(Cr) Metal-Organic Frameworks: The Impact of Surface Functional Groups. *Energy Fuels* **2016**, *30*, 5593–5600. [CrossRef]
114. Qin, L.; Zhou, Y.; Li, D.; Zhang, L.; Zhao, Z.; Zuhra, Z.; Mu, C. Highly Dispersed HKUST-1 on Milimeter-Sized Mesoporous γ-Al_2O_3 Beads for Highly Effective Adsorptive Desulfurization. *Ind. Eng. Chem. Res.* **2016**, *55*, 7249–7258. [CrossRef]

115. Ahmed, I.; Khan, N.A.; Zubair Hasan, Z.; Sung Hwa Jhung, A.H. Adsorptive denitrogenation of model fuels with porous metal-organic framework (MOF) MIL-101 impregnated with phosphotungstic acid: Effect of acid site inclusion. *J. Hazard. Mater.* **2013**, *250–251*, 37–44. [CrossRef] [PubMed]
116. Khan, N.A.; Jhung, S.H. Adsorptive removal of benzothiophene using porous copper- benzenetricarboxylate loaded with phosphotungstic acid. *Fuel Process. Technol.* **2012**, *100*, 49–54. [CrossRef]
117. Ahmed, I.; Khan, N.A.; Jhung, S.H. Graphite oxide/metal-organic framework (MIL-101): Remarkable performance in the adsorptive denitrogenation of model fuels. *Inorg. Chem.* **2013**, *52*, 14155–14161. [CrossRef] [PubMed]
118. Radwan, D.R.; Matloob, A.; Mikhail, S.; Saad, L.; Guirguis, D. Metal organic framework-graphene nano-composites for high adsorption removal of DBT as hazard material in liquid fuel. *J. Hazard. Mater.* **2019**, *373*, 447–458.
119. Emam, H.E.; Ahmed, H.B.; El-Deib, H.R.; El-Dars, F.M.S.E.; Abdelhameed, R.M. Non-invasive route for desulfurization of fuel using infrared-assisted MIL-53(Al)-NH$_2$ containing fabric. *J. Colloid Interface Sci.* **2019**, *556*, 193–205. [CrossRef] [PubMed]
120. Hasan, Z.; Jhung, S.H. Removal of hazardous organics from water using metal-organic frameworks (MOFs): Plausible mechanisms for selective adsorptions. *J. Hazard. Mater.* **2015**, *283*, 329–339. [CrossRef] [PubMed]
121. Wei, K.; Ni, J.; Cui, Y.; Han, H.; Xie, Y.; Liu, Y. Desulfurization by liquid phase adsorption: Role of exposed metal sites in metal-organic frameworks. *J. Mol. Struct.* **2019**, *1184*, 163–167. [CrossRef]
122. Maes, M.; Trekels, M.; Boulhout, M.; Schouteden, S.; Vermoortele, F.; Alaerts, L.; Heurtaux, D.; Seo, Y.K.; Hwang, Y.K.; Chang, J.S.; et al. Selective removal of N-heterocyclic aromatic contaminants from fuels by lewis acidic metal-organic frameworks. *Angew. Chem. Int. Ed.* **2011**, *50*, 4210–4214. [CrossRef]
123. Grajciar, L.; Bludský, O.; Nachtigall, P. Water adsorption on coordinatively unsaturated sites in CuBTC MOF. *J. Phys. Chem. Lett.* **2010**, *1*, 3354–3359. [CrossRef]
124. Dietzel, P.D.C.; Besikiotis, V.; Blom, R. Application of metal-organic frameworks with coordinatively unsaturated metal sites in storage and separation of methane and carbon dioxide. *J. Mater. Chem.* **2009**, *19*, 7362–7370. [CrossRef]
125. Yoon, J.W.; Seo, Y.K.; Hwang, Y.K.; Chang, J.S.; Leclerc, H.; Wuttke, S.; Bazin, P.; Vimont, A.; Daturi, M.; Bloch, E.; et al. Controlled reducibility of a metal-organic framework with coordinatively unsaturated sites for preferential gas sorption. *Angew. Chem. Int. Ed.* **2010**, *49*, 5949–5952. [CrossRef] [PubMed]
126. Hwang, Y.K.; Hong, D.Y.; Chang, J.S.; Jhung, S.H.; Seo, Y.K.; Kim, J.; Vimont, A.; Daturi, M.; Serre, C.; Férey, G. Amine grafting on coordinatively unsaturated metal centers of MOFs: Consequences for catalysis and metal encapsulation. *Angew. Chem. Int. Ed.* **2008**, *47*, 4144–4148. [CrossRef] [PubMed]
127. Khan, N.A.; Jhung, S.H. Adsorptive removal and separation of chemicals with metal-organic frameworks: Contribution of π-complexation. *J. Hazard. Mater.* **2017**, *325*, 198–213. [CrossRef] [PubMed]
128. Furukawa, H.; Go, Y.B.; Ko, N.; Park, Y.K.; Uribe-Romo, F.J.; Kim, J.; O'Keeffe, M.; Yaghi, O.M. Isoreticular expansion of metal-organic frameworks with triangular and square building units and the lowest calculated density for porous crystals. *Inorg. Chem.* **2011**, *50*, 9147–9152. [CrossRef] [PubMed]
129. Bae, Y.S.; Dubbeldam, D.; Nelson, A.; Walton, K.S.; Hupp, J.T.; Snurr, R.Q. Strategies for characterization of large-pore metal-organic frameworks by combined experimental and computational methods. *Chem. Mater.* **2009**, *21*, 4768–4777. [CrossRef]
130. Jiao, L.; Seow, J.Y.R.; Skinner, W.S.; Wang, Z.U.; Jiang, H.-L. Metal–organic frameworks: Structures and functional applications. *Mater. Today* **2019**, *27*, 43–68. [CrossRef]
131. Xue, M.; Liu, Y.; Schaffino, R.M.; Xiang, S.; Zhao, X.; Zhu, G.S.; Qiu, S.L.; Chen, B. New prototype isoreticular metal - Organic framework Zn$_4$O(FMA)$_3$ for gas storage. *Inorg. Chem.* **2009**, *48*, 4649–4651. [CrossRef] [PubMed]
132. Farha, O.K.; Eryazici, I.; Jeong, N.C.; Hauser, B.G.; Wilmer, C.E.; Sarjeant, A.A.; Snurr, R.Q.; Nguyen, S.T.; Yazaydın, A.Ö.; Hupp, J.T. Metal-organic framework materials with ultrahigh surface areas: Is the sky the limit? *J. Am. Chem. Soc.* **2012**, *134*, 15016–15021. [CrossRef]
133. Petit, C.; Bandosz, T.J. Engineering the surface of a new class of adsorbents: Metal-organic framework/graphite oxide composites. *J. Colloid Interface Sci.* **2015**, *447*, 139–151. [CrossRef]
134. Bandosz, T.J.; Petit, C. MOF/graphite oxide hybrid materials: Exploring the new concept of adsorbents and catalysts. *Adsorption* **2011**, *17*, 5–16. [CrossRef]

135. Petit, C.; Bandosz, T.J. MOF-graphite oxide composites: Combining the uniqueness of graphene layers and metal-organic frameworks. *Adv. Mater.* **2009**, *21*, 4753–4757. [CrossRef]
136. Ebrahim, A.M.; Jagiello, J.; Bandosz, T.J. Enhanced reactive adsorption of H_2S on Cu–BTC/ S- and N-doped GO composites. *J. Mater. Chem. A* **2015**, *3*, 8194–8204. [CrossRef]
137. Giannakoudakis, D.A.; Bandosz, T.J. Graphite Oxide Nanocomposites for Air Stream Desulfurization. In *Composite Nanoadsorbents*; Elsevier: Amsterdam, The Netherlands, 2019; pp. 1–24.
138. Giannakoudakis, D.A.; Jiang, M.; Bandosz, T.J. Highly efficient air desulfurization on self-assembled bundles of copper hydroxide nanorods. *ACS Appl. Mater. Interfaces* **2016**, *8*, 31986–31994. [CrossRef] [PubMed]
139. Chen, M.; Ding, Y.; Liu, Y.; Wang, N.; Yang, B.; Ma, L. Adsorptive desulfurization of thiophene from the model fuels onto graphite oxide/metal-organic framework composites. *Pet. Sci. Technol.* **2018**, *36*, 141–147. [CrossRef]
140. Hong, J.; Chen, C.; Bedoya, F.E.; Kelsall, G.H.; O'Hare, D.; Petit, C. Carbon nitride nanosheet/metal–organic framework nanocomposites with synergistic photocatalytic activities. *Catal. Sci. Technol.* **2016**, *6*, 5042–5051. [CrossRef]
141. Wang, H.; Yuan, X.; Wu, Y.; Zeng, G.; Chen, X.; Leng, L.; Li, H. Synthesis and applications of novel graphitic carbon nitride/metal-organic frameworks mesoporous photocatalyst for dyes removal. *Appl. Catal. B Env.* **2015**, *174–175*, 445–454. [CrossRef]
142. Huang, J.; Zhang, X.; Song, H.; Chen, C.; Han, F.; Wen, C. *Protonated graphitic carbon nitride coated metal-organic frameworks with enhanced visible-light photocatalytic activity for contaminants degradation*; Elsevier: Amsterdam, The Netherlands, 2018.
143. Giannakoudakis, D.A.; Travlou, N.A.; Secor, J.; Bandosz, T.J. Oxidized g-C 3 N 4 Nanospheres as Catalytically Photoactive Linkers in MOF/g-C_3N_4 Composite of Hierarchical Pore Structure. *Small* **2017**, *13*, 1601758. [CrossRef]
144. Giannakoudakis, D.A.; Hu, Y.; Florent, M.; Bandosz, T.J. Smart textiles of MOF/g-C 3 N 4 nanospheres for the rapid detection/detoxification of chemical warfare agents. *Nanoscale Horiz.* **2017**, *2*, 356–364. [CrossRef]
145. Giannakoudakis, D.A.; Bandosz, T.J. Defectous UiO-66 MOF Nanocomposites as Reactive Media of Superior Protection against Toxic Vapors. *ACS Appl. Mater. Interfaces* **2019**. [CrossRef]

© 2019 by the authors. Licensee MDPI, Basel, Switzerland. This article is an open access article distributed under the terms and conditions of the Creative Commons Attribution (CC BY) license (http://creativecommons.org/licenses/by/4.0/).

Review

Applications of Metal-Organic Frameworks in Food Sample Preparation

Natalia Manousi [1], George A. Zachariadis [1], Eleni A. Deliyanni [2] and Victoria F. Samanidou [1,*]

[1] Laboratory of Analytical Chemistry, Department of Chemistry, Aristotle University of Thessaloniki, Thessaloniki 54124, Greece; nmanousi@chem.auth.gr (N.M.); zacharia@chem.auth.gr (G.A.Z.)
[2] Division of Chemical Technology, Department of Chemistry, Aristotle University of Thessaloniki, Thessaloniki 54124, Greece; lenadj@chem.auth.gr
* Correspondence: samanidu@chem.auth.gr; Tel.: +30-2310-342-507

Received: 21 October 2018; Accepted: 5 November 2018; Published: 6 November 2018

Abstract: Food samples such as milk, beverages, meat and chicken products, fish, etc. are complex and demanding matrices. Various novel materials such as molecular imprinted polymers (MIPs), carbon-based nanomaterials carbon nanotubes, graphene oxide and metal-organic frameworks (MOFs) have been recently introduced in sample preparation to improve clean up as well as to achieve better recoveries, all complying with green analytical chemistry demands. Metal-organic frameworks are hybrid organic inorganic materials, which have been used for gas storage, separation, catalysis and drug delivery. The last few years MOFs have been used for sample preparation of pharmaceutical, environmental samples and food matrices. Due to their high surface area MOFs can be used as adsorbents for the development of sample preparation techniques of food matrices prior to their analysis with chromatographic and spectrometric techniques with great performance characteristics.

Keywords: metal-organic frameworks; MOF; sample preparation; HPLC; GC; food samples

1. Introduction

Sample preparation is the most challenging step of the analytical procedure for the analysis of most samples. An appropriate sample preparation technique should not only be simple, fast and economical, but it should also be in regard with the main principles of green chemistry [1,2]. Solid-phase extraction (SPE) is a well-established sample preparation technique; which however shows some fundamental disadvantages, such as including complicated and time-consuming steps, as well as requiring large amounts of sample and organic solvents. As a result, many novel techniques have been developed [1–5]. Nowadays, a trend in analytical chemistry is to develop new sorbents either for the well-established SPE procedure or for the novel microextraction procedures, which have been gaining more and more attention [1]. New sorbents such as molecular imprinted polymers (MIPs), carbon-based nanomaterials, carbon tubes, graphene based materials, or metal-organic frameworks (MOFs) are becoming more and more popular [6–8]. Metal-organic frameworks are a new class of hybrid organic inorganic supramolecular materials, which are based on the coordination of metal ions or clusters with bi- or multidentate organic linkers [9,10]. Metal-organic frameworks became popular in 1995, when Yangi and Li reported the synthesis of a metal-organic framework containing large rectangular channels [11] What makes the use of MOF materials so promising is the fact that they bare great physical and chemical properties, such as their high surface areas (up to 10,000 m^2/g), in addition with their tunable pore size and functionality, and can act as hosts for a variety of guest molecules. Some of MOFs great properties are luminosity, flexibility of their structure, charge transfer ability from the ligand to the metal or from the metal to the ligand, thermal stability, properties that include electronic and conducting effects and pH-sensitive stability [11,12].

For the synthesis of MOF materials many alternative ways have been proposed. The most famous method is the solvothermal method, which is normally performed in an autoclave with high temperature and pressure and with the use of an organic solvent at its boing point (typically dialkyl formamides, alcohols and pyridine) [13]. Other synthetic methods that have been applied for MOF materials include microwave, electrochemical, mechanochemical, ultrasonic, high-throughput syntheses and more novel techniques include post-synthetic deprotection [12].

Therefore, MOFs have been applied in many different scientific fields and their most famous application is for storage of gas fuels such as hydrogen and methane [14]. Other applications of MOFs include gas separation proton, electron, and ion conduction, capture of carbon dioxide and organic reaction catalysis applications [15] Biomedical applications of MOFs include biomedical imaging, disease diagnosing, drug delivery, biosensing and magnetic resonance imaging [15–19]. In the field of analytical chemistry many different applications of MOF materials have been reported. In 2006, Chen et al. used for the first time MOF-508 material as stationary phase in a packed column in gas chromatography (GC) [20]. After that, some other MOFs have been used in packed GC columns [21–23]. Moreover, MOFs have been used as stationary phases in HPLC columns both for normal-phase and for reversed-phase high performance liquid chromatography (HPLC) applications [24–26]. However, the most popular applications of the use of MOFs in analytical chemistry are in the field of sample preparation as absorbents for the extraction of a wide range of analytes in different matrices [27].

In the last few years the very promising properties of MOF materials, such as the high surface area, made MOF ideal materials to be used as absorbents for sample preparation to meet various separation needs for many different compounds including either organic compounds or metal compounds from a wide range of matrices, such as environmental samples, food samples, drinking water etc. Typical examples of MOF materials that have been used as absorbents for sample preparation are MOF-199, MOF-5(Zn), ZIF-8, and MIL-53(Al). Most of the times, the mechanism of absorption may be due to the π–π stacking interaction between the MOF material and the analytes because of the presence of sp^2 hybridized carbons [15].

Another interesting category of materials are metal organic frameworks derived nanoporous carbons, which are also useful materials for sample preparation. These materials have properties similar to MOFs and therefore they can form π-interactions between them and benzene rings of the target analytes. Direct carbonization or carbonization/polymerization after impregnation of MOF carbon precursors with furfuryl alcohol can lead to the formation of those materials. As a result, MOF derived nanoporous carbons are also considered as useful adsorbents for sample preparation [28].

Herein, we aim to point out the applications of MOFs, which are reported in the literature which include the use of metal-organic compounds and their derived carbons, as absorbents in combination with dispersive sample preparation techniques, magnetic sample preparation techniques, in-tube sample preparation techniques and on-line sample preparation techniques for the analysis of complex food samples, such as milk, tea and beverages, fruits and vegetables, meat, chicken, fish etc.

2. Food Matrices

Metal-organic frameworks have been used for many different food matrices (Figure 1).

2.1. Milk Samples

Milk is an important and well-studied matrix because its quality depends directly on the nutrition and medication that is given to the animals that produce it. A wide range of antibiotics have been examined in milk samples and many analytical methods have been developed for the determination of those compounds in milk samples based on the legislation and the maximum residue limits. MOF materials have been used as absorbents for the extraction of different kind of analytes such as sulfonamides, penicillins, tetracyclines, etc. The wide sulfonamide class of antibacterial compounds has been widely examined due to their excessive use in veterinary practice [29,30]. In 2017 Jia et al., synthesized a novel hybrid MOF/graphene oxide (GO) material for the dispersive micro-solid phase extraction (d-μSPE) prior to the determination of trace sulfonamides in milk with ultra-high pressure liquid chromatography-tandem mass spectrometry (UHPLC-MS/MS). For this purpose, GO was synthesized and then MIL-101(Cr)@GO material was formed with the hydrothermal method by mixing the graphene oxide with hydrofluoric acid, chromium(III) nitrate nonahydrate and terephthalic acid. For milk sample preparation, ethyl acetate was added to the sample, and the mixture was centrifuged. Accordingly, the supernatant was evaporated, re-dissolved in deionized water and pH was adjusted to 5. Then 5 mg of the material was dispersed in the sample and vortex mixing took place for 20 min. When extraction was finished, the material was separated from the liquid with centrifugation and desorption of the analytes was achieved with 5% ammonia-methanol in ultrasonic bath for 10 min. Detection limits ranged between 0.012 and 0.145 µg/L and recoveries ranged between 79.83% and 103.8%. The developed method was rapid and easy, and the composite MOF material can be implemented for milk analysis and its use can be extended for other non-volatile analytes [31]. Table 1 summarizes the use of different MOFs in various food matrices as well as some analytical characteristics of the novel developed methods.

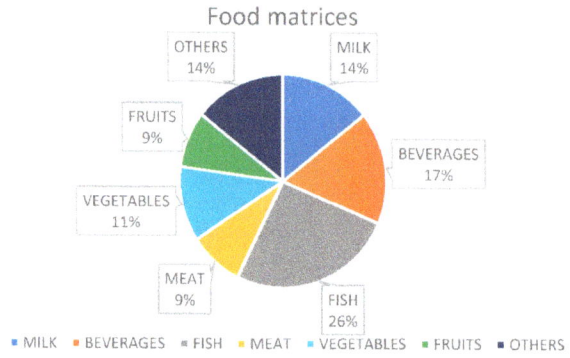

Figure 1. Food matrices treated with MOFs for analytical purposes.

Penicillins are another class of antibiotics widely used for animals. In 2015, Lirio et al. developed an aluminum-based MOF-polymer (MIL-53) monolithic column for the in tube solid phase micro-extraction (SPME) of penicillins from river water and milk samples. The material was synthesized by mixing aluminum nitrate nonahydrate and 1,4-benzenedicarboxylic acid. A 0.8 mm inside diameter (I.D.) capillary tube was pretreated with NaOH, water, methanol and a mixture of 3-trimethoxysilyl-propyl methacrylate/methanol, washed and dried prior to the filling. Then, the Al-MOF material was suspended in a mixture of methacrylate-based monomers and azo-bis-isobutyronitrile prior to vortex mixing, sonication and degassing and the column was filled. Subsequently, the column was suspended in water and the filling was polymerized in situ with microwave assistance. Milk samples were treated with acetonitrile prior to their loading to the column and for the solid phase extraction procedure the optimized parameters were: sample matrix at pH 3, 200 µL desorption volume using methanol, 37.5% of MOF in a 4-cm column length, flow rate of 0.100 mL min^{-1}, column conditioning with 0.5 mL methanol (MeOH) and 0.5 mL pH 3 phosphate buffer saline solution, sample volume 2 mL, column washing with 0.5 mL pH 2 phosphate buffer saline. With this procedure high extraction efficiency was succeed not only due to the π-π interactions between the absorbent and the analytes but also due to the breathing ability of MIL-53. In addition, three other aluminum-based MOFs; DUT-5, CYCU-4, MIL-68, were prepared according to the optimized condition for MIL-53-polymer in order to find out which one is more suitable for the extraction procedure. The method efficiency was satisfactory, with recoveries ranging from 80.8% to 90.9% and a limit of detection between 0.06–0.26 µg/L [32].

Table 1. Applications of MOF use for food sample preparation.

Matrix	Analytes	Analytical Technique	MOF Material	Sample Preparation Technique	Recovery	LODs	Reference
Milk	Sulfonamides	UHPLC-MS/MS	MIL-101(Cr)@GO	d-µSPE	79.83–103.8%	0.012–0.145 µg/L	[31]
Milk	Penicillins	UHPLC-TUV	MIL-53	In tube SPME	80.8–90.9%	0.06–0.26 µg/L	[32]
Milk	Tetracyclines	HPLC-PDA	ZIF-8	on-line SPE	70.3–107.4%	1.5–8.0 µg/L	[33]
Milk	Estrogens	HPLC-UV	MOF-5	SPME	73.1–96.7%	0.17–0.56 ng/mL	[34]
Fruit tea	Polycyclic aromatic hydrocarbons	UHPLC-FLD	Fe_3O_4@HKUST-1	D-µSPE	On average 75%	0.8 ng/L	[35]
Tea samples	Pyrethroids	GC-ECD	MIL-101(Cr)	MSPE-DLLME-SFO	>0.015 ng/mL	78.3–103.6%	[36]
Chrysanthemum tea	Luteolin	Square wave anodic stripping voltammetry	$Cu_3(BTC)_2$/GO	SPE	7.9×10^{-10} mol/L	99.4–101.0%	[37]
In tea and mushroom	Hg(II)	AFS	JUC-62	SPE	>0.58 mg/kg	On average 93.3%	[38]
Fish	Polychlorinated biphenyls	GC-MS	Fe_3O_4-MOF-5(Fe)	SBSE	0.061–0.096 ng/g	>80%	[39]
Fish	Polychlorinated biphenyls	GC-MS	MOF-5	SBSE	0.003–0.004 ng/mL	>80%	[40]
Fish	Aromatic hydrocarbons and gibberellic acids	GC-MS LC-MS/MS	MOF-5	MSPE	0.91–1.96 ng/L for PAHs and 0.006–0.08 µg/L for GAs	66.4–120.0% for PAHs and 90.5–127.4% for GAs	[41]
Fish	Triphenylmethane dyes	HPLC-MS/MS	MOF-5	MSPE	0.30–0.80 ng/mL	83.15–96.53	[42]
Fish	Cd(II) and Pb(II)	FAAS	MOF-199	MSPE	0.2–1.1 µg/L	92.8–117%	[43]
Fish	Cd(II), Zn(II), Ni(II), and Pb(II)	FAAS	MOF-199	MSPE	0.12–1.2 ng/mL	>90%	[44]
Fish	Hg(II)	Cold Vapor AAS	MOF-199	MSPE	10 ng/L	95–102%	[45]
Fish and shrimps	Cd(II), Pb(II), and Ni(II)	FAAS	Fe_3O_4@TAR	MSPE	0.15–0.8 ng/mL	NA	[46]
Shrimp samples, chicken and pork meat	Sulfonamides	HPLC-DAD	Fe_3O_4@JUC-48	MSPE	1.73–5.23 ng/g	76.1–102.6%	[47]
Chicken breast	Drug traces	LC-MS/MS	MIL-101(Cr)@GO	d-µSPE	0.08 and 1.02 ng/kg	88.9–102.3%	[48]
Lettuce	Pesticides	GC-MS	$\infty\{[La_{0.9}Eu_{0.1})_2(DPA)_3(H_2O)_3]\}$	MSPD	0.02–0.05 mg/kg	78–107%	[49]
Fruits and vegetables	Phytohormones	HPLC-FLD	UiO-66	Pipette Tip SPE	0.01–0.02 ng/mL	88.3–105.2%	[50]
Fruits	Plant growth regulator	HPLC-FLD	UiO-67	d-SPE	89.3–102.3%	0.21–0.57 ng/mL	[51]

Table 1. *Cont.*

Matrix	Analytes	Analytical Technique	MOF Material	Sample Preparation Technique	Recovery	LODs	Reference
Fruits	Phytohormones	HPLC-UV	Zeolitic imidazolate framework-8	SBSE	82.7–111%	0.11–0.51 µg/L	[52]
Fruits and vegetables	of insecticides	HPLC-UV	$Fe_3O_4@SiO_2$-GO MOF	MSPE	81.2–105.8%	0.30–1.58 µg/L	[53]
Shellfish	Shellfish poisoning toxin	LC-MS/MS	$Fe_3O_4@SiO_2$@UiO-66	MSPE	93.1% and 107.3%	1.45 pg/mL	[54]
Rice	Herbicides	HPLC-UV	MIL-101(Cr)	MSPE	83.9–103.5%	0.010–0.080 µg/kg	[55]
Tomato sauce	Sudan dyes	HPLC-DAD	Fe_3O_4-NH_2@MIL-101	MSPE	69.6–92.9%	0.5–2.5 µg/kg	[56]
Peanuts	Herbicides	HPLC-DAD	MIL-101(Cr)	d-SPE	89.5–102.7%	0.98–1.9 µg/kg	[57]
In cereal, beverages and water samples	Lead	FAAS	MOF-545	Vortex Assisted SPE	91–96%	1.78 µg/L	[58]

A novel on-line solid-phase extraction application of MOF material ZIF-8, which is a zeolite imidazole framework was proposed by Yang et al., for the determination of tetracyclines in milk samples by HPLC. For this purpose, 390 mg of the material were packed into a stainless-steel column (3 cm × 4.6 mm I.D.) which was coupled on the HPLC injector valve, in order to replace the sample loop. The extraction was achieved at a flow rate of 3 mL min^{-1} for 10 min with the use of a flow-injection system. The milk was treated with McIlvane/ethylenediaminetetraacetic acid (EDTA) buffer and the mixture was centrifuged. For the preconcentration "load" valve position was used, while unwanted sample water was going to waste. Then, with the use of "inject" position the HPLC mobile phase (10% MeOH-20% acetonitrile (ACN)-70% of 0.02 M oxalic acid solution) was pumped in the backflush mode to elute the analytes from the SPE column into the analytical column for the HPLC analysis. The proposed method was the first online SPE method that used MOF material as absorbent for milk samples. Enhancement factors of 35–61 were obtained, recoveries were between 70.3% and 107.4% and limits of detection ranged from 1.5–8.0 µg/L. Good validation results were obtained which indicates that the developed method can be used for preconcentration of multiple analytes from complex samples [33].

Lan et al. in 2016 published an interesting approach for the determination of estrogens in milk by solid phase micro extraction including a novel fiber coating synthesis. For this purpose, cathodic electrodeposition (CED) was used for the in situ synthesis of a MOF-5 coating material for an SPME fiber (average thickness 12.5 µm). The fiber was used for the extraction of estrogens from milk samples that had been previously treated with acetonitrile for ultrasound-assisted extraction, dried and been reconstituted in n-hexane. The fiber was immersed in 5 mL of the extract for 30 min under mixing at 1000 rpm for the extraction and then it was rinsed with hexane for 10 s and immersed in an SPME-HPLC coupling device for 10 min to desorb the analytes in 60 mL of methanol prior to the injection to the HPLC system. After the whole procedure the fiber was conditioned with methanol for 10 min. For the estrogens low LODs (0.17–0.56 ng/mL) and recoveries ranged between 73.1% and 96.7%. Moreover, good method validation results were obtained, which shows that the fiber could be industrialized [34].

2.2. Beverages

A composite HKUST-1 MOF was used for the preconcentration of ultra-high-performance liquid chromatography with fluorescence detection (FLD), for the determination of polycyclic aromatic hydrocarbons (PAHs) in water and fruit tea infusions [35]. Tea beverages contain caffeine, as well as other xanthine derivatives like theobromine and theophylline. Phenolic compounds including phenols, phenolic acids, phenylpropanoids, flavonoids, flavones, flavonones, isoflavones, xanthones, aurones, quinines, and tannins can also be present as they can be found in tea leaves and extracted into the infusion [59]. Water was boiled, and tea was placed in water for 10 min prior to filtration and dispersive micro-extraction with the composite magnetic Fe_3O_4@HKUST-1 material. The interaction between the MOF material and the iron (II, III) oxide (Fe_3O_4) magnetic nanoparticles was achieved with the application of vortex mixing for 30 s. Afterwards, the material was placed together with the tea sample for the ultra-sound assisted extraction of the analytes for 5 min and a strong external magnetic field was used to separate the material. For the elution an aliquot of 0.5 mL of acetonitrile was used together with vortex mixing, and the procedure was repeated thrice. The eluent was filtered, dried under nitrogen and reconstituted in the mobile phase prior to the injection to the UHPLC system. LODs were 0.8 ng/L and recoveries for fruit tea were on average 75% [35].

A combination of two microextraction techniques was implemented by Lu et al. for the extraction and preconcentration of pyrethroids in water and two different tea samples prior to gas chromatography-electron capture detection (GC-ECD) detection. Therefore, a magnetic solid phase extraction coupled with dispersive liquid-liquid microextraction, with solidification of a floating organic drop (MSPE-DLLME-SFO) procedure was developed that included the MOF material MIL-101(Cr)-based for the MSPE step. Firstly, the tea was added in boiling water for 10 min and the extract was filtered. For the MSPE procedure, 10 mg of the MOF material was dispersed in a conical flask containing 50 mL of the sample together with ultrasonic irradiation for 10 min. A magnet was used to transfer the pyrethroid-absorbed magnetic material into a centrifuge tube and 600 µL of methanol was added during ultrasonic mixing for 2 min to desorb the analytes. Magnetic separation took place and the eluate was injected rapidly in a tube containing 5 mL of water. Methanol was used as the elution solvent for the MSPE procedure and the disperser solvent for the DLLME-SFO. Immediately, 50 µL of 1-dodecanol was injected into the solution and mixed in vortex for 10 s to form a cloudy solution with entire dispersion of 1-dodecanol droplets which extracted the analytes within seconds. The alcohol was separated from the mixture by centrifugation and solidification of the floating organic drop in an ice bath. As the last step, the 1-dodecanol became liquid again in room temperature and was injected in the gas chromatography system. The developed composite extraction procedure showed high sensitivity (LODs less than 0.015 ng/mL, LOQs less than 0.050 ng/mL), satisfactory precision and recovery (78.3–103.6%) [36].

Wang et al., synthesized a MOF and GO hybrid composite for solid-phase extraction and preconcentration of luteolin (i.e., a common flavonoid) from tablets and chrysanthemum tea samples. The $Cu_3(BTC)_2$/GO material was made using the solvothermal method by mixing copper nitrate trihydrate and benzene-1,3,5-tricarboxylic acid in N,N-dimethylformamide (DMF) and by adding the GO powder. Tea samples were transferred into a beaker together with 10 mL ethanol for 20 min with ultrasonic, filtration took place and the extract was diluted to 50 mL of ethanol. For the luteoline extraction 10 mL sample solution was transferred to a beaker and the pH of the solution was adjusted to 6. Then, 15 mg of sorbent was added, and the solution was stirred for 20 min for the adsorption. After that the suspension was separated and the sorbent was shaken with 2.5 mL ethanol and phosphate buffer solution mixture to elute the analytes. The eluent was placed into an electrochemical cell for subsequent detection by square wave anodic stripping voltammetry. Limit of detections were of 7.9×10^{-10} mol/L and recovery values for chrysanthemum tea were 99.4–101.0%. Moreover, good adsorption capacity was obtained by the novel material for the extraction procedure and the method was efficient for the enrichment of the sample [37].

In 2016, Wu et al. published a method for the determination of Hg(II) in tea and mushroom samples based on MOF as solid phase extraction sorbents. The MOF material which was used in this study was JUC-62 and it was made of 3,3′5,5′-azobenzenetetracarboxylic acid and copper nitrate trihydrate. Tea samples were dried and digested with nitric acid prior to dilution to 25 mL with deionized water and pH value was adjusted to 6–7. Both static and kinetic adsorption conditions were studied. For the static adsorption experiment 5 mg of the material were added to 5 mL of sample solution. The suspension was then shaken at room temperature for one hour, and then centrifuged. Accordingly, for the dynamic adsorption study, shaking was separately applied for a period of time. After adsorption, the suspension was centrifuged and the crystals were dispersed in 5 mL of acetate buffer together by shaking at room temperature for 10 min. Then, the suspension was centrifuged and the concentration of desorbed mercury was measured by atomic fluorescence spectrometry (AFS). Recovery values for tea samples were in average 93.3%. The static adsorption isotherm exhibited excellent adsorption capacity. The obtained results indicate that the material is promising for the sample preparation for the determination of Hg^{2+} [38].

2.3. Fish

Lipids and proteins constitute the main components of fish tissues. The exact chemical composition of fish depends on the fish species as well as on age, season, sex and environment [60]. For fish sample preparation, many MOF materials are reported in the literature. Lin et al. developed a Fe_3O_4-MOF-5(Fe) composite magnetic material, which was used as a coating for a Nd-Fe-B permanent magnet for stir bar sorptive extraction (SBSE) of six polychlorinated biphenyls, a class of toxic persistent organic pollutants. SBSE is a sensitive equilibrium technique with good reproducibility, which is generally classified as a "green analytical technique", because it is considered to be solvent-free, or uses very low volumes of organic solvents. Due to the high coating amount, SBSE shows good recovery and extraction capacity and it has been used for the analysis of different complex matrices and a wide class of analytes. Lin et al. synthesized four different MOF materials (MIL-101(Cr), MOF-5(Zn), ZIF-8, and MOF-5(Fe)) and used them as coating of SBSE bar and found that Fe_3O_4-MOF-5(Fe) material had the best extraction efficiency. The material was synthesized by mixing amine-functionalized Fe_3O_4 nanoparticles with terephthalic acid and ferric nitrate nonahydrate and 40 mg of it were used as the coating for the SBSE procedure. For the fish sample preparation, the samples were homogenized, extracted with n-hexane under sonication, followed by filtration. Then the filtrate was loaded onto the cartridge, it was dried and finally dispersed in deionized water. Afterwards, 20 mL aqueous sample was placed into a vial into which the stir bar was immersed, and extraction took place at 700 rpm for 30 min. Desorption of the analytes was succeed with 2.5 mL of n-hexane in an ultrasonic bath within 3 min. The elution solution was dried under nitrogen atmosphere and was further diluted into 1 mL of isooctane for the Gas Chromatography-Mass Spectrometry (GC-MS) analysis. During extraction process optimization, it was found that the best efficiency is achieved at pH 7 and with a sodium chloride (NaCl) concentration of 10% m/v. The developed method was simple and sensitive and showed good linearity, low detection limits (0.061–0.096 ng/g) for the six studied polychlorinated biphenyls and recovery values more than 80% [39].

The same working group published in 2016 another analytical process for the selective enrichment and determination of polychlorinated biphenyls in fish samples using aptamer-functionalized stir bar sorptive extraction prior to GC-MS analysis. Therefore, immobilization of aptamer, which could recognize two analytes took place on a MOF-5 material that was fabricated by electro-deposition. For the extraction, the bar was placed into sample solution for 1.0 h. Desorption was performed in 5 mL of dichloromethane/glycine-hydrochloric acid (HCl) buffer (v/v, 1/10) under stirring. Limit of detections ranged from 0.003 to 0.004 ng/mL and recoveries were higher than 80%. Because of the high surface area and high selective recognition of aptamer towards the biphenyls, the prepared MOF based coating showed high selectivity and it can be used for other target analytes by changing the aptamer [40].

In 2013, Hu et al. used a Fe_3O_4-MOF-5(Zn) composite material for the determination of polycyclic aromatic hydrocarbons and gibberellic acids (GAs) in environmental, plant and food samples prior to GC-MS and liquid chromatography-tandem mass spectrometry (LC-MS/MS) analysis. For the synthesis of the MOF-5 material terephthalic acid and zinc diacetate hydrate was used together with amine-functionalized Fe_3O_4 nanoparticles. For the preparation of fish samples, the fish were ground and then extracted with Florisil and a mixture of n-hexane and dichloromethane (1:1, v/v) was added. For the MSPE procedure for the determination of the gibberellic acids, a quantity of 30 mg of the composite material was added to 10 mL of extracted fish sample dissolved in n-hexane, which was placed into ultrasonic bath for 30 s and then shaken on a rotator. The MOF material was collected by applying a magnet to the outer wall of the vial and elution of the analytes took place with 1.0 mL of acetonitrile containing 1% formic acid under ultrasound. Afterwards, the supernatant was evaporated under N_2 atmosphere and re-dissolved in 100 µL of formic acid in water (0.1%, v/v) prior to LC-MS/MS analysis. Accordingly, for the enrichment of the polycyclic aromatic hydrocarbons, a portion of 50 mg of the MOF material was added to 25 mL of the extracted solution, the same procedure was followed, thus desorption was performed with 0.25 mL of acetone prior to GC-MS analysis. This method was proved to be ideal both for polar and for non-polar analytes and it showed good sensitivity, linearity, repeatability, low detection limits (0.91–1.96 ng/L for PAHs and 0.006–0.08 µg/L for GAs) and satisfactory recoveries (66.4–120.0% for PAHs and 90.5–127.4% for GAs) [41].

In 2018, Zhou et al. used a magnetic mesoporous metal-organic framework-5 for the effective enrichment of malachite green and crystal violet; two triphenylmethane dyes in fish samples. For the synthesis of the material polyethyleneimine functionalized Fe_3O_4 nanoparticles were mixed with zinc acetate dihydrate and terephthalic acid. Fish samples were treated with acetonitrile together and the mixture was sonicated for 10 min followed by centrifugation. The extract was dried and dissolved in 1 mL of ethanol and 10 mg of the magnetic MOF composite material was added for the MSPE procedure combined with shaking for 40 min. A magnet was used for separation of the material and analytes were extracted in methanol containing 1% formic acid, and the desorption time was 20 min prior to UHPLC-MS/MS analysis. Detection limits were 0.30 ng/mL for malachite green and 0.08 ng/mL for crystal violet, while recoveries ranged from 83.15% to 96.53%. The developed MOF material can be further studied for the adsorption of these compounds from various complex matrices [42].

HKUST-1 (MOF-199) material have been extensively studied for the determination of heavy metals from fish samples prior detection with atomic absorption spectroscopy (AAS). In 2012, Sohrabi et al. published an analytical method for the determination of Cd(II) and Pb(II) with flame atomic absorption spectroscopy (FAAS) using this magnetic MOF. For this purpose, Fe_3O_4-pyridine conjugate was prepared and mixed with copper(II) nitrate trihydrate and trimesic acid. For the sample preparation the fish samples were digested with nitric acid. For the MSPE procedure, 30 mg of the magnetic sorbent was added into the solutions (pH 6.3) and the mixture was stirred for 14 min to extract heavy metal ions completely. As final step 6.0 mL of 0.01 mol L^{-1} NaOH in EDTA solution was used for the elution. The elution step required 16.5 min. For Cd(II) limit of detection were 0.2 µg/L and recoveries in real samples ranged between 95–117%, while for Pb(II) 1.1 µg/L and recoveries in real samples ranged between 92.8–103.3% [43].

In 2013, Taghizadeh et al. developed a method for the determination of Cd(II), Zn(II), Ni(II), and Pb(II) ions with FAAS using the same MOF material as absorbent. For the preparation of the material, Fe_3O_4 nanoparticles were modified with dithizone and a copper-(benzene-1,3,5-tricarboxylate) MOF made after the reaction of trimesic acid and copper(II) nitrate trihydrate. For the sample preparation fish samples were digested with nitric acid. For the MSPE procedure, 25 mg of the magnetic sorbent was added into the solutions (pH 6.4) and the mixture was stirred for 13 min to extract the metal ions completely. As final step 7.8 mL of 0.9 mol L^{-1} thiourea in 0.01 mol L^{-1} NaOH solution were used for the elution that required 19 min. Limits of detection were found to be 0.12 ng/mL for Cd(II), 0.39 ng/mL for Zn(II), 0.98 ng/mL for Ni(II), and 1.2 ng/mL Pb(II) and recovery values were more than 90%. Potentially interfering ions does not affect the determination of the Cd(II), Zn(II), Ni(II) and Pb(II) ions [44].

In 2016 Tadjarodi and Abbaszadeh developed a method for the determination of Hg(II) with cold vapor AAS. For this purpose, Fe_3O_4 nanoparticles were modified with 4-(5)-imidazole-dithiocarboxylic acid and then reacted with trimesic acid and Cu(II) acetate to form the metal-organic framework HKUST-1 (MOF-199). The material was used as the adsorbent for the determination Hg(II) with MSPE. Therefore, after digestion with HNO_3 a portion of 24 mg of the MOF material was added to the aqueous sample the pH of which was adjusted to 6.0 and the mixture was stirred for 8 min. An external magnetic field was applied in order to collect the material and elution took place with the use of 3.5 mL of 1.1 mol L^{-1} of thiourea solution under shaking for 11 min. With the developed method LODs was 10 ng/L and LOQs was 40 ng/L and satisfactory recovery (95–102%) was obtained [45].

In 2016, Ghorbani-Kalhor used HKUST-1 that was modified with magnetic nanoparticles carrying covalently immobilized 4-(thiazolylazo) resorcinol (Fe_3O_4@TAR) for the determination of Cd(II), Pb(II), and Ni(II) ions with FAAS from seafood (fish and shrimps) and agricultural samples. Fish samples were digested with nitric acid and then 50 mg of the material were added to the sample (pH 6.2, 10 min), for the MSPE procedure. Elution was achieved with 5 mL of a 0.6 mol/L EDTA solution for 15.2 min. LOD values ranged from 0.15 to 0.8 ng/mL [46].

All the four above mentioned methods, which included metal ions determination, were found to be characterized as simple, fast, reproducible, and selective method and the developed sorbent shows high sorption capacity, low limit of detection and high enrichment factor. Moreover, their breakthrough volume was 1000 mL and potentially interfering ions did not affect the determination of the examined metal ions. The LOD values were 0.15, 0.40, and 0.8 ng/mL for Cd(II), Ni(II) and Pb(II) ions, accordingly [42–46].

2.4. Meat, Chicken and Shrimps

Shrimp tissues contain proteins in high concentration. It also contains fatty acids (unsaturated) as well as minerals like calcium. The final shrimp tissue composition depends on the feed [61]. Shrimp samples were treated with composite HKUST-1 material for the determination of Cd(II), Pb(II), and Ni(II) ions by FAAS after digestion with nitric acid as mentioned above [46]. Sulfonamides have been also examined in shrimp samples, together with chicken and pork meat. Meat besides water contains protein and amino acids, minerals, fats and fatty acids, vitamins and other bioactive components, and small quantities of carbohydrates. Percentage composition varies according to animal species (beef, porcine, chicken, etc.) [62]. Sample preparation was achieved with a magnetic and mesoporous metal-organic framework and determination was performed by high-performance liquid chromatography. For this purpose, Fe_3O_4@JUC-48 material was prepared by mixing mercaptoacetic acid functionalized Fe_3O_4 nanoparticles, cadmium nitrate tetrahydrate and 1,4-biphenyldicarboxylic acid. Shrimps, pork and chicken samples were homogenized and placed in a centrifuge tube with acetonitrile under vortex mixing for 5 min, followed by ultra-sound assisted extraction for 30 min. Then, for the MSPE procedure a portion of 25 mg of the composite MOF material was added to the extracts and the mixture was shaken for 8 min. An external magnetic field was used to collect the material and 0.8 mL of methanol with 5% acetic acid was added into the tube and ultrasonic elution of the analytes took place in 10 min. Limit of detection ranged from 1.73 to 5.23 ng/g, recovery valued were between 76.1% and 102.6%. The developed method was successfully applied to real samples [47].

In 2017, Wang et al. published a dispersive micro-solid-phase extraction (d-μ-SPE) of three different kinds of traces of drugs in chicken breast using MIL-101(Cr)@GO composite material in microwave-assisted extraction coupled with HPLC–MS/MS detection. GO was dispersed in water and mixed with terephthalic acid and chromium nitrate nonahydrate. Then, acetonitrile was used with the following extraction conditions: 5 min extraction time at 50 °C with microwave power at 500 W. Then, 8 mg of MIL-101(Cr)@GO was used for the extraction of analytes in combination with vortex mixing for 10 min and centrifugation for 5 min. Elution was achieved with 1 mL of methanol and sonication in 15 min. As last step centrifugation took place for 5 min to separate the material. The liquid was evaporated and re-dissolved in the mobile phase for the HPLC analysis. The process reduced the consumption of organic solvent and was simple to operate. Good precision results were obtained, LODs were between 0.08 and 1.02 ng/kg and recoveries ranged from 88.9% to 102.3% [48].

2.5. Fruits and Vegetables

The main chemical components of fruit and vegetables include carbohydrates, dietary fiber, enzymes, protein, fat, minerals, vitamins, phenolic acids and carotenoids. However, the exact chemical composition of different fruit and vegetables depend greatly on the ripening stage, the cultivation conditions as well as the postharvest conditions [63].

In 2010, Barreto et al. indicated that lettuce samples can also be treated with MOF materials for sample preparation. A three dimensional $\infty[(La_{0.9}Eu_{0.1})_2(DPA)_3(H_2O)_3]$ material was used for the matrix solid phase dispersion (MSPD) of pesticides from lettuce prior to GC-MS determination. The material was prepared by mixing La_2O_3, Eu_2O_3, pyridine-2,6-dicarboxylic acid and water under pressure at 180 °C for three days. Lettuce samples were diced with a knife and placed into a glass mortar with 0.5 g of the material and the pestle was used for homogenous mixing. After that the mixture was placed in a 100 × 20 mm i.d. polypropylene column packed with glass wool together with anhydrous magnesium sulfate and activated carbon. Then a volume of 30 mL of acetonitrile was introduced into the column to elute the analytes. The eluent was collected and concentrated in a vacuum evaporator to a volume of 1 mL and 1 μL of it was directly analyzed with gas chromatography. LOD values were found to be 0.02–0.05 mg/kg, LOQ values were found to be 0.05–0.10 mg/kg, while recoveries ranged from 78% to 107%. Good validation results were obtained, and the developed method can be useful in screening protocols for the determination of pesticides by GC [49].

In 2018, Yan et al. synthesized electrospun UiO-66/polyacrylonitrile nanofibers and used them as adsorbent for pipette tip solid phase extraction of phytohormones in watermelon and mung bean sprouts. The vegetables were cut and ground to form fine powder with liquid nitrogen, followed by extraction with methanol with the use of ultrasonic radiation for 40 min. After centrifugation the obtained liquid could be treated with the MOF material. UiO-66 material was made by mixing terephthalic acid and zirconium(IV) chloride and polyacrylonitrile was added to the spinning solution to obtain the composite material. The nanofibers (5 mg) was placed in a 200 µL pipette-tip that was inserted into a 1.0 mL pipette-tip. The tip was activated with methanol and water, and 1.0 mL of the sample solution was loaded into it to adsorb the analytes. The pipette tip cartridge was washed with 15% methanol-water and a solution of 90% acetonitrile-ammonia was used for the elution. The eluent was dried under nitrogen and re-dissolved in the mobile phase for HPLC analysis. For the four phytohormones recoveries ranged from 88.3% to 105.2%. Limit of detection values were low (0.01 ng/mL to 0.02 ng/mL) and the method was found to be reliable, which indicates that it can be used for real samples analysis [50].

In 2016, Liu et al. developed a zirconium(IV)-based metal-organic framework (UIO-67) and used it as sorbent in dispersive solid phase extraction of plant growth regulator from fruits (pear, apple, grapefruit, orange and grape) prior to HPLC fluorescence detection. Zirconium tetrachloride and 4,4-biphenyldicarboxylic acid were used to make the material, while fruit samples were homogenized and centrifuged to collect the liquid. Then, 15 mg of the material was added to the sample solution and mixed with vortex for 8 min. Centrifugation was used to separate the material from the liquid. Subsequently, 0.6 mL of methanol with 5% formic acid was added for elution with sonication, followed by fluorescence labeling of the analytes with 1-(9H-carbazol-9-yl) propan-2-yl-methane-sulfonate. With this method good analytical performance was achieved in combination with low detection limits (0.21–0.57 ng/mL), high recoveries (89.3–102.3%) and short extraction times [51].

A one-pot synthesis of zeolitic imidazolate framework-8/poly (methyl methacrylate-ethylene-glycol dimethacrylate) monolith coating for stir bar sorptive extraction of phytohormones from fruit samples followed by high performance liquid chromatography-ultraviolet detection has been reported on the literature. The monolithic coating was made by mixing methyl methacrylate, ethylene-glycol dimethacrylate, methanol, azo-di-isobutyronitrile, zinc nitrate hexahydrate and 2-methylimidazole. The mixture was injected in a polytetrafluoroethylene (PTFE) mold with a prepared bar inserted inside under N_2 and remained at 60 °C for 24 h, before being taken out and aged for 8 h at 60 °C. The bar was used to extract the analytes from pear and apple samples, which were previously cut and extracted with methanol/water (85/15, v/v) with vortex mixing for 1 min and sonication for 10 min. Under the optimum parameters, the stir bar was placed into a glass vial containing the sample pH value 3.0 under stirring for 50 min. After washing the bar was immersed in a small vial with 120 µL 30 mM NaOH (dissolved in methanol) for elution followed by neutralization. The stir coating was found to be sensitive and selective towards the examined analytes. Low detection limits were obtained (0.11–0.51 µg/L), as well as good recoveries (82.7–111%) [52].

In 2018, hybrid magnetic nanocomposites based on Cu-MOFs embedded with graphene oxide (GO) were used for the sample preparation of apples, plums, grapes, cucumbers and spinach for the determination of insecticides by HPLC. Firstly, 3.0 g copper(II) nitrate trihydrate was mixed with terephthalic acid in ethanol. Then, amino-functionalized Fe_3O_4 microspheres were prepared and mixed with the MOF material and graphene oxide. The $Fe_3O_4@SiO_2$-GO MOF material was prepared and 10 mg of it were added into the sample solutions obtained from the fruit samples with shaking at 15 °C for 50 min. Subsequently the material was removed with a magnet and 200 µL of methanol were added for elution in 5 min with shaking prior to HPLC-UV analysis. Preconcentration was achieved and good analytical method performance was obtained. LODs ranged from 0.30 to 1.58 µg/L, LOQs ranged from 1.0 to 5.2 µg/L and recovery values were found to be 81.2–105.8% [53].

2.6. Other Food Samples

Another complex food matrix category; shellfish have also been treated with MOF materials prior to LC-MS/MS determination of domoic acid, the primary amnesic shellfish poisoning toxin. For this purpose, magnetic Fe_3O_4@SiO_2 microspheres were synthesized with the solvothermal method and became carboxylate terminated after treatment with glutaric acid anhydride. Then Fe_3O_4@SiO_2@UiO-66 core-shell microspheres were formed with the addition of terephthalic acid and zirconium(IV) chloride. The material was used for MPSE treatment of shellfish tissue, which was previously homogenized and extracted with methanol: water (1:1, v/v). For the magnetic solid phase extraction procedure 1 mg of the magnetic MOF material was added to the extraction solution and the mixture was vortexed for 6 min. Then, the material was removed with the use of a magnet and 0.5 mL of acetonitrile containing 20% acetic acid was used for the elution in combination with vortex mixing for 5 min. The elution procedure was performed three times, followed by evaporation and re-dissolving of the eluent in the mobile phase. LOD values were 1.45 pg/mL and recovery ranged between 93.1% and 107.3%. During the optimization procedure it was found that no pH adjusting, or salt addition was needed, and the developed method was fast and efficient for the extraction of polar analytes from complex matrices [54].

In 2018, Liang et al., used in situ synthesized MOF MIL-101(Cr) functionalized magnetic particles for the determination of seven triazine herbicides in rice. Firstly, Fe_3O_4 nanoparticles were prepared and used for the synthesis of Fe_3O_4@SiO_2-NH_2. Then the material was treated with graphene oxide and the resulting Fe_3O_4@SiO_2-GO were added to the n-hexane extract obtained from the crushed rice powder together with MIL 101(Cr) and the mixture was sonicated for 25 min to obtain the Fe_3O_4@SiO_2-GO/MIL-101(Cr) composite and extract analytical targets from sample. With the use of a magnet the material was removed, and the analytes were eluted with acetonitrile prior to HPLC analysis. The MSPE procedure was optimized and good recoveries (83.9–103.5%) and low limits of detection (0.010–0.080 µg/kg) were achieved for the pesticides [55].

In 2018, Shi et al. used a magnetic Fe_3O_4-NH_2@MIL-101 material for the extraction of six Sudan dyes in tomato sauce. Magnetic nanoparticles were amino-functionalized and mixed with previously prepared MOL-101. Tomato sauce was treated with acetonitrile for 1 min followed by centrifugation at 4 °C for 10 min, twice. The resulting solutions were dried and re-dissolved in 1 mL of MeOH/water (1/1, v/v) in a vial, where 3 mg of the composite MOF material were added. The mixture was shaken for 2 min and then a magnet was used to separate the material for elution with 1 mL of ethyl acetate for 10 min by vortexing, twice. The two fractions of eluent were dried and reconstituted in acetonitrile prior to HPLC analysis. The method was efficient rapid and easy to apply, low LODs were obtained (0.5–2.5 µg/kg) as well as good recovery values (69.6–92.9%) [56].

MIL-101(Cr) was also used as sorbent for the dispersive solid phase extraction of herbicides in peanuts. The herbicides were ultrasonically extracted from peanut using ethyl acetate with ultrasound radiation. The resulting solution was evaporated and reconstituted into n-hexane and 7 mg of the material were added with shaking for 5 min. Separation of the material was achieved with centrifugation and elution was achieved with acetonitrile followed by evaporation and reconstitution in methanol prior to HPLC analysis. The above-mentioned developed method was efficient for the analysis of high in fat matrices. Low LOD values were obtained (0.98–1.9 µg/kg), as well as satisfactory recoveries (89.5–102.7%) [57].

Lead has been determined in cereal, beverages and water samples, using the zirconium-based highly porous metal-organic framework MOF-545 as adsorbent for vortex assisted-solid phase extraction prior to determination with FAAS. For the material preparation, zirconyl chloride octahydrate was mixed with *meso*-tetra(4-carboxyphenyl) porphyrin in dimethylformamide (DMF). Cereal and legume samples (chickpea, bean, wheat and lentil) were dried at 80 °C and digested firstly with HNO_3 (65%, w/w) and then with hydrogen peroxide (30%, w/w) and diluted after evaporation in water. Cherry juice was digested with the same method and mineral water was used without digestion. For the extraction procedure 10 mg of MOF-545 was added to sample solution and the mixture was vortexed for 15 min. Centrifugation was used to separate the material and elution was achieved with 2 mL of 1 mol/L HCl solution by vortex mixing for 15 min and the eluent was separated by centrifuging. High adsorption capacity achieved, in combination with low detection limit 1.78 µg/L (with a preconcentration factor of 125) and recovery ranged from 91–96% [58].

3. MOF-Derived Carbon Materials

Recently, the use of magnetic nanoporous carbons derived from metal-organic framework as adsorbent for sample preparation is gaining more and more attention. Since MOFs are known for their high surface area and in combination with their mesoporous properties and the high carbon content, these materials consist a useful template to synthesize porous carbons with many potential uses, such as hydrogen storage, toxic aromatic compounds sensing, electrocatalysis, etc. [28]. As Lim et al. have found, even non-porous MOFs could result in highly nanoporous carbons [64]. In general, there are two different techniques to construct a MOF-derived nanoporous carbon. The first attempt includes impregnation of MOF carbon precursors with furfuryl alcohol as carbon source and then polymerization/carbonization as a second step. Another simpler attempt is a single-step direct carbonization [28]. MOF-5 is the most common MOF material that has been used for the synthesis of MOF derived nanoporous carbons. Figure 2 shows the structure of MOF-5 [64,65].

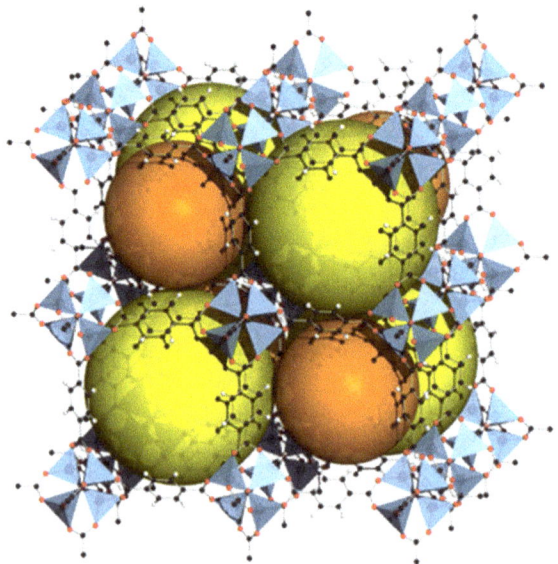

Figure 2. The structure of MOF-5 with orange and yellow spheres showing the pores. [Credit: Tony Boehle].

In 2008, Liu et al. first used A MOF as a template to make porous carbon and subsequently many applications have been reported on the literature. Those materials as well as MOFs show high surface area, large porous volume and due to their thermal stability and great electrochemical performance can be used for the same purposes as metal-organic frameworks [66,67]. Moreover, due to the presence of sp^2 hybridized carbons they are able to form π-stacking interaction with benzene ring and aromatic compounds. As a result, those materials can be used for the adsorption of these kind of chemical compounds. When combined with magnetic precursors, MOFs can form magnetic nanoporous carbons that combine the great adsorption ability of porous carbons and handling convenience of magnetic materials [28]. Table 2 summarizes the applications of MF derived carbons for food sample preparation and some analytical details about the novel developed method.

Table 2. Applications of MOF-derived carbons for food sample preparation.

Matrix	Analytes	Analytical Technique	Precursor MOF Material	Sample Preparation Technique	Recovery	LODs	Reference
Apples	Carbamates	HPLC-UV	MOF-5	MSPE	89.3–109.7%	0.1–0.2 ng/g	[67]
Grapes and bitter gourd	Herbicides	HPLC-UV	ZIF-67	MSPE	88.9–105.1% for grapes, 89.6–105.0% for bitter gourd	0.17–0.4 ng/g for grapes, 0.23–0.46 ng/g for bitter gourd	[68]
Fatmelon	Neonicotinoid insecticides	HPLC-UV	ZIF-67	MSPE	93.0–99.3%	0.2–0.5 ng/g	[69]
Honey tea	Chlorophenols	HPLC-UV	ZIF-8	MSPE	83.0–114.0%	0.1–0.2 ng/mL	[70]
Fruit juice and milk	Endocrine disrupting compounds	UHPLC-FLD	MIL-53	MSPE	92.2–108.3%	0.05–0.10 ng/mL	[71]
Mushrooms	Chlorophenols	HPLC-UV	MOF-5	MSPE	0.25–0.30 ng/g	85.4–97.5%	[72]
Chicken	Fluoroquinolones	HPLC-UV	Cu based MOF	DSPE	0.18–0.58 ng/g	81.3–104.3%	[73]

In 2015, Liu et al., used the well-known material MOF-5 as template to form a magnetic porous material. For this purpose, MOF-5 was loaded in quartz boat and transferred in tube furnace at 80 °C for one hour in order to be carbonized under argon atmosphere at 900 °C for 6 h. Afterwards, iron(III) chloride hexahydrate and iron(II) sulfate heptahydrate was added to the porous carbon under nitrogen atmosphere at 50 °C for one hour in combination with mechanical stirring. The material was used for the determination of carbamates in apple samples with HPLC. For the sample preparation a quantity of 25.0 g of homogenized apple samples was placed in a 50 mL centrifugal tube and was centrifuged at 4000 rpm for 10 min. The supernatant was collected and filtered, and the procedure was repeated after the addition of 10 mL of water to the sediment and vortex mixing. Then, the whole extract was transferred in a conical flask for the MSPE process, where 60 mg of the MOF derived magnetic porous carbon was added and mechanical shaking was implemented.

After 25 min, the material was collected to the bottom of the flask with the use of external field (magnet) and the liquid was discarded. Elution was achieved with 200 µL of methanol and the procedure was repeated three times prior to HPLC analysis. During extraction method optimization it was found that pH value should not be higher than 6 and no salt addition is needed. The developed method showed good repeatability, linearity, precision, recovery values (89.3–109.7%) and low LODs (0.1–0.2 ng/g). Moreover, the material can be used 13 times without any loss in functionality [67].

The same research group published in 2015 an analytical method for simultaneous determination of phenylurea herbicides in grapes and bitter gourd samples, using magnetic carbon as adsorbent material for the sample preparation. The magnetic nanoporous carbon was synthesized by direct carbonization of Co-based metal-organic framework, ZIF-67. For the fabrication of ZIF-67, cobalt(II) nitrate hexahydrate and 2-methylimidazole were used. For the carbonization, ZIF-67 was heated at 150 °C for 1 h, and after that it was heated at 700 °C for 6 h under nitrogen to pyrolyze the organic species. For the sample preparation of grapes and gourd sample homogenization, centrifugation and collection of the supernatant and filtration was carried out with the same procedure as in apple sample preparation. [65,66]. For the MSPE process, a quantity of 10 mg of the magnetic porous carbon material was placed in a 100 mL flask that contained the sample solution. Shaking of the mixture took place for 25 min for the extraction and then the magnetic material was gathered to the bottom of the flask with the use of a magnet. After discarding the supernatant 0.1 mL of acetone was added to desorb the analytes for the HPLC analysis. It was found that no pH adjusting, or salt addition was required for the optimum extraction procedure. No significant loss of adsorption capacity was observed when the material was used 15 times. The developed method showed good repeatability, linearity and precision, LODs were 0.17–0.4 ng/g for the grape samples and 0.23–0.46 ng/g for the bitter gourd samples, while recoveries were 88.9–105.1% for grape sample and 89.6–104.0% for bitter gourd sample [68].

The same MOF derived magnetic nanoporous carbon was used for the determination of neonicotinoid insecticides from water and fat-melon samples by high-performance liquid chromatography ultraviolet detection (HPLC-UV). The samples were cut, homogenized and centrifuged and the MSPE procedure was similar to the above- mentioned procedure for grapes and gourd samples. Same amount of material and volume of extraction solvent was used, however shaking for the extraction took place for 20 min. Separation of the phase was carried out with the use of a magnet and after discarding the liquid phase, a volume of 0.2 mL acetone was added into the isolated MOF and vortexed for 1 min to desorb the chemical compounds. Desorption procedure was repeated one more time before HPLC analysis. During method optimization it was found that pH 6 was the ideal value for extraction and no salt addition was necessary. The MOF derived material can be used at least 15 times without functionality loss. Linearity, repeatability and method precision were good. Moreover, and LOD for the analytes in fat melon samples were 0.2–0.5 ng/g and recoveries ranged from 93.0% to 99.3% [69].

In 2016, Li et al. synthesized a Zn/Co bimetallic metal–organic framework by introducing cobalt into ZIF-8 and by direct carbonization of the resulting Zn/Co-ZIF-8 and used it as an adsorbent for the extraction of chlorophenols from water and honey tea samples prior to their determination by

HPLC-UV. The MOF material was prepared by mixing cobalt(II) nitrate hexahydrate, zinc nitrate hexahydrate and 2-methylimidazole. Different molar ratios of Zn and Co complex compounds were examined and finally the ratio of Zn:Co 7:1 was chosen. Carbonization of the material took place at 900 °C for 6 h under nitrogen. For the honey tea sample preparation, the samples were diluted in a volume of 1:1 with distilled water and filtered and 15 mg of the material was added in 100 mL of the solution and the mixture was shaken for 20 min for the MSPE procedure. The material was separated from the mixture with the use of a magnet and desorption took place with 2 × 0.2 mL alkaline methanol solution and pH was neutralized with HCl solution prior to the injection to the HPLC system. As a result, a rapid, convenient, and efficient MSPE method was developed with low LOD values (0.1–0.2 ng/mL) and good recoveries (83.0–114.0%) [70].

The same year Liu et al. developed a nanoporous carbon/iron composite material MIL-53-C by one-step carbonization of the MOF material MIL-53. The novel material was used as an adsorbent for MSPE for the determination of endocrine disrupting compounds (EDCs) in fruit juices and milk by HPLC. Firstly, MIL-53 (Fe) was fabricated by mixing terephthalic acid and iron(III) chloride hexahydrate at high temperature and pressure and then carbonization was achieved by heating the material at 700 °C for 6 h under nitrogen atmosphere to pyrolyze the organic species. After juice samples were filtrated and milk samples were deproteinized and extracted with acetone, a portion of 12 mg of the material was added to the solutions for the MSPE procedure. Under optimum conditions extraction lasted for 20 min with mixing, adsorption was achieved with 0.2 mL alkaline acetone thrice and no pH adjusting, or salt addition was needed. LODs were 0.05–0.10 ng/mL for fruit juice and 0.10–0.20 ng/mL, while recovery values ranged from 92.2% to 108.3%. The method showed high adsorption capability for trace levels of EDCs and could be a promising extraction method for preconcentration of other organic compounds [71].

In 2016, Hao et al. used a metal-organic framework-derived nanoporous carbon (MOF-5-C) modified with Fe_3O_4 magnetic nanoparticles for the extraction of chlorophenols from mushroom samples prior to HPLC-UV determination. Excellent adsorption capacity was achieved. The carbonization of the MOF-5 nanoparticles was performed at 900 °C for 6 h under Ar. For the MSPE, 8.0 mg of Fe_3O_4@MOF-5-C was added to 50 mL sample solution obtained from homogenization and centrifugation of mushroom samples. The mixture was shaken on a slow-moving platform shaker for 10 min. Subsequently, the material was separated from the sample solution by putting an external magnet and 0.4 mL (0.2 mL × 2) of alkaline methanol was used for elution. The developed method was characterized as simple, fast and sensitive. Limit of detection ranged between of 0.25–0.30 ng/g, while recovery values were 85.4–97.5% [72].

In 2017, Wang et al. synthesized three-dimensional porous Cu@graphitic octahedron carbon cages that were constructed by rapid room-temperature synthesis of a Cu-based metal–organic framework (MOF) followed by further pyrolysis at 700 °C under nitrogen for the dispersive solid phase extraction of four fluoroquinolones (FQs) from chicken muscle and fish tissue prior to their determination with HPLC. The material was synthesized by the reaction of 1,3,5-benzenetricarboxylic acid and copper(II) nitrate trihydrate. Chicken and fish samples were homogenized and treated with methanol with sonication for 10 min to extract the analytes. The resulting solution was filtered, and 36.0 mg of the porous Cu@graphitic carbon cages was added into it. The mixture was vibrated for 30 min followed by centrifugation to separate the material. Elution was performed with ethanol (EtOH)/NaOH 1 mol L^{-1} (7/1, v/v) and the liquid was evaporated under nitrogen. Finally, acidic methanol was added for HPLC analysis. Low detection limits (0.18–0.58 ng/g) were obtained in combination with satisfying recoveries (81.3–104.3%). Good method performance was obtained showing great potential to further increase the applications of this novel material [73].

4. Conclusions

MOFs are novel composite organic-inorganic materials that have been successfully used for sample preparation of different food samples. Most applications include modification of the MOF material with magnetic nanoparticles such as Fe_3O_4 for the magnetic solid phase extraction of different analytes prior to their determination. In other applications the material is introduced into a column either online or offline for the extraction and pre-concentration of organic compounds in food sample matrices. However, more research has to be carried out and many factors have to be investigated for the possible automation of MOF use in sample preparation. Moreover, sensitivity of sample preparation techniques can be improved and limits of quantification (LOQs) can be obtained with the use of more sensitive analytical technique. These techniques could be inductively coupled plasma-atomic emission spectroscopy (ICP-AES) and inductively coupled plasma-mass spectroscopy (ICP-MS) for metal ions or GC-MS and LC-MS for organic compounds.

Carbonization of MOFs for the formation of MOF derived porous carbons has been also used for the sample preparation of different matrices because of the promising properties of those materials. Since there is a great variety of metal ions or clusters and organic linkers suitable to build MOF materials, many materials can be synthesized. The use of these materials or the sample preparation of food samples tend to be very promising in order to simplify the analytical procedure, to reduce the analysis cost and the organic solvent consumption.

Author Contributions: The authors have equally contributed to the manuscript.

Funding: This research received no external funding.

Conflicts of Interest: The authors declare no conflict of interest.

References

1. Manousi, N.; Raber, G. Recent advances in microextraction techniques of antipsychotics in biological fluids prior to liquid chromatography analysis. *Separations* **2017**, *4*, 18. [CrossRef]
2. Green Chemistry. Available online: http://www.epa.gov/greenchemistry (accessed on 30 September 2018).
3. Arthur, L.C.; Pawliszyn, J. Solid phase microextraction with thermal desorption using fused silica optical fibers. *Anal. Chem.* **1990**, *62*, 2145–2148. [CrossRef]
4. Anthemidis, A.N.; Ioannou, K.I. On-line sequential injection dispersive liquid-liquid microextraction system for flame atomic absorption spectrometric determination of copper and lead in water samples. *Talanta* **2009**, *30*, 86–91. [CrossRef] [PubMed]
5. Karageorgou, E.; Manousi, N.; Samanidou, V.; Kabir, A.; Furton, K.G. Fabric phase sorptive extraction for the fast isolation of sulfonamides residues from raw milk followed by high performance liquid chromatography with ultraviolet detection. *Food Chem.* **2016**, *196*, 428–436. [CrossRef] [PubMed]
6. Samanidou, F.V.; Karageorgou, E. Carbon nanotubes in sample preparation. *Curr. Org. Chem.* **2012**, *16*, 1645–16699. [CrossRef]
7. Kyzas, G.Z.; Deliyanni, E.A.; Matis, K.A. Graphene oxide and its application as an adsorbent for wastewater treatment. *J. Chem. Technol. Biotechnol.* **2014**, *89*, 196–205. [CrossRef]
8. Maya, F.; Cabello, C.P.; Frizzarin, R.M.; Estela, J.M.; Palomino, G.T.; Cerda, V. Magnetic solid-phase extraction using metal-organic frameworks (MOFs) and their derived carbons. *Trends Anal. Chem.* **2017**, *90*, 142–152. [CrossRef]
9. Rostami, S.; Pour, A.N.; Salimi, A.; Abolghasempour, A. Hydrogen adsorption in metal-organic frameworks (MOFs): Effects of adsorbent architecture. *Int. J. Hydrog. Energy* **2018**, *43*, 7072–7080. [CrossRef]
10. Meek, S.T.; Greathouse, J.A.; Allendorf, M.D. Metal-organic frameworks: A rapidly growing class of versatile nanoporous materials. *Adv. Mater.* **2011**, *23*, 249–267. [CrossRef] [PubMed]
11. Yaghi, O.M.; Li, H. Hydrothermal synthesis of a metal-organic framework containing large rectangular channels. *J. Am. Chem. Soc.* **1995**, *117*, 10401–10402. [CrossRef]
12. Zhou, H.; Long, J.R.; Yaghi, O.M. Introduction to metal-organic frameworks. *Chem. Rev.* **2012**, *112*, 673–674. [CrossRef] [PubMed]
13. Hashemi, B.; Zohrabi, P.; Raza, N.; Kim, K. Metal-organic frameworks as advanced sorbents for the extraction and determination of pollutants from environmental, biological, and food media. *Trends Anal. Chem.* **2017**, *97*, 65–82. [CrossRef]
14. Furukawa, H.; Cordova, K.E.; O'Keeffe, M.; Yagh, O.M. The chemistry and applications of metal-organic frameworks. *Science* **2013**, *341*. [CrossRef] [PubMed]

15. Chen, Y.; Zhang, W.; Zhang, Y.; Deng, Z.; Zhao, W.; Du, H.; Ma, X.; Yin, D.; Xie, F.; Chen, W.; et al. In situ preparation of core-shell magnetic porous aromatic framework nanoparticles for mixed-mode solid-phase extraction of trace multitarget analytes. *J. Chromatogr. A* **2018**, *1556*, 1–9. [CrossRef] [PubMed]
16. Wu, H.B.; Lou, X.W. Metal-organic frameworks and their derived materials for electrochemical energy storage and conversion: Promises and challenges. *Sci. Adv.* **2017**, *13*. [CrossRef] [PubMed]
17. Chen, W.; Wu, C. Synthesis, functionalization, and applications of metal-organic frameworks in biomedicine. *Dalton Trans.* **2018**, *47*, 2114–2133. [CrossRef] [PubMed]
18. Chen, D.; Yang, D.; Dougherty, C.A.; Lu, W.; Wu, H.; He, X.; Cai, T.; Van Dort, M.E.; Ross, B.D.; Hong, H. In vivo targeting and positron emission tomography imaging of tumor with intrinsically radioactive metal-organic frameworks nanomaterials. *ACS Nano* **2017**, *25*, 4315–4327. [CrossRef] [PubMed]
19. Peller, M.; Böll, K.; Zimpel, A.; Wuttke, S. Metal-organic framework nanoparticles for magnetic resonance imaging. *Inorg. Chem. Front.* **2018**, *5*, 1760–1779. [CrossRef]
20. Chen, B.; Liang, C.; Yang, J.; Contreras, D.S.; Clancy, Y.L.; Lobkovsky, E.B.; Yaghi, O.M.; Dai, S. A microporous metal-organic framework for gas-chromatographic separation of alkanes. *Angew. Chem.* **2006**, *45*, 1390–1393. [CrossRef] [PubMed]
21. Gu, Z.-Y.; Yan, X.-P. Metal-organic framework MIL-101 for high-resolution gas chromatographic separation of xylene isomers and ethylbenzene. *Angew. Chem.* **2010**, *49*, 1477–1480. [CrossRef] [PubMed]
22. Gu, Z.-Y.; Jiang, J.-Q.; Yan, X.-P. Fabrication of isoreticular metal-organic framework coated capillary columns for high-resolution gas chromatographic separation of persistent organic pollutants. *Anal. Chem.* **2011**, *83*, 5093–5100. [CrossRef] [PubMed]
23. Chang, N.; Gu, Z.-Y.; Wang, H.-F.; Yan, X.-P. Metal-organic frameworks-based tandem molecular sieves as a dual platform for selective microextraction and high resolution gas chromatographic separation of n-alkanes in complex matrixes. *Anal. Chem.* **2011**, *83*, 7094–7101. [CrossRef] [PubMed]
24. Yang, C.-X.; Yan, X.-P. Metal-organic framework MIL-101(Cr) for high-performance liquid chromatographic separation of substituted aromatics. *Anal. Chem.* **2011**, *83*, 7144–7150. [CrossRef] [PubMed]
25. Liu, S.-S.; Yang, C.-X.; Wang, S.-W.; Yan, X.-P. Metal-organic frameworks for reverse-phase high-performance liquid chromatography. *Analyst* **2012**, *137*, 816–818. [CrossRef] [PubMed]
26. Yang, C.-X.; Chen, Y.-J.; Wang, H.-F.; Yan, X.-P. High-performance separation of fullerenes on metal-organic framework MIL-101(Cr). *Chem. Eur. J.* **2011**, *17*, 11734–11737. [CrossRef] [PubMed]
27. Gu, Z.-Y.; Yang, C.X. Metal-organic frameworks for analytical chemistry: From sample collection to chromatographic separation. *Acc. Chem. Res.* **2012**, *45*, 734–745. [CrossRef] [PubMed]
28. Chaikittisilp, W.; Ariga, K.; Yamauchi, Y. A new family of carbon materials: Synthesis of MOF-derived nanoporous carbons and their promising applications. *J. Mater. Chem. A* **2013**, *1*, 14–19. [CrossRef]
29. Tolika, E.P.; Samanidou, V.F.; Papadoyannis, I.N. Development and validation of an HPLC method for the determination of ten sulfonamide residues in milk according to 2002/657/EC. *J. Sep. Sci.* **2011**, *34*, 1627–1635. [CrossRef] [PubMed]
30. Bitas, D.; Samanidou, V. Molecularly imprinted polymers as extracting media for the chromatographic determination of antibiotics in milk. *Molecules* **2018**, *23*, 316. [CrossRef] [PubMed]
31. Jia, X.; Zhao, P.; Ye, X.; Zhang, L.; Wang, T.; Chen, Q.; Hou, X. A novel metal-organic framework composite MIL-101(Cr)@GO as an efficient sorbent in dispersive micro-solid phase extraction coupling with UHPLC-MS/MS for the determination of sulfonamides in milk samples. *Talanta* **2017**, *169*, 227–238. [CrossRef] [PubMed]
32. Lirio, S.; Liu, W.-L.; Lin, C.-L.; Lin, C.-H.; Huang, H.Y. Aluminum based metal-organic framework-polymer monolith in solid-phase microextraction of penicillins in river water and milk samples. *J. Chromatogr. A* **2016**, *1428*, 236–245. [CrossRef] [PubMed]
33. Yang, X.-Q.; Yang, C.-X.; Yan, X.-P. Zeolite imidazolate framework-8 as sorbent for on-line solid-phase extraction coupled with high-performance liquid chromatography for the determination of tetracyclines in water and milk samples. *J. Chromatogr. A* **2013**, *1304*, 28–33. [CrossRef] [PubMed]
34. Lan, H.; Pan, D.; Sun, Y.; Guo, Y.; Wu, Z. Thin metal-organic frameworks coatings by cathodic electrodeposition for solid-phase microextraction and analysis of trace exogenous estrogens in milk. *Anal. Chim. Acta* **2016**, *937*, 53–60. [CrossRef] [PubMed]

35. Rocío-Bautista, P.; Pino, V.; Ayala, J.H.; Pasán, J.; Ruiz-Pérez, C.; Afonso, A.M. A magnetic-based dispersive micro-solid-phase extraction method using the metal-organic framework HKUST-1 and ultra-high-performance liquid chromatography with fluorescence detection for determining polycyclic aromatic hydrocarbons in waters and fruit tea infusions. *J. Chromatogr. A* **2016**, *1436*, 42–50. [CrossRef] [PubMed]
36. Lu, N.; He, X.; Wang, T.; Liu, S.; Hou, X. Magnetic solid-phase extraction using MIL-101(Cr)-based composite combined with dispersive liquid-liquid microextraction based on solidification of a floating organic droplet for the determination of pyrethroids in environmental water and tea samples. *Microchem. J.* **2018**, *137*, 449–455. [CrossRef]
37. Wang, Y.; Wu, Y.; Ge, H.; Chen, H.; Ye, G.; Hu, X. Fabrication of metal-organic frameworks and graphite oxide hybrid composites for solid-phase extraction and preconcentration of luteolin. *Talanta* **2014**, *122*, 91–96. [CrossRef] [PubMed]
38. Wu, Y.; Xu, G.; Wei, F.; Song, Q.; Tang, T.; Wang, X.; Hu, Q. Determination of Hg (II) in tea and mushroom samples based on metal-organic frameworks as solid phase extraction sorbents. *Microporous Mesoporous Mater.* **2016**, *235*, 204–210. [CrossRef]
39. Lin, S.; Gan, N.; Qiao, L.; Zhang, J.; Cao, Y.; Chen, Y. Magnetic metal-organic frameworks coated stir bar sorptive extraction coupled with GC-MS for determination of polychlorinated biphenyls in fish samples. *Talanta* **2015**, *144*, 1139–1145. [CrossRef] [PubMed]
40. Lin, S.; Gan, N.; Zhang, J.; Qiao, L.; Chen, Y.; Cao, Y. Aptamer-functionalized stir bar sorptive extraction coupled with gas chromatography–mass spectrometry for selective enrichment and determination of polychlorinated biphenyls in fish samples. *Talanta* **2016**, *149*, 266–274. [CrossRef] [PubMed]
41. Hu, Y.; Huang, Z.; Liao, J.; Li, G. Chemical bonding approach for fabrication of hybrid magnetic metal-organic framework-5: High efficient adsorbents for magnetic enrichment of trace analytes. *Anal. Chem.* **2013**, *85*, 6885–6893. [CrossRef] [PubMed]
42. Zhou, Z.; Fu, Y.; Qin, Q.; Lu, X.; Shia, X.; Zhao, C.; Xu, G. Synthesis of magnetic mesoporous metal-organic framework-5 for the effective enrichment of malachite green and crystal violet in fish samples. *J. Chromatogr. A* **2018**, *1560*, 19–25. [CrossRef] [PubMed]
43. Sohrabi, M.R.; Matbouie, Z.; Asgharinezhad, A.A.; Dehghani, A. Solid phase extraction of Cd(II) and Pb(II) using a magnetic metal-organic framework, and their determination by FAAS. *Microchim. Acta* **2013**, *180*, 589–597. [CrossRef]
44. Taghizadeh, M.; Asgharinezhad, A.A.; Pooladi, M.; Barzin, M.; Abbaszadeh, A.; Tadjarodi, A. A novel magnetic metal-organic framework nanocomposite for extraction and preconcentration of heavy metal ions, and its optimization via experimental design methodology. *Microchim. Acta* **2013**, *180*, 1073–1084. [CrossRef]
45. Tadjarodi, A.; Abbaszadeh, A. A magnetic nanocomposite prepared from chelator-modified magnetite (Fe$_3$O$_4$) and HKUST-1 (MOF-199) for separation and preconcentration of mercury(II). *Microchim. Acta* **2016**, *183*, 1391–1399. [CrossRef]
46. Ghorbani-Kalhor, E. A metal-organic framework nanocomposite made from functionalized magnetite nanoparticles and HKUST-1 (MOF-199) for preconcentration of Cd(II), Pb(II), and Ni(II). *Microchim. Acta* **2016**, *183*, 2639–2647. [CrossRef]
47. Xia, L.; Liu, L.; Lv, X.; Qu, F.; Li, G.; You, J. Towards the determination of sulfonamides in meat samples: A magnetic and mesoporous metal-organic framework as an efficient sorbent for magnetic solid phase extraction combined with high-performance liquid chromatography. *J. Chromatogr. A* **2017**, *1500*, 24–31. [CrossRef] [PubMed]
48. Wang, Y.; Dai, X.; He, X.; Chen, L.; Hou, X. MIL-101(Cr)@GO for dispersive micro-solid-phase extraction of pharmaceutical residue in chicken breast used in microwave-assisted coupling with HPLC-MS/MS detection. *J. Pharm. Biomed. Anal.* **2017**, *145*, 440–446. [CrossRef] [PubMed]
49. Barreto, A.S.; da Silva, R.L.; Dos Santos Silva, S.C.; Rodrigues, M.O.; de Simone, C.A.; de Sá, G.F.; Júnior, S.A.; Navickiene, S.; de Mesquita, M.E. Potential of a metal-organic framework as a new material for solid-phase extraction of pesticides from lettuce (*Lactuca sativa*), with analysis by gas chromatography-mass spectrometry. *J. Sep. Sci.* **2010**, *33*, 3811–3816. [CrossRef] [PubMed]
50. Yan, Z.; Wu, M.; Hu, B.; Yao, M.; Zhang, L.; Lu, Q.; Pang, J. Electrospun UiO-66/polyacrylonitrile nanofibers as efficient sorbent for pipette tip solid phase extraction of phytohormones in vegetable samples. *J. Chromatogr. A* **2018**, *1542*, 19–27. [CrossRef] [PubMed]

51. Liu, L.; Xia, L.; Wu, C.; Qu, F.; Li, G.; Sun, Z.; You, J. Zirconium(IV)-based metal-organic framework(UIO-67) as efficient sorbent in dispersive solid phase extraction of plant growth regulator from fruits coupled with HPLC fluorescence detection. *Talanta* **2016**, *154*, 23–30. [CrossRef] [PubMed]
52. You, L.; He, M.; Chen, B.; Hu, B. One-pot synthesis of zeolitic imidazolate framework-8/poly (methylmethacrylate-ethyleneglycol dimethacrylate) monolith coating for stir bar sorptive extraction of phytohormones from fruit samples followed by high performance liquid chromatography-ultraviolet detection. *J. Chromatogr. A* **2017**, *1524*, 57–65. [CrossRef] [PubMed]
53. Wang, X.; Ma, X.; Huang, P.; Wang, J.; Du, T.; Dua, X.; Lu, X. Magnetic Cu-MOFs embedded within graphene oxide nanocomposites for enhanced preconcentration of benzenoid-containing insecticides. *Talanta* **2018**, *181*, 112–117. [CrossRef] [PubMed]
54. Zhang, W.; Yan, Z.; Gao, J.; Tong, P.; Liu, W.; Zhang, L. Metal-organic framework UiO-66 modified magnetite@silica core-shell magnetic microspheres for magnetic solid-phase extraction of domoic acid from shellfish sample. *J. Chromatogr. A* **2015**, *1400*, 10–18. [CrossRef] [PubMed]
55. Liang, L.; Wang, X.; Sun, Y.; Ma, P.; Li, X.; Piao, H.; Jiang, Y.; Song, D. Magnetic solid-phase extraction of triazine herbicides from rice using metal-organic framework MIL-101(Cr) functionalized magnetic particles. *Talanta* **2019**, *179*, 512–519. [CrossRef] [PubMed]
56. Shi, X.-R.; Chen, X.-L.; Hao, Y.-L.; Li, L.; Xu, H.-J.; Wang, M.-M. Magnetic metal-organic frameworks for fast and efficient solid-phase extraction of six Sudan dyes in tomato sauce. *J. Chromatogr. B* **2018**, *1086*, 146–152. [CrossRef] [PubMed]
57. Li, N.; Wang, Z.; Zhang, L.; Nian, L.; Lei, L.; Yang, X.; Zhang, H.; Yu, A. Liquid-phase extraction coupled with metal-organic frameworks-based dispersive solid phase extraction of herbicides in peanuts. *Talanta* **2014**, *128*, 345–353. [CrossRef] [PubMed]
58. Tokalıoglu, S.; Yavuz, E.; Demir, S.; Patat, S. Zirconium-based highly porous metal-organic framework (MOF-545) as an efficient adsorbent for vortex assisted-solid phase extraction of lead from cereal, beverage and water samples. *Food Chem.* **2017**, *237*, 707–715. [CrossRef] [PubMed]
59. Costa, D.; Costa, H.; Albuquuerque, T.; Ramos, F.; Castilho, M.; Sanches-Silva, A. Advances in phenolic compounds analysis of aromatic plants and their potential applications. *Trends Food Sci. Technol.* **2015**, *45*, 336–354. [CrossRef]
60. Petricorena, Z.C. Chemical Composition of Fish and Fishery Products. In *Handbook of Food Chemistry*; Cheung, P., Ed.; Springer: Heidelberg/Berlin, Germany, 2014; pp. 1–28.
61. Samanidou, V.; Charitonos, S.; Papadoyannis, I.; Bitas, D. On the extraction of antibiotics from shrimps prior to chromatographic analysis. *Separations* **2016**, *3*, 8. [CrossRef]
62. Sources of Meat. Available online: www.fao.org/ag/againfo/themes/en/meat/backgr_composition.html (accessed on 30 October 2018).
63. 2.2 Chemical Composition. Available online: http://www.fao.org/docrep/V5030E/V5030E06.htm (accessed on 30 October 2018).
64. Introduction to Inorganic Chemistry/Coordination Chemistry and Crystal Field Theory. Available online: https://en.wikibooks.org/wiki/Introduction_to_Inorganic_Chemistry/Coordination_Chemistry_and_Crystal_Field_Theory (accessed on 27 September 2018).
65. An Underused Framework for Simpler Sample Prep? Available online: https://theanalyticalscientist.com/issues/0618/an-underused-framework-for-simpler-sample-prep/ (accessed on 1 October 2018).
66. Liu, B.; Shioyama, H.; Akita, T.; Xu, Q. Metal-organic framework as a template for porous carbon synthesis. *J. Am. Chem. Soc.* **2008**, *30*, 5390–5391. [CrossRef] [PubMed]
67. Liu, X.; Wang, C.; Wu, Q.; Wang, Z. Magnetic porous carbon-based solid-phase extraction of carbamates prior to HPLC analysis. *Microchim. Acta* **2015**, *183*, 415–421. [CrossRef]
68. Liu, X.; Wang, C.; Wu, Q.; Wang, Z. Metal-organic framework-templated synthesis of magnetic nanoporous carbon as an efficient absorbent for enrichment of phenylurea herbicides. *Anal. Chim. Acta* **2015**, *870*, 67–70. [CrossRef] [PubMed]
69. Hao, L.; Wang, C.; Wu, Q.; Li, Z.; Zang, X.; Wang, Z. Metal-organic framework derived magnetic nanoporous carbon: Novel adsorbent for magnetic solid-phase extraction. *Anal. Chem.* **2014**, *86*, 12199–12205. [CrossRef] [PubMed]

70. Li, M.; Wang, J.; Jiao, C.; Wang, C.; Wu, Q.; Wang, Z. Magnetic porous carbon derived from a Zn/Co bimetallic metal-organic framework as an adsorbent for the extraction of chlorophenols from water and honey tea samples. *J. Sep. Sci.* **2016**, *39*, 1884–1891. [CrossRef] [PubMed]
71. Liu, X.; Feng, T.; Wang, C.; Hao, L.; Wang, C.; Wu, Q.; Wang, C. A metal-organic framework-derived nanoporous carbon/iron composite for enrichment of endocrine disrupting compounds from fruit juices and milk samples. *Anal. Methods* **2016**, *8*, 3528. [CrossRef]
72. Hao, L.; Liu, X.-L.; Wang, J.-T.; Wang, C.; Wu, Q.-H.; Wang, Z. Metal-organic framework derived magnetic nanoporous carbon as an adsorbent for the magnetic solid-phase extraction of chlorophenols from mushroom sample. *Chin. Chem. Lett.* **2016**, *27*, 783–788. [CrossRef]
73. Wang, Y.; Tong, Y.; Xu, X.; Zhang, L. Metal-organic framework-derived three-dimensional porous graphitic octahedron carbon cages-encapsulated copper nanoparticles hybrids as highly efficient enrichment material for simultaneous determination of four fluoroquinolones. *J. Chromatogr. A* **2018**, *1533*, 1–9. [CrossRef] [PubMed]

© 2018 by the authors. Licensee MDPI, Basel, Switzerland. This article is an open access article distributed under the terms and conditions of the Creative Commons Attribution (CC BY) license (http://creativecommons.org/licenses/by/4.0/).

MDPI
St. Alban-Anlage 66
4052 Basel
Switzerland
Tel. +41 61 683 77 34
Fax +41 61 302 89 18
www.mdpi.com

Molecules Editorial Office
E-mail: molecules@mdpi.com
www.mdpi.com/journal/molecules

www.ingramcontent.com/pod-product-compliance
Lightning Source LLC
LaVergne TN
LVHW071947080526
838202LV00064B/6700